M. A. Shubin
Pseudodifferential Operators and Spectral Theory

M. A. Shubin
Pseudodifferential Operators and Spectral Theory

Springer

Berlin
Heidelberg
New York
Barcelona
Hong Kong
London
Milan
Paris
Singapore
Tokyo

M. A. Shubin

Pseudodifferential Operators and Spectral Theory

Second Edition

Translated from the Russian
by Stig I. Andersson

Springer

Mikhail A. Shubin
Department of Mathematics
Northeastern University
Boston, MA 02115, USA
e-mail: shubin@neu.edu

Stig I. Andersson
Institute of Theoretical Physics
University of Göteborg
41296 Göteborg, Sweden

Title of the Russian original edition:
Psevdodifferentsialnye operatory i spektralnaya teoria
Publisher Nauka, Moscow 1978

The first edition was published in 1987 as part of the *Springer Series in Soviet Mathematics*
Advisers: L. D. Faddeev (Leningrad), R. V. Gamkrelidze (Moscow)

Mathematics Subject Classification (2000): 35S05, 35S30, 35P20, 47G30, 58J40

Library of Congress Cataloging-in-Publication Data

Shubin, M. A. (Mikhail Aleksandrovich), 1944-
 [Psevdodifferentsial'nye operatory i spektral'naia teoriia. English]
 Pseudodifferential operators and spectral theory / M.A. Shubin ; translated from the
Russian by Stig I. Andersson.-- 2nd ed.
 p. cm.
 ISBN 354041195X (softcover : alk. paper)
 1. Pseudodifferential operators. 2. Spectral theory (Mathematics)

QA381 .S4813 2001
515'.7242--dc21

                                                                              2001020695

ISBN 3-540-41195-X Springer-Verlag Berlin Heidelberg New York
ISBN 3-540-13621-5 1st edition Springer-Verlag New York Berlin Heidelberg

Springer-Verlag Berlin Heidelberg New York
a member of BertelsmannSpringer Science+Business Media GmbH
http://www.springer.de
© Springer-Verlag Berlin Heidelberg 1987, 2001
Printed in Germany

Typesetting: Daten- und Lichtsatz-Service, Würzburg
Cover design: *design & production* GmbH, Heidelberg

Printed on acid-free paper   SPIN 10786064   44/3142LK – 5 4 3 2 1 0

Preface to the Second Edition

I had mixed feelings when I thought how I should prepare the book for the second edition. It was clear to me that I had to correct all mistakes and misprints that were found in the book during the life of the first edition. This was easy to do because the mistakes were mostly minor and easy to correct, and the misprints were not many.

It was more difficult to decide whether I should update the book (or at least its bibliography) somehow. I decided that it did not need much of an updating. The main value of any good mathematical book is that it teaches its reader some language and some skills. It can not exhaust any substantial topic no matter how hard the author tried.

Pseudodifferential operators became a language and a tool of analysis of partial differential equations long ago. Therefore it is meaningless to try to exhaust this topic. Here is an easy proof. As of July 3, 2000, MathSciNet (the database of the American Mathematical Society) in a few seconds found 3695 sources, among them 363 books, during its search for "pseudodifferential operator". (The search also led to finding 963 sources for "pseudo-differential operator" but I was unable to check how much the results of these two searches intersected). This means that the corresponding words appear either in the title or in the review published in Mathematical Reviews. On the other major topics of the book the results were as follows:

Fourier Integral operator: 1022 hits (105 books),
Microlocal analysis: 500 hits (82 books),
Spectral asymptotic: 367 hits (56 books),
Eigenvalue asymptotic: 127 hits (21 books),
Pseudodifferential operator AND spectral theory: 142 hits (36 books).

Similar results were obtained by searching the Zentralblatt database.

And there were only 132 references (total) in the original book. So I decided to quote here additionally only three books which I can not resist quoting (in chronological order):

1. J. Brüning, V. Guillemin (eds.), *Mathematics Past and Present. Fourier Integral Operators. Selected Classical Articles by J.J. Duistermaat, V.W. Guillemin and L. Hörmander.*, Springer-Verlag, 1994.

2. Yu. Safarov, D. Vassiliev, *The Asymptotic Distribution of Eigenvalues of Partial Differential Operators*, Amer. Math. Soc., 1997.

3. V. Ivrii, *Microlocal Analysis and Precise Spectral Asymptotics*. Springer-Verlag, 1998.

These books fill what I felt was missing already in the first edition: treatment of more advanced spectral asymptotics by more advanced microlocal analysis (in particular, by Fourier Integral operators).

By the reasons quoted above I did not add anything to the old bibliography at the end of the book, but I made the references more precise whenever this was possible. In case of books I added some references to English translations and also switched the references to the newest editions when I was aware of the existence of such editions.

I made some clarifying changes to the text in some places where I felt these changes to be warranted. I am very grateful to the readers of the book who informed me about the places which need clarifying. Unfortunately, I did not make the list of those readers and I beg forgiveness of those whom I do not mention. However, I decided to mention Pablo Ramacher who was among the most recent and most thorough readers. His comments helped me a lot.

I am also very grateful to Eugenia Soboleva for her selfless work which she generously put in helping me with the proofreading of the second edition.

I hope that my book still has a chance to perform its main function: to teach its readers beautiful and important mathematics.

March 21, 2001   *Mikhail Shubin*

Preface to the Russian Edition

The theory of pseudodifferential operators (abbreviated ΨDO) is comparatively young; in its modern form it was created in the mid-sixties. The progress achieved with its help, however, has been so essential that without ΨDO it would indeed be difficult to picture modern analysis and mathematical physics. ΨDO are of particular importance in the study of elliptic equations. Even the simplest operations on elliptic operators (e.g. taking the inverse or the square root) lead out of the class of differential operators but will, under reasonable assumptions, preserve the class of ΨDO. A significant role is played by ΨDO in the index theory for elliptic operators, where ΨDO are needed to extend the class of possible deformations of an operator. ΨDO appear naturally in the reduction to the boundary for any elliptic boundary problem. In this way, ΨDO arise not as an end-in-themselves, but as a powerful and natural tool for the study of partial differential operators (first and foremost elliptic and hypoelliptic ones). In many cases, ΨDO allow us not only to establish new theorems but also to have a fresh look at old ones and thereby obtain simpler and more transparent formulations of already known facts. This is, for instance, the case in the theory of Sobolev spaces.

A natural generalization of ΨDO are the Fourier integral operators (abbreviated FIO), the first version of which was the Maslov canonical operator. The solution operator to the Cauchy problem for a hyperbolic operator provides an example of a FIO. In this way, FIO play the same role in the theory of hyperbolic equations as ΨDO play in the theory of elliptic equations.

One of the most significant areas for applications of ΨDO and FIO is the spectral theory of elliptic operators. The possibility of describing the structure of various nontrivial functions of an operator (resolvents, complex powers, exponents, approximate spectral projection) is of importance here. By means of ΨDO and FIO one gets the theorem on analytic continuation of the ζ-function of an operator and a number of essential theorems on the asymptotic behaviour of the eigenvalues.

This book contains a slightly elaborated and extended version of a course on ΨDO and spectral theory which I gave at the Department of Mechanics and Mathematics of Moscow State University. The aim of the course was a complete presentation of the theory of ΨDO and FIO in connection with the spectral theory of elliptic and hypoelliptic operators. I have therefore sought to make the presentation accessible to students familiar with the standard Analysis course (including the elementary theory of distributions) and, at the same time, tried to lead the reader to the level of modern journal articles. All this has required a fairly restrictive selection of the material, which was naturally influenced by my personal interests.

The most essential material of an instructional educational nature is in Chapter I and Appendix 1, which also uses theorems from §17 and §18 of Chapter III (note that §17 is not based at all on any foregoing material and §18 is based only on Chapter I). We unite all of this conventionally as the first theme, which constitutes a self-contained introduction to the theory of ΨDO and wave fronts of distributions. In my opinion, this theme is useful to all mathematicians specializing in functional analysis and partial differential equations.

Let me emphasize once more that the first theme can be studied independently of the rest.

Chapters II and III constitute the second and third themes, respectively. From Chapter II the reader will learn about the theory of complex powers and the ζ-function of an elliptic operator. Apparently the theorem on the poles of the ζ-function is one of the most remarkable applications of ΨDO. The derivation of a rough form of the asymptotic behaviour of the eigenvalues is also shown in this chapter. In Chapter III there is a more precise form of the theorem on the asymptotic behaviour of the eigenvalues. This theorem makes use of a number of essential facts from the theory of FIO, also presented here. Let us note that it is in exactly this way that further essential progress in spectral theory was achieved, using, however, a more complete theory of FIO which falls outside the framework of this book (see the section "Short Guide to the Literature").

Finally, Chapter IV together with Appendices 2 and 3 constitute the final (fourth) theme. (Appendix 3 contains auxiliary material from functional analysis which is used in Chapter IV and is singled out in an appendix only for convenience. Advanced readers or those familiar with the material need not look at Appendix 3 or may use it only for reference, whereas it is advisable for a beginner to read it through.) Here we present the theory of ΨDO in $\mathbb{R}^n$ which arises in connection with some mathematical questions in quantum mechanics.

It is necessary to say a few words about the exercises and problems in this book. The exercises, inserted into the text, are closely connected with it and are an integral part of the text. As a rule the results in these exercises are used in what follows. All these results are readily verified and are not proved in the text only because it is easier to understand them by yourself than to simply read them through. The problems are usually more difficult than the exercises and are not used in the text although they develop the basic material in useful directions. The problems can be used to check your understanding of what you have read and solving them is useful for a better assimilation of concepts and methods. It is, however, hardly worthwhile solving all the problems one after another, since this might strongly slow down the reading of the book. At a first reading the reader should probably solve those problems which seem of most interest to him. In the problems, as well as in the basic text, apart from the already presented material, we do not use any information falling outside the framework of an ordinary university course.

I hope that this book will be useful for beginners as well as for the more experienced mathematicians who wish to familiarize themselves quickly with ΨDO and their important applications and also to all who use or take an interest in spectral theory.

In conclusion, I would like to thank V.I. Bezyaev, T.E. Bogorodskaya, T.I. Girya, A.I. Gusev, V.Yu. Kiselev, S.M. Kozlov, M.D. Missarov and A.G. Sergeev who helped to record and perfect the lectures; V.N. Tulovskij who communicated to me his proof of the theorem on propagation of singularities and allowed me to include it in this book; V.L. Roitburd who on my request has written Appendix 2; V.Ya. Ivriĭ and V.P. Palamodov who have read the manuscript through and made a number of useful comments and also all those who have in any way helped me in the work.

M.A. Shubin

Interdependence of the parts of the book

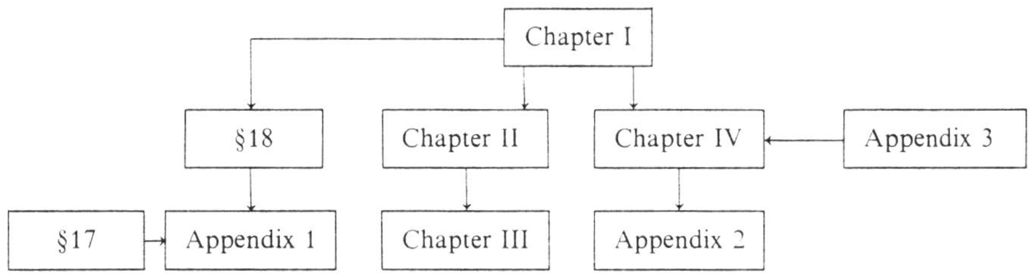

Preface to the English Edition

There are so many books on pseudodifferential operators (which was not the case when the Russian edition of this book appeared) that one naturally questions the need for one more. I hope, nevertheless, that this book can be useful because of its selfcontained approach aimed directly at the spectral theory applications. In addition it contains some ideas which have not been described in any other monograph in English. (I should mention, for instance, the approximate spectral projection method which is a universal method of investigating the asymptotic behaviour of the spectrum – see Chapter IV and also a review paper of Levendorskii in Acta Applicandae Mathematicae *.)

Certainly many new developments have taken place since the Russian edition of the book appeared. The most important ones can be found in the monographs listed below.

September 3, 1985   M. A. Shubin

References

Egorov, Yu. V.: Linear partial differential equations of principal type. Mir Publishers, Moscow 1984 (in Russian). Also: Consultants Bureau, New York, 1986.

Hörmander, L.: The analysis of linear partial differential operators, v. I–IV. Springer-Verlag, Berlin e.a. 1983–1985

Helffer, B.: Théorie spectrale pour des opérateurs globalement elliptiques. Société Math. de France, Astérisque **112**, 1984

Ivrii, V.: Precise spectral asymptotics for elliptic operators. Lecture Notes in Math. **1100**; Springer-Verlag, Berlin e.a. 1984

Kumano-go,H.: Pseudo-differential operators. MIT Press, Cambridge Mass. 1981

Reed, M., Simon, B.: Methods of modern mathematical physics, v. I–IV. Academic Press, New York e.a. 1972–1979

Rempel, S., Schulze, B.-W.: Index theory of elliptic boundary problems. Akademie-Verlag, Berlin, 1982

Taylor, M.: Pseudodifferential operators. Princeton Univ. Press, Princeton, N.J. 1981

Treves, F.: Introduction to pseudodifferential and Fourier integral operators, v. I, II. Plenum Press, New York e.a. 1980

* See also: Levendorskii, S.: Asymptotic distribution of eigenvalues of differential operators. Kluwer Academic Publishers, 1990.

Table of Contents

Chapter I
Foundations of ΨDO Theory

§1. Oscillatory Integrals

1.1 The Fourier transformation. The simplest example of an oscillatory integral is provided by the Fourier transform of a function (or distribution) of tempered growth. Let $S(\mathbb{R}^n)$ be the Schwartz space of functions $u(x) \in C^\infty(\mathbb{R}^n)$ all derivatives of which decrease faster than any power of $|x|$ as $|x| \to \infty$, i.e. for arbitrary α, β

$$\sup_{x \in \mathbb{R}^n} |x^\alpha (\partial^\beta u)(x)| < +\infty. \tag{1.1}$$

As usual x here stands for $(x_1, \ldots, x_n)$; α and β are multiindices, so for example $\alpha = (\alpha_1, \ldots, \alpha_n)$ and α_j is a non-negative integer; $x^\alpha = x_1^{\alpha_1} \ldots x_n^{\alpha_n}$; $\partial = (\partial_1, \ldots, \partial_n)$ where $\partial_j = \dfrac{\partial}{\partial x_j}$; $\partial^\beta = \partial_1^{\beta_1} \ldots \partial_n^{\beta_n} = \dfrac{\partial^{|\beta|}}{\partial x_1^{\beta_1} \ldots \partial x_n^{\beta_n}}$ with $|\beta| = \beta_1 + \ldots + \beta_n$. The left hand sides of (1.1) define a collection of semi-norms in $S(\mathbb{R}^n)$ which turn $S(\mathbb{R}^n)$ into a Fréchet space.

The Fourier transform of a function $u(x) \in S(\mathbb{R}^n)$ is given by the formula

$$(Fu)(\xi) = \hat{u}(\xi) = \int e^{-ix \cdot \xi} u(x)\, dx, \tag{1.2}$$

where $\xi \in \mathbb{R}^n$, $x \cdot \xi = x_1 \xi_1 + \ldots + x_n \xi_n$, $i = \sqrt{-1}$ and $dx = dx_1 \ldots dx_n$ is Lebesgue measure on $\mathbb{R}^n$. The integral in (1.2) is taken over the whole of $\mathbb{R}^n$, which will always be the case unless a domain of integration is explicitly indicated.

It is well known that the operator F defines a linear topological isomorphism

$$F: S(\mathbb{R}^n) \to S(\mathbb{R}^n)$$

and that the inverse operator (the inverse Fourier transformation) is given by the inversion formula

$$(F^{-1} \hat{u})(x) = u(x) = \int e^{ix \cdot \xi} \hat{u}(\xi)\, d\xi, \tag{1.3}$$

where $d\xi = (2\pi)^{-n} d\xi_1 \ldots d\xi_n$.

Now we are going to show how to extend the Fourier transformation (1.2) to continuous functions $u(x)$ satisfying the following condition: there exist constants $C > 0$ and $N > 0$ such that

$$|u(x)| \leqq C\langle x\rangle^N, \tag{1.4}$$

where $\langle x\rangle$ stands for $(1 + |x|^2)^{1/2}$ and $|x|^2 = x_1^2 + \ldots + x_n^2$. We will define $\hat{u}(\xi) \in S'(\mathbb{R}^n)$, the dual space of $S(\mathbb{R}^n)$, i.e. the space of all continuous linear functionals on $S(\mathbb{R}^n)$. So we want to regularize the integral

$$\langle \hat{u}, \psi\rangle = \int\int e^{-ix\cdot\xi} u(x)\, \psi(\xi)\, dx\, d\xi, \tag{1.5}$$

with $\psi(\xi) \in S(\mathbb{R}^n)$, an integral which we will also regard as the value of the functional $\hat{u}$ at the element $\psi(\xi)$. If $u(x) \in S(\mathbb{R}^n)$ it is obvious that

$$\langle \hat{u}, \psi\rangle = \int \hat{u}(\xi)\, \psi(\xi)\, d\xi,$$

since in this case (1.5) converges absolutely.

We give two equivalent means of regularizing (1.5) both differing from the well-known method, based on the Parseval identity, and both extendable to considerably more general situations.

First method. Put $D_j = \dfrac{1}{i}\dfrac{\partial}{\partial x_j}$, $D = (D_1, \ldots, D_n)$ and $\langle D\rangle = (1 + D_1^2 + \ldots + D_n^2)^{1/2}$ (usually we will make use of $\langle D\rangle^k$ with k a non negative even number so that $\langle D\rangle^k$ becomes a differential operator). The vector D will also be used to indicate differentiation in the ξ variable. To avoid confusion we then denote by D_x the just described vector D and by D_ξ the same vector but acting on the ξ variable. We have

$$e^{-ix\cdot\xi} = \langle x\rangle^{-k}\langle D_\xi\rangle^k e^{-ix\cdot\xi}. \tag{1.6}$$

To begin with, suppose that $u(x) \in S(\mathbb{R}^n)$. Then inserting this expression for $e^{-ix\cdot\xi}$ in (1.5) and integrating by parts, we obtain

$$\langle \hat{u}, \psi\rangle = \int\int e^{-ix\cdot\xi} u(x)\langle x\rangle^{-k}\langle D_\xi\rangle^k \psi(\xi)\, dx\, d\xi. \tag{1.7}$$

This integral is now defined not just for $u(x) \in S(\mathbb{R}^n)$ or for absolutely integrable $u(x)$. Indeed, if $u(x)$ satisfies (1.4) and $k > N + n$, then (1.7) converges absolutely and we can consider it as the required regularization of (1.5).

Exercise 1.1. Verify that formula (1.7) defines a continuous linear functional $\hat{u} \in S'(\mathbb{R}^n)$ for $k > N + n$.

Second method. Suppose that $\varphi(x) \in C_0^\infty(\mathbb{R}^n)$ (the space of compactly supported infinitely differentiable functions on $\mathbb{R}^n$) and that $\varphi(0) = 1$. Put

$$I_\varepsilon = \int\int e^{-ix\cdot\xi}\,\varphi\,(\varepsilon x)\,u\,(x)\,\psi\,(\xi)\,dx\,d\xi, \qquad \varepsilon > 0. \tag{1.8}$$

This integral converges absolutely. It turns out that there is a limit $I = \lim\limits_{\varepsilon\to 0} I_\varepsilon$ independent of the choice of $\varphi\,(x)$. Indeed, carrying out in (1.8) the same integration by parts as before, we get

$$I_\varepsilon = \int\int e^{-ix\cdot\xi}\,\varphi\,(\varepsilon x)\,u\,(x)\langle x\rangle^{-k}\langle D_\xi\rangle^k\,\psi\,(\xi)\,dx\,d\xi,$$

and if $k > N + n$, by the Lebesgue dominated convergence theorem, the limit as $\varepsilon \to 0$ exist, and equals $\langle \hat{u}, \psi\rangle$ as defined by formula (1.7).

Exercise 1.2. Verify that for different values of k formula (1.7) leads to the same functional $\hat{u}$.

1.2 Definition of the oscillatory integral and its regularization. Now consider an integral more general than (1.5)

$$I_\Phi\,(au) = \int\int e^{i\Phi(x,\,\theta)}\,a\,(x,\theta)\,u\,(x)\,dx\,d\theta. \tag{1.9}$$

Here $\theta \in \mathbb{R}^N$, $x \in X$, where X is an open set in $\mathbb{R}^n$ and $u\,(x)\in C_0^\infty(X)$, i.e. $u\,(x)\in C^\infty(X)$ and there is a compact set $K \subset X$ such that $u|_{X\setminus K} = 0$. To describe $a\,(x,\theta)$ and $\Phi\,(x,\theta)$ we introduce a number of definitions.

Definition 1.1. Let m, ϱ and δ be real numbers; $0 \leqq \delta \leqq 1$, $0 \leqq \varrho \leqq 1$. The class $S^m_{\varrho,\delta}(X \times \mathbb{R}^N)$ consists of functions $a\,(x,\theta)\in C^\infty(X\times\mathbb{R}^N)$ such that for any multi-indices α, β and any compact set $K \subset X$ a constant $C_{\alpha,\beta,K}$ exists for which

$$|\partial^\alpha_\theta\partial^\beta_x a\,(x,\theta)| \leqq C_{\alpha,\beta,K}\langle\theta\rangle^{m-\varrho|\alpha|+\delta|\beta|}. \tag{1.10}$$

where $x \in K$ and $\theta \in \mathbb{R}^N$.

Instead of $S^m_{1,0}(X\times\mathbb{R}^N)$ we simply write $S^m(X\times\mathbb{R}^N)$. Furthermore, instead of $S^m_{\varrho,\delta}(X\times\mathbb{R}^N)$ we will sometimes simply write $S^m_{\varrho,\delta}$. We also put $S^{-\infty} = \bigcap\limits_m S^m$.

Definition 1.2. We call $\Phi\,(x,\theta)$ a *phase function* if $\Phi\,(x,\theta)\in C^\infty(X\times(\mathbb{R}^N\setminus 0))$, $\Phi\,(x,\theta)$ is real valued and positively homogeneous of degree 1 in θ (i.e. $\Phi\,(x,t\theta) = t\Phi\,(x,\theta)$ for any $x \in X$, $\theta \in \mathbb{R}^N$ and $t > 0$) and $\Phi\,(x,\theta)$ does not have critical points for $\theta \neq 0$, i.e. $\Phi'_{x,\theta}(x,\theta) \neq 0$ for $x \in X$ and $\theta \in \mathbb{R}^N\setminus 0$ ($\Phi'_{x,\theta}$ denotes the gradient of $\Phi\,(x,\theta)$ with respect to x and θ).

Definition 1.3. An integral (1.9) in which $a\,(x,\theta)\in S^m_{\varrho,\delta}(X\times\mathbb{R}^N)$ and $\Phi\,(x,\theta)$ is a phase function is called an *oscillatory integral*.

Exercise 1.3. Verify that if $a\,(x,\theta)\in S^m_{\varrho,\delta}(X\times\mathbb{R}^N)$ then $\partial^\alpha_\theta\partial^\beta_x a\,(x,\theta) \in S^{m-\varrho|\alpha|+\delta|\beta|}_{\varrho,\delta}(X\times\mathbb{R}^N)$. Verify also that for $b\,(x,\theta)\in S^{m'}_{\varrho,\delta}(X\times\mathbb{R}^N)$ we have $a\,(x,\theta)\cdot b\,(x,\theta)\in S^{m+m'}_{\varrho,\delta}(X\times\mathbb{R}^N)$.

Our immediate goal is the regularization of the oscillatory integral (1.9) which is not, generally speaking, absolutely convergent.

The following lemma allows us to write down an equality of the type (1.6) in the general case.

Lemma 1.1. *There exists on $X \times \mathbb{R}^N$, an operator*

$$L = \sum_{j=1}^{N} a_j(x, \theta) \, \frac{\partial}{\partial \theta_j} + \sum_{k=1}^{n} b_k(x, \theta) \, \frac{\partial}{\partial x_k} + c(x, \theta), \qquad (1.11)$$

such that $a_j(x, \theta) \in S^0(X \times \mathbb{R}^N)$, $b_k(x, \theta) \in S^{-1}(X \times \mathbb{R}^N)$, $c(x, \theta) \in S^{-1}(X \times \mathbb{R}^N)$ and defining the formal adjoint ${}^tL$ by the formula

$$ {}^tLu(x, \theta) = - \sum_{j=1}^{N} \frac{\partial}{\partial \theta_j}(a_j u) - \sum_{k=1}^{n} \frac{\partial}{\partial x_k}(b_k u) + cu, \qquad (1.12)$$

we have

$$ {}^tL \, e^{i\Phi} = e^{i\Phi} \qquad (1.13)$$

Exercise 1.4. The operator ${}^tL$ may also be written in the form (1.11) with other a_j, b_k and c, still belonging to the same classes as stated in the definition of L.

Exercise 1.5. Show that if $M = {}^tL$ then $L = {}^tM$.

Proof of Lemma 1.1. We have

$$\frac{\partial}{\partial \theta_j} e^{i\Phi} = i \, \frac{\partial \Phi}{\partial \theta_j} \, e^{i\Phi}, \qquad \frac{\partial}{\partial x_k} e^{i\Phi} = i \, \frac{\partial \Phi}{\partial x_k} \, e^{i\Phi},$$

therefore

$$\left(\sum_{i=1}^{N} -i \, \frac{\partial \Phi}{\partial \theta_j} \, |\theta|^2 \, \frac{\partial}{\partial \theta_j} + \sum_{k=1}^{n} -i \, \frac{\partial \Phi}{\partial x_k} \, \frac{\partial}{\partial x_k} \right) e^{i\Phi}$$

$$= \left(\sum_{j=1}^{N} |\theta|^2 \left| \frac{\partial \Phi}{\partial \theta_j} \right|^2 + \sum_{k=1}^{n} \left| \frac{\partial \Phi}{\partial x_k} \right|^2 \right) e^{i\Phi} = \frac{1}{\psi} \, e^{i\Phi},$$

where $\psi(x, \theta) \in C^\infty(X \times (\mathbb{R}^N \setminus 0))$ is positively homogeneous of degree -2 in θ. Therefore

$$-i\psi \left\{ \sum_{j=1}^{N} |\theta|^2 \, \frac{\partial \Phi}{\partial \theta_j} \, \frac{\partial}{\partial \theta_j} + \sum_{k=1}^{n} \frac{\partial \Phi}{\partial x_k} \, \frac{\partial}{\partial x_k} \right\} e^{i\Phi} = e^{i\Phi}$$

and it remains only to get rid of the singularity at $\theta = 0$. Let $\chi(\theta) \in C_0^\infty(\mathbb{R}^N)$, $\chi(\theta) = 1$ for $|\theta| < \frac{1}{4}$ and $\chi(\theta) = 0$ for $|\theta| > \frac{1}{2}$. Let us put

$$M = - \sum_{j=1}^{N} i(1 - \chi) \, \psi \, |\theta|^2 \, \frac{\partial \Phi}{\partial \theta_j} \, \frac{\partial}{\partial \theta_j} - \sum_{k=1}^{n} i(1 - \chi) \, \psi \, \frac{\partial \Phi}{\partial x_k} \, \frac{\partial}{\partial x_k} + \chi \, .$$

It is obvious that $Me^{i\Phi} = e^{i\Phi}$ and one can easily verify that all the coefficients of the operator M have the required properties. The same is also true for $L = {}^t M$ (cf. Exercise 1.4). It only remains to note that ${}^t L = M$ in view of Exercise 1.5. □

We will achieve the regularization of (1.9) using two different methods.

First method. To begin with, let $m < -N$ so that (1.9) converges absolutely. Utilizing (1.13) write in this integral $({}^t L)^k e^{i\Phi}$ instead of $e^{i\Phi}$ and integrate by parts k times. In this way we get

$$I_\Phi(au) = \iint e^{i\Phi(x,\theta)} L^k(a(x,\theta)u(x))\,dx\,d\theta. \tag{1.14}$$

Exercise 1.6. Verify that this operation is well defined.

Putting $s = \min(\varrho, 1-\delta)$, from Exercise 1.3 we deduce $L^k(au) \in S_{\varrho,\delta}^{m-ks}$ $(X \times \mathbb{R}^N)$. If $\varrho > 0$ and $\delta < 1$ (so that $s > 0$), which will always be assumed in the sequel, then formula (1.14) already allows us to define the integral $I_\Phi(au)$ for an arbitrary m, if we select k so that $m - ks < -N$. This, of course, makes the integral (1.14) absolutely convergent.

Exercise 1.7. Demonstrate that for fixed a and Φ, $I_\Phi(au)$ considered as a functional of $u \in C_0^\infty(X)$, defines a distribution on X, i.e. is an element of $\mathscr{D}'(X)$ (the dual of $C_0^\infty(X)$).

Second method. Picking $\chi(\theta) \in C_0^\infty(\mathbb{R}^N)$ such that $\chi(\theta) = 1$ in a neighbourhood of $0 \in \mathbb{R}^N$, we put

$$I_{\Phi,\varepsilon}(au) = \iint \chi(\varepsilon\theta)\,e^{i\Phi(x,\theta)}\,a(x,\theta)\,u(x)\,dx\,d\theta, \qquad \varepsilon > 0. \tag{1.15}$$

Integrating by parts as in the first method, we get

$$I_{\Phi,\varepsilon}(au) = \iint e^{i\Phi(x,\theta)} L^k(\chi(\varepsilon\theta)\,a(x,\theta)\,u(x))\,dx\,d\theta. \tag{1.16}$$

Note that

$$|\partial_\theta^\gamma \chi(\varepsilon\theta)| \leqq C_\gamma \langle\theta\rangle^{-|\gamma|}, \tag{1.17}$$

where C_γ does not depend on ε for $0 < \varepsilon \leqq 1$.

Exercise 1.8. Verify the estimate (1.17).

We now see that using the dominated convergence theorem we may pass to the limit as $\varepsilon \to 0$ in (1.16). In this way $\lim_{\varepsilon \to 0} I_{\Phi,\varepsilon}(au)$ exists and is equal to $I_\Phi(au)$ in the sense of formula (1.14). In particular, it follows that the integral (1.14) does not depend on k (provided k is sufficiently large) and also that the limit of (1.15) as $\varepsilon \to 0$ is independent of the choice of cut-off function χ.

In what follows we will more or less freely deal with oscillatory integrals, assuming that they could be regularized by one of the above methods.

Exercise 1.9. Prove that the oscillatory integral $I_\Phi(au)$, for fixed Φ and u, represents a continuous linear functional on the Fréchet space $S^m_{\varrho,\delta}(X\times\mathbb{R}^N)$. Here the topology is given by the semi-norms equal to the infima of the constants $C_{\alpha,\beta,K}$ in (1.10). Verify that the closure of $S^{-\infty}(X\times\mathbb{R}^N)$ in $S^m_{\varrho,\delta}(X\times\mathbb{R}^N)$ contains $S^{m'}_{\varrho,\delta}(X\times\mathbb{R}^N)$ for arbitrary $m' < m$. In this way we can view regularization as an extension by continuity of a linear functional.

1.3 Smoothness of distributions defined by oscillatory integrals. Let us introduce the following important notation

$$C_\Phi = \{(x,\theta)\colon x\in X,\ \theta\in\mathbb{R}^N\setminus 0,\ \Phi'_\theta(x,\theta)=0\} \tag{1.18}$$

$\left(\text{Here }\Phi'_\theta\text{ denotes the gradient of }\Phi\text{ w.r.t. }\theta,\text{ i.e. the vector }\left(\dfrac{\partial\Phi}{\partial\theta_1},\ldots,\dfrac{\partial\Phi}{\partial\theta_N}\right)\right).$

The set C_Φ is a conic subset of $X\times(\mathbb{R}^N\setminus 0)$, i.e. together with the point (x_0,θ_0) it also contains all points of the form $(x_0,t\theta_0)$ with $t>0$.

Denoting the natural projection by $\pi\colon X\times(\mathbb{R}^N\setminus 0)\to X$ we set

$$S_\Phi = \pi\, C_\Phi,\qquad R_\Phi = X\setminus S_\Phi. \tag{1.19}$$

Consider the distribution $A\in\mathscr{D}'(X)$ defined via the oscillatory integral $I_\Phi(au)$ by

$$\langle A,u\rangle = I_\Phi(au).$$

Theorem 1.1. *sing supp $A \subset S_\Phi$ or, equivalently, $A\in C^\infty(R_\Phi)$.*

Proof. The assertion of the theorem is equivalent to the existence of $A(x)\in C^\infty(R_\Phi)$ such that if $u\in C^\infty_0(R_\Phi)$ then

$$I_\Phi(au) = \int A(x)\,u(x)\,dx, \tag{1.20}$$

Put

$$A(x) = \int e^{i\Phi(x,\theta)}a(x,\theta)\,d\theta. \tag{1.21}$$

The last integral is itself an oscillatory integral for $x\in R_\Phi$, depending on the parameter x. Differentiating w.r.t this parameter we obtain integrals of the same kind. Essentially here we speak about differentiating w.r.t the parameter of a convergent integral obtained from (1.21) by the above transformation. Therefore $A(x)\in C^\infty(R_\Phi)$ and (1.20) is straightforward. $\square$

Theorem 1.2. *If $a\in S^m_{\varrho,\delta}(X\times\mathbb{R}^N)$ and $a=0$ in a conical neighbourhood of C_Φ, then $A\in C^\infty(X)$.*

Proof. Similar to the proof of Theorem 1.1, in that, since $\Phi'_\theta(x,\theta)\neq 0$ on the support of $a(x,\theta)$, it amounts to a study of the oscillatory integral (1.21). $\square$

Definition 1.4. The phase function $\Phi(x, \theta)$ is called *non-degenerate* if the differentials $d\left(\dfrac{\partial\Phi}{\partial\theta_j}\right)$, $j = 1, \ldots, N$ are all linearly independent on C_Φ or, equivalently rank $\|\Phi''_{\theta\theta}\,\Phi''_{\theta x}\| = N$ (in detail;

$$\|\Phi''_{\theta\theta}\,\Phi''_{\theta x}\| = \left\|\begin{array}{cccccccc} \dfrac{\partial^2\Phi}{\partial\theta_1\partial\theta_1} & \cdots & \dfrac{\partial^2\Phi}{\partial\theta_1\partial\theta_N} & \dfrac{\partial^2\Phi}{\partial\theta_1\partial x_1} & \cdots & \dfrac{\partial^2\Phi}{\partial\theta_1\partial x_n} \\ \cdots & \cdots & \cdots & \cdots & \cdots & \cdots \\ \dfrac{\partial^2\Phi}{\partial\theta_N\partial\theta_1} & \cdots & \dfrac{\partial^2\Phi}{\partial\theta_N\partial\theta_N} & \dfrac{\partial^2\Phi}{\partial\theta_N\partial x_1} & \cdots & \dfrac{\partial^2\Phi}{\partial\theta_N\partial x_n} \end{array}\right\|$$

).

Proposition 1.1. *If Φ is a non-degenerate phase function, then C_Φ is an n-dimensional submanifold in $X \times (\mathbb{R}^N \backslash 0)$.*

Proof. A trivial consequence of the implicit function theorem. $\quad\square$

The following theorem makes theorem 1.2 more precise in the case of a non-degenerate phase function.

Theorem 1.3. *Let Φ be non-degenerate and let $a \in S^m_{\varrho,\delta}(X \times \mathbb{R}^N)$ with the condition:*

$$\text{"either } \varrho > \delta \text{ and } \varrho + \delta = 1 \text{ or } \varrho > \delta \text{ and } \Phi \text{ is linear in } \theta\text{"} \qquad (1.22).$$

Then

1) if a has a zero on C_Φ of infinite order then $A(x) \in C^\infty(X)$;

2) if $a = 0$ on C_Φ, we can find $b \in S^{m-(\varrho-\delta)}_{\varrho,\delta}(X \times \mathbb{R}^N)$ such that $I_\Phi(au) = I_\Phi(bu)$ for arbitrary $u \in C_0^\infty(X)$.

Remark. The latter statement shows, that if $a\,|_{C_\Phi} = 0$, then the distribution A may also be defined by substituting $b(x, \theta)$ for $a(x, \theta)$ and keeping the phase function. The function $b(x, \theta)$ has a lower degree of growth in θ, meaning higher regularity $A(x)$.

To prove theorem 1.3 we need a series of lemmas. The first of these concerns the change of variables in functions of the class $S^m_{\varrho,\delta}$. First of all note, that it makes sense to say that $a(x, \theta) \in S^m_{\varrho,\delta}(U)$, where U is an arbitrary region in $\mathbb{R}^n \times \mathbb{R}^N$, which is conic with respect to θ. Indeed, we will write that $a(x, \theta) \in S^m_{\varrho,\delta}(U)$, if for any compact set $K \subset (\mathbb{R}^n \times S^{N-1}) \cap U$ (S^{N-1} is the unit sphere in $\mathbb{R}^N$) and for arbitrary multi-indices α, β there is a constant $C_{\alpha,\beta,K} > 0$ such that (1.10) is satisfied for $(x, \theta/|\theta|) \in K$ and $|\theta| \geq 1$. Now assume, that we are given a diffeomorphism from a conical region $V \subset \mathbb{R}^{n_1} \times \mathbb{R}^{N_1}$ onto the conical region $U \subset \mathbb{R}^n \times \mathbb{R}^N$, commuting with the natural action of the multiplicative group $\mathbb{R}_+$ of positive numbers, i.e. the diffeomorphism maps a point $(y, \eta) \in V$ to a point $(x(y, \eta), \theta(y, \eta)) \in U$, where $x(y, \eta)$ and $\theta(y, \eta)$ are positively homogeneous in η of degree 0 and 1 respectively. Change the variables in $a(x, \theta)$:

$$b(y, \eta) = a(x(y, \eta), \theta(y, \eta)). \qquad (1.23)$$

Lemma 1.2. *Let $a(x, \xi) \in S^m_{\varrho,\delta}(U)$ and assume that one of the following three assumptions hold*:

a) $\varrho + \delta = 1$;

b) $\varrho + \delta \geq 1$ *and* $x = x(y)$ *does not depend on* η;

c) $x = x(y)$, $\xi = \xi(\eta)$.

   Then $b(y, \eta) \in S^m_{\varrho,\delta}(V)$.

Proof. Differentiating $b(y, \eta)$, we obtain from (1.23)

$$\frac{\partial b}{\partial \eta_l} = \sum_j b^{(j)} \frac{\partial \theta_j}{\partial \eta_l} + \sum_k b_{(k)} \frac{\partial x_k}{\partial \eta_l}, \tag{1.24}$$

$$\frac{\partial b}{\partial y_r} = \sum_j b^{(j)} \frac{\partial \theta_j}{\partial y_r} + \sum_k b_{(k)} \frac{\partial x_k}{\partial y_r}, \tag{1.25}$$

where $b^{(j)} = \dfrac{\partial a}{\partial \theta_j}(x(y, \eta), \theta(y, \eta))$, $b_{(k)} = \dfrac{\partial a}{\partial x_k}(x(y, \eta), \theta(y, \eta))$. The functions $\dfrac{\partial \theta_j}{\partial \eta_l}$, $\dfrac{\partial x_k}{\partial \eta_l}$, $\dfrac{\partial \theta_j}{\partial y_r}$ and $\dfrac{\partial x_k}{\partial y_r}$ are positively homogeneous in η of degrees 0, -1, 1 and 0 respectively. They belong therefore to the classes S^0, S^{-1}, S^1 and S^0 respectively (in V). Estimating the derivatives of a, we easily obtain for $|\eta| \geq 1$

$$\left| \frac{\partial b}{\partial \eta_l} \right| \leq C_K (|\eta|^{m-\varrho} + |\eta|^{m+\delta-1}), \qquad \left(y, \frac{\eta}{|\eta|} \right) \in K,$$

$$\left| \frac{\partial b}{\partial y_r} \right| \leq C_K (|\eta|^{m-\varrho+1} + |\eta|^{m+\delta}), \qquad \left(y, \frac{\eta}{|\eta|} \right) \in K,$$

where K is some compact set in V. If $m + \delta - 1 \leq m - \varrho$, i.e. $\varrho + \delta \leq 1$, then from (1.24) we obtain the estimate $\left| \dfrac{\partial b}{\partial \eta_l} \right| \leq 2 C_K \langle \eta \rangle^{m-\varrho}$. If $x = x(y)$ then $\dfrac{\partial x_k}{\partial \eta_l} = 0$ and we obtain this estimate from (1.24) without assuming $\varrho + \delta \leq 1$.

   Similarly, if $m - \varrho + 1 \leq m + \delta$, i.e. $\varrho + \delta \geq 1$, it follows from (1.25) that $\left| \dfrac{\partial b}{\partial y_r} \right| \leq 2 C_K \langle \eta \rangle^{m+\delta}$, and the same estimate is obtained without the extra assumption $\varrho + \delta \geq 1$ if $\theta = \theta(\eta)$.

   The necessary estimates of the form (1.10) are thus verified when $|\alpha + \beta| \leq 1$ for an arbitrary function $a(x, \theta) \in S^m_{\varrho,\delta}(U)$ in all the three cases a), b), c). Now, inductively, assume that the estimates hold for $|\alpha + \beta| \leq k$ and arbitrary $a \in S^m_{\varrho,\delta}(U)$. In particular, we then obtain that for the derivatives of order $\leq k$ of $b^{(j)}$ and $b_{(j)}$ the estimates of the classes $S^{m-\varrho}(V)$ and $S^{m+\delta}(V)$ respectively hold. But then we obtain from (1.24), (1.25) by analogous reasoning, that these estimates hold for derivatives of order $\leq (k+1)$ and for arbitrary $a \in S^m_{\varrho,\delta}(U)$. $\square$

Lemma 1.3 (variant of the Hadamard lemma). *Let the functions $\Phi_1(x, \theta), \ldots$ $\Phi_k(x, \theta)$ belong to $C^\infty(U)$, with U a conic region in $\mathbb{R}^n \times (\mathbb{R}^N \setminus 0)$, and assume that they are positively homogeneous in θ of degree 0; $d\Phi_1, \ldots, d\Phi_k$ being linearly independent at points of the set*

$$C = \{(x, \theta) \in U, \; \Phi_j(x, \theta) = 0, \; j = 1, \ldots, k\}.$$

Let $a \in S^m_{\varrho, \delta}(U)$, $a|_C = 0$ and $\varrho + \delta = 1$. Then there is a representation

$$a = \sum_{i=1}^{k} a_j \Phi_j, \tag{1.26}$$

where $a_j(x, \theta) \in S^{m+\delta}_{\varrho, \delta}(U)$, $j = 1, \ldots, k$. If the function $a(x, \theta)$ here has a zero of infinite order on C, then all the functions $a_j(x, \theta), j = 1, \ldots, k$, also have a zero of infinite order on C.

Proof. Note, that (1.26) is a linear equation in the functions a_j. It is therefore enough to be able to find the functions a_j locally (for $(x, \theta/|\theta|)$ close to $(x_0, \theta_0/|\theta_0|)$, (x_0, θ_0) being a fixed point in U). A global solution could then be glued together in U using a partition of the unity in U consisting of functions homogeneous of degree zero supported in conical regions in which the required functions a_j have already been constructed.

Thus let $(x_0, \theta_0) \in U$. If $(x_0, \theta_0) \notin C$, there exists a j_0 such that $\Phi_{j_0}(x_0, \theta_0) \neq 0$. For $(x, \theta/|\theta|)$ close to $(x_0, \theta_0/|\theta_0|)$ we can put $a_{j_0} = a/\Phi_{j_0}$ and $a_j = 0$ for $j \neq j_0$. It remains to verify the existence of a_j locally for $(x, \theta/|\theta|)$ close to $(x_0, \theta_0/|\theta_0|) \in C$. By the implicit function theorem $\Phi_1, \ldots, \Phi_k$ can be supplemented by functions $\Phi_{k+1}, \ldots, \Phi_l (l = N + n - 1)$, homogeneous of degree zero, to form a local coordinate system $\Phi_1, \ldots, \Phi_l$ on the manifold $\{(x, \theta): |\theta| = 1\}$ in a neighbourhood of $(x_0, \theta_0/|\theta_0|)$. Therefore, the transformation

$$(x, \theta) \to (\Phi_1(x, \theta), \ldots, \Phi_l(x, \theta), |\theta|) \in \mathbb{R}^l \times \mathbb{R}_+$$

is a diffeomorphism of a conical neighbourhood of the ray $(x_0, t\theta_0)$ onto the conical set $B \times \mathbb{R}_+$, B a ball in $\mathbb{R}^l$, and the image of C has the form $\{(\Phi, |\theta|): \Phi_1 = \ldots = \Phi_k = 0\}$. Let us now consider the symbol

$$\tilde{a}(\Phi, |\theta|) = a(x(\Phi, |\theta|), \theta(\Phi, |\theta|)),$$

obtained from $a(x, \theta)$ under this diffeomorphism. It follows from lemma 1.2 that $\tilde{a}(\Phi, |\theta|) \in S^m_{\varrho, \delta}(B \times \mathbb{R}_+)$. But then, by the Newton-Leibnitz formula

$$\tilde{a}(\Phi, |\theta|) = \sum_{j=1}^{k} \Phi_j \int_0^1 \tilde{a}_{(j)}(t\Phi_1, \ldots, t\Phi_k, \Phi_{k+1}, \ldots, \Phi_l, |\theta|)\, dt,$$

where $\tilde{a}_{(j)} = \dfrac{\partial \tilde{a}}{\partial \Phi_j} \in S^{m+\delta}_{\varrho, \delta}(B \times \mathbb{R}_+)$. It remains to carry out the inverse substitution.

Proof of Theorem 1.3. Assume that $\varrho + \delta = 1$. Then, if $a|_{C_\Phi} = 0$, by lemma 1.3 with $\Phi_j = \dfrac{\partial \Phi}{\partial \theta_j}$, we may represent $a(x, \theta)$ in the form

$$a = \sum_{j=1}^{N} a_j \frac{\partial \Phi}{\partial \theta_j}, \qquad a_j \in S_{\varrho, \delta}^{m+\delta}(U). \tag{1.27}$$

However, taking into account that $\dfrac{\partial \Phi}{\partial \theta_j} e^{i\Phi} = -i \dfrac{\partial}{\partial \theta_j} e^{i\Phi}$, we obtain, on integrating by parts, that

$$I_\Phi(au) = \sum_{i=1}^{N} I_\Phi\left(i \frac{\partial a_j}{\partial \theta_j} u\right).$$

But $\dfrac{\partial a_j}{\partial \theta_j} \in S_{\varrho, \delta}^{m+\delta-\varrho}(U)$, demonstrating the second statement of the theorem. From this proof it is obvious, that if $a(x, \theta)$ had a zero of infinite order in C, then $b(x, \theta)$ could also be chosen to possess this property. So in proving the first statement we can assume $a(x, \theta) \in S_{\varrho, \delta}^{-M}(X \times \mathbb{R}^N)$, M as large as desired. But then the integral (1.21) converges absolutely and uniformly in x as do the integrals obtained from it by differentiation of degree $\leq l(M)$, where $l(M) \to +\infty$ as $M \to +\infty$, and hence the smoothness of $A(x)$ follows. $\square$

Exercise 1.10. Prove theorem 1.3 when the second of the assumptions (1.22) is fulfilled ($\Phi(x, \theta)$ linear in θ).

Hint. It amounts to applying of part c) of Lemma 1.2.

§2. Fourier Integral Operators (Preliminaries)

2.1 Definition of the Fourier integral operator and its kernel. Let X, Y be open sets in $\mathbb{R}^{n_X}$ and $\mathbb{R}^{n_Y}$. Consider the expression

$$Au(x) = \int e^{i\Phi(x, y, \theta)} a(x, y, \theta) u(y) \, dy \, d\theta, \tag{2.1}$$

where $u(y) \in C_0^\infty(Y)$, $x \in X$, $\Phi(x, y, \theta)$ is a phase function on $X \times Y \times \mathbb{R}^N$ and $a(x, y, \theta) \in S^m(X \times Y \times \mathbb{R}^N)$ with $\varrho > 0$ and $\delta < 1$.

Under these conditions the integral

$$\langle Au, v \rangle = \iiint e^{i\Phi(x, y, \theta)} a(x, y, \theta) u(y) v(x) \, dx \, dy \, d\theta, \qquad v \in C_0^\infty(X), \tag{2.2}$$

is defined and is an ordinary oscillatory integral. It is easily verified, that for fixed u the expression (2.2) viewed as a functional of v, defines a distribution $Au \in \mathscr{D}'(X)$. Therefore a linear operator

$$A: C_0^\infty(Y) \to \mathscr{D}'(X), \tag{2.3}$$

is defined. We will formally write it as the integral (2.1)

Definition 2.1. An operator A of the form (2.1) is called a *Fourier integral operator* (*abbreviated FIO*) with phase function $\varphi(x, y, \theta)$.

Definition 2.2. The distribution $K_A \in \mathscr{D}'(X \times Y)$ defined by the oscillatory integral

$$\langle K_A, w \rangle = \int\!\int\!\int e^{i\Phi(x, y, \theta)}\, a(x, y, \theta)\, w(x, y)\, dx\, dy\, d\theta, \qquad w \in (X \times Y), \tag{2.4}$$

is called the *kernel* of A.

Proposition 2.1. a) $K_A \in C^\infty(R_\Phi)$ where $R_\Phi = \{(x, y):\ \Phi'_\theta(x, y, \theta) \neq 0,\ \theta \in \mathbb{R}^N \setminus 0\}$

b) *If $a(x, y, \theta) = 0$ in a conical neighbourhood of the set*

$$C_\Phi = \{(x, y, \theta):\ \Phi'_\theta(x, y, \theta) = 0\},$$

then $K_A \in C^\infty(X \times Y)$.

Proof. This follows immediately from Theorems 1.1 and 1.2. $\square$

In view of the obvious formula

$$\langle Au, v \rangle = \langle K_A, u(y)\, v(x) \rangle, \qquad u \in C_0^\infty(Y), \qquad v \in C_0^\infty(X), \tag{2.5}$$

the kernel K_A is the usual kernel of A in the sense of L. Schwartz.

Exercise 2.1. Verify that the kernel K_A is uniquely defined by the map (2.3) given by A and, conversely, uniquely determines this map.

Remark. One can easily construct two different pairs consisting of a phase function and a function $a(x, y, \theta)$, both pairs giving rise to the same operator (2.3). Furthermore, as a rule, $a(x, y, \theta)$ is not uniquely defined by A, even with the same phase function Φ.

2.2 Operator phase functions.

Definition 2.3. A phase function $\Phi(x, y, \theta)$ is called an *operator phase function*, if the following two conditions are fulfilled;

$$\Phi'_{y, \theta}(x, y, \theta) \neq 0 \ \text{ for }\ \theta \neq 0, \qquad x \in X, \qquad y \in Y \tag{2.6}$$

$$\Phi'_{x, \theta}(x, y, \theta) \neq 0 \ \text{ for }\ \theta \neq 0 \qquad x \in X, \qquad y \in Y \tag{2.7}$$

The role of these two conditions is brought out by the following two propositions.

Proposition 2.2. *Under condition (2.6) the operator (2.1) continuously maps* $C_0^\infty(Y)$ *into* $C^\infty(X)$.

Proof. The integral (2.1) is already defined as an oscillatory integral, depending on the parameter x. Its x- derivatives are of the same form. $\square$

In what follows, $\mathscr{E}'(Y)$ denotes the dual to $C^\infty(Y)$ (and is the set of compactly supported distributions in Y).

Proposition 2.3. *Under condition (2.7) the map (2.3) defined via (2.1) extends by continuity to a continuous map*:

$$A\colon \mathscr{E}'(Y) \to \mathscr{D}'(X). \tag{2.8}$$

Here the continuity is understood in the sense of the weak topologies on $\mathscr{E}'(Y)$ and $\mathscr{D}'(X)$. Recall that the weak topology on the space E', consisting of the linear functionals on E, is defined by the family of semi-norms $p_\varphi(f) = |\langle f, \varphi \rangle|$, where $f \in E'$ and φ is any fixed element of E.

Proof. The transposed operator

$${}^t\!Av(y) = \iint e^{i\Phi(x, y, \theta)}\, a(x, y, \theta)\, v(x)\, dx\, d\theta \tag{2.9}$$

defines, by proposition 2.2, a map

$${}^t\!A\colon C_0^\infty(X) \to C^\infty(Y).$$

Thus, defining A by

$$\langle Au, v \rangle = \langle u, {}^t\!Av \rangle$$

with $u \in \mathscr{E}'(Y)$, $v \in C_0^\infty(X)$ we are done. $\square$

Exercise 2.2. Verify that the operator A, defined in this way, is indeed an extension by continuity of the map (2.3).

So an FIO A with operator phase function Φ maps $C_0^\infty(Y)$ into $C^\infty(X)$ and $\mathscr{E}'(Y)$ into $\mathscr{D}'(X)$. We now study the change in the singular support under the action of A.

Let us settle a notation. If X and Y are two sets, S a subset of $X \times Y$ and K a subset of Y, then $S \circ K$ is the subset of X consisting of the points $x \in X$, for which there exists a $y \in K$ with $(x, y) \in S$.

Theorem 2.1. *The following inclusion holds*:

$$\text{sing supp } Au \subset S_\Phi \circ \text{sing supp } u \tag{2.10}$$

where $S_\Phi = (X \times Y) \backslash R_\Phi$ *consists of those pairs* (x, y) *for which there exists a* $\theta \in \mathbb{R}^N \backslash 0$ *with* $\Phi'_\theta(x, y, \theta) = 0$.

Proof. Splitting $u \in \mathscr{E}'(Y)$ into a sum of a function in $C_0^\infty(Y)$ and a distribution with the support in a neighbourhood of sing supp u we see that it suffices to demonstrate that sing supp $(Au) \subset S_\Phi \circ \mathrm{supp}\, u$.

Let $K = \mathrm{supp}\, u$ and K' an arbitrary compact set in X not intersecting $S_\Phi \circ K$ and so that $K' \times K \subset R_\Phi$. Since R_Φ is open, there are open neighbourhoods Ω and Ω' of the compact sets K and K' respectively such that $\Omega' \times \Omega \subset R_\Phi$. So it suffices to verify that $Au \in C^\infty(\Omega')$. But this is evident, since $K_A(x, y) \in C^\infty(R_\Phi)$ and in particular, $K_A(x, y) \in C^\infty(\Omega' \times \Omega)$. $\square$

Exercise 2.3. Verify the statement used above that if $K_A \in C^\infty(\Omega' \times \Omega)$ then A maps $\mathscr{E}'(\Omega)$ into $C^\infty(\Omega')$.

2.3 Example 1: The Cauchy problem for the wave equation. Consider the Cauchy problem

$$\frac{\partial^2 f}{\partial t^2} = \Delta f \tag{2.11}$$

$$f|_{t=0} = 0 \qquad f_t'|_{t=0} = u(x) \tag{2.12}$$

where $x \in \mathbb{R}^n$, $f = f(t, x)$, Δ is the Laplacian in x and $-$ to begin with $-$ $u(x) \in C_0^\infty(\mathbb{R}^n)$. We solve (2.11)–(2.12) with the help of the Fourier transformation in x, putting

$$\tilde{f}(t, \xi) = \int e^{-iy \cdot \xi} f(t, y)\, dy .$$

In this way, we have

$$\frac{\partial^2 \tilde{f}}{\partial t^2} = -|\xi|^2\, \tilde{f}(t, \xi) \tag{2.13}$$

$$\tilde{f}|_{t=0} = 0, \qquad \tilde{f}_t'|_{t=0} = \tilde{u}(\xi) \tag{2.14}$$

where $\tilde{u}(\xi)$ is the Fourier transform of $u(x)$.

From (2.13) and (2.14) we easily obtain that

$$\tilde{f}(t, \xi) = \tilde{u}(\xi)\frac{\sin t|\xi|}{|\xi|} .$$

Therefore by the Fourier inversion formula

$$\begin{aligned}
f(t, x) &= \iint e^{i(x-y)\cdot\xi} |\xi|^{-1} \sin t|\xi| u(y)\, dy\, d\xi \\
&= \iint e^{i(x-y)\cdot\xi} (2i|\xi|)^{-1} (e^{it|\xi|} - e^{-it|\xi|}) u(y)\, dy\, d\xi .
\end{aligned}$$

We would like to split the last integral into two parts separating the exponents $e^{it|\xi|}$ and $e^{-it|\xi|}$. However this would lead to a singularity at $\xi = 0$. To avoid this singularity, let us again use a cut-off function $\chi = \chi(\xi) \in C_0^\infty(\mathbb{R}^n)$, such that $\chi(\xi) = 1$ near 0 and split the integral into three parts:

$$f(t, x) = f_+(t, x) - f_-(t, x) + r(t, x)$$

$$f_+(t, x) = \int\!\!\int e^{i[(x-y)\cdot\xi+t|\xi|]}(1 - \chi(\xi))(2i|\xi|)^{-1}dy\,d\xi,$$
$$f_-(t, x) = \int\!\!\int e^{i[(x-y)\cdot\xi-t|\xi|]}(1 - \chi(\xi))(2i|\xi|)^{-1}dy\,d\xi,$$
$$r(t, x) = \int\!\!\int e^{i(x-y)\cdot\xi}\chi(\xi)|\xi|^{-1}\sin t|\xi|\,dy\,d\xi.$$

It is clear e.g. that $f_+ = Au$ where A is a FIO with the phase function

$$\Phi(t, x, y, \xi) = (x - y)\cdot\xi + t|\xi|.$$

This is an operator phase function. Since $\Phi'_\xi = x - y + t\xi/|\xi|$, we have

$$C_\Phi = \{(t, x, y, \xi) : y - x = t\xi/|\xi|\},$$
$$S_\Phi = \{(t, x, y) : |x - y|^2 = t^2\}.$$

The second term $f_-(t, x)$ can be similarly presented as $f_- = \tilde{A}u$, where $\tilde{A}$ is a FIO with the phase function

$$\tilde{\Phi}(t, x, y, \xi) = (x - y)\cdot\xi - t|\xi|.$$

It has the same set $S_{\tilde{\Phi}} = S_\Phi$.

For the third term we clearly have $r = Ru$ where R has a C^∞ Schwartz kernel $K_R(t, x, y)$. In fact, it is easy to see that any such operator R can be also presented as a FIO in the form (2.1) with an arbitrary choice of the phase function and with an amplitude $a \in S^{-\infty}$. (See also Exercise 2.4 below.)

So we see that each of the terms $f_\pm(t, x), r(t, x)$ can be presented as a result of the application of a FIO to the initial condition u. In particular, by Proposition 2.3 we can define $f(t, x)$ for any $u \in e'(\mathbb{R}^n)$. By Theorem 2.1 the singularities of $f(t, x)$ belong to

$$\{(t, x) : \exists u \in \operatorname{sing\,supp} u, \ |x - y|^2 = t^2\}.$$

This is the classical statement that singularities propagate with the speed of light (which is equal to 1 in this case). In particular, the singularities of the fundamental solution (the case $u(y) = \delta(y)$) belong to the light cone $|x|^2 = t^2$.

Note also that if we fix the time t_0, then $\Phi_{t_0}(x, y, \xi) = \Phi(t_0, x, y, \xi)$ remains an operator phase function. Therefore the mapping of u into $f(t_0, \cdot)$ is a FIO.

2.4 Example 2: Linear differential operators. Let

$$A = \sum_{|\alpha| \leqq m} a_\alpha(x) D^\alpha, \tag{2.16}$$

where $a_\alpha(x) \in C^\infty(X)$, X an open set in $\mathbb{R}^n$ and $D = i^{-1}\dfrac{\partial}{\partial x}$.

Using the Fourier transformation, we may write

$$D^\alpha u(x) = \int\!\!\int \xi^\alpha e^{i(x-y)\cdot\xi}u(y)\,dy\,d\xi,$$

hence

$$Au(x) = \iint e^{i(x-y)\cdot\xi}\, \sigma_A(x,\xi)\, u(y)\, dy\, d\xi, \tag{2.17}$$

where $\sigma_A(x,\xi) = \sum\limits_{|\alpha|\le m} a_\alpha(x)\,\xi^\alpha$ is called the *symbol* of the operator A. Since $\sigma_A(x,\xi)\in S^m(X\times\mathbb{R}^n)$, we see from (2.17) that A is an FIO with phase function $\Phi(x,y,\xi) = (x-y)\cdot\xi$.

2.5 Example 3: Pseudodifferential operators

Definition 2.4. Let $n_X = n_Y = N = n$ and $X = Y$. Then an FIO with the phase function $\Phi(x,y,\xi) = (x-y)\cdot\xi$ is called a *pseudodifferential operator* (briefly: ΨDO). The class of ΨDO, defined by $a(x,y,\xi) \in S^m_{\varrho,\delta}(X\times X\times\mathbb{R}^n)$ is denoted by $L^m_{\varrho,\delta}(X)$ or simply by $L^m_{\varrho,\delta}$. We also put L^m instead of $L^m_{1,0}$ and write $L^{-\infty} = \bigcap\limits_m L^m$.

As demonstrated in the previous example, any linear partial differential operator is a ΨDO.

We now display the properties of ΨDO, which follow from the already shown properties of FIO.

Proposition 2.4. *Let A be a ΨDO given by the formula*

$$Au(x) = \iint e^{i(x-y)\cdot\xi}\, a(x,y,\xi)\, u(y)\, dy\, d\xi \tag{2.18}$$

Let K_A be the kernel of A and Δ the diagonal in $X\times X$.
Then a) $K_A \in C^\infty((X\times X)\setminus\Delta))$;
b) *A defines continuous linear maps*

$$A:\ C_0^\infty(X) \to C^\infty(X) \tag{2.19}$$

$$A:\ \mathscr{E}'(X) \to \mathscr{D}'(X) \tag{2.20}$$

and

$$\text{sing supp } Au \subset \text{sing supp } u \tag{2.21}$$

for $u\in\mathscr{E}'(X)$ (this property is called pseudolocality of A);
c) *if the function $a(x,y,\xi)\in S^m_{\varrho,\delta}(X\times X\times\mathbb{R}^n)$ vanishes for $x=y$ and $\delta<\varrho$ then we can write A in the form (2.18) with $b(x,y,\xi)\in S^{m-(\varrho-\delta)}_{\varrho,\delta}(X\times X\times\mathbb{R}^n)$ instead of $a(x,y,\xi)$.*
d) *if $a(x,y,\xi)$ has a zero of infinite order at $x=y$, then $K_A\in C^\infty(X\times X)$ and the operator A transforms $\mathscr{E}'(X)$ into $C^\infty(X)$.*

Proof. Left to the reader as an exercise. $\square$

Exercise 2.4. Let $K(x,y)\in C^\infty(X\times X)$ and A be an operator from $C_0^\infty(X)$ to $C^\infty(X)$ with kernel $K(x,y)$. Prove that A is a ΨDO and that in the representation (2.18) we can take $a(x,y,\xi)\in S^{-\infty}(X\times X\times\mathbb{R}^n)$.

Hint. If $\chi(\xi)\in C_0^\infty(\mathbb{R}^n)$, $\chi(\xi)\ge 0$ and $\int\chi(\xi)\,d\xi = 1$ we can take

$$a(x,y,\xi) = e^{-i(x-y)\cdot\xi}\, K(x,y)\, \chi(\xi).$$

Note that linear differential operators enjoy the *locality* property:

$$\operatorname{supp}(Au) \subset \operatorname{supp} u, \quad u \in C_0^\infty(X) \tag{2.22}$$

The following exercise shows that in general this is not the case for ΨDO.

Exercise 2.5. Show that an operator whose kernel $K(x,y) \in C^\infty(X \times X)$ is not identically zero does not obey (2.22)

Problem 2.1. Given a linear continuous operator

$$A: \; C_0^\infty(X) \to C_0^\infty(X)$$

satisfying (2.22), then for any subdomain $X' \subset X$, whose closure is compact in X, we get a linear differential operator by restricting A to $C_0^\infty(X')$.

Hint. Verify that for any fixed $x_0 \in X$ the linear functional given by $(A\varphi)(x_0)$ for $\varphi \in C_0^\infty(X)$, is supported at x_0 and thus can be written as

$$(A\varphi)(x_0) = \sum_\alpha a_\alpha(x_0) (D^\alpha \varphi)(x_0).$$

Derive from the continuity of A the local finiteness of this sum and the smoothness (in x_0) of the coefficients $a_\alpha(x_0)$.

§3. The Algebra of Pseudodifferential Operators and Their Symbols

3.1 Properly supported pseudodifferential operators. Let A be a ΨDO with kernel K_A and let supp K_A denote the support of K_A (the smallest closed subset $Z \subset X \times X$ such that $K_A|_{(X \times X) \setminus Z} = 0$. Consider the canonical projections $\Pi_1, \Pi_2 : \operatorname{supp} K_A \to X$, obtained by restricting the corresponding projections of the direct product $X \times X$. Recall that a continuous map $f : M \to N$ between topological spaces M, N is called *proper* if for any compact $K \subset N$ the inverse image $f^{-1}(K)$ is a compact in M.

Definition 3.1. A ΨDO A is called *properly* supported if both projections $\Pi_1, \Pi_2 : \operatorname{supp} K_A \to X$ are proper maps.

Example. Linear differential operators (cf. item 2.4) are properly supported ΨDO (in this case supp $K_A = \Delta$, the diagonal in $X \times X$).

Proposition 3.1. *Let A be a properly supported ΨDO. Then A defines a map*

$$A: \; C_0^\infty(X) \to C_0^\infty(X) \tag{3.1}$$

which extends to continuous maps

$$A: \; \mathscr{E}'(X) \to \mathscr{E}'(X) \tag{3.2}$$

$$A: \quad C^\infty(X) \to C^\infty(X) \tag{3.3}$$

$$A: \quad \mathcal{D}'(X) \to \mathcal{D}'(X) \tag{3.4}$$

Proof. For $u(y) \in C_0^\infty(X)$, we have the inclusion

$$\operatorname{supp}(Au) \subset (\operatorname{supp} K_A) \circ (\operatorname{supp} u) \tag{3.5}$$

Indeed, if $v \in C_0^\infty(X)$ is such that $\operatorname{supp} v \cap (\operatorname{supp} K_A) \circ (\operatorname{supp} u) = \emptyset$, then $\operatorname{supp} K_A \cap \operatorname{supp}[u(y)\,v(x)] = 0$ and therefore $\langle Au, v \rangle = 0$ by (2.5). Furthermore, from the obvious formula

$$(\operatorname{supp} K_A) \circ (\operatorname{supp} u) = \Pi_1 (\operatorname{supp} K_A \cap \Pi_2^{-1}(\operatorname{supp} u)) \tag{3.6}$$

it follows that the right hand side of (3.5) is compact, so that $Au \in C_0^\infty(X)$, which establishes (3.1). The continuity is easily verified.

Since the kernel $K_{{}^tA}$ of the transposed operator ${}^tA$ is obtained from K_A by permuting x and y (more precisely: $\langle K_{{}^tA}, w(x, y) \rangle = \langle K_A, w(y, x) \rangle$): then ${}^tA$ also defines a continuous map

$$^tA: \quad C_0^\infty(X) \to C_0^\infty(X),$$

which yields (3.4) by duality. Finally, as is easily verified, formula (3.5) also applies for $u \in \mathscr{E}'(X)$ which gives (3.2). Since this can also be said of ${}^tA$, by duality we obtain (3.3). $\quad \square$

Exercise 3.1. Let X_n be a sequence of open subsets of X such that
1) $X_1 \subset X_2 \subset \ldots \subset X_n \subset \ldots$,
2) $\bigcup_n X_n = X$
3) the closure $\overline{X}_n$ of X_n in X is compact in X.

Let $\chi_n(\mathrm{x}) \in C_0^\infty(X)$ and $\chi_n(x) = 1$ for $x \in X_n$. Finally let A be a properly supported ΨDO in X. Show that if $u \in \mathcal{D}'(X)$, then for arbitrary m one can find $N = N(m)$ such that the distribution $[A(\chi_n u)]|_{X_m}$ does not depend on n for $n \geq N$. In this way, we can define $Au \in \mathcal{D}'(X)$ by the formula

$$Au = \lim_{n \to \infty} A(\chi_n u) \tag{3.7}$$

Show that this definition coincides with the one given above in proving Proposition 3.1.

Exercise 3.2. Show that all the three definitions of a properly supported ΨDO given above, coincide on $C^\infty(X)$:
   a) by duality from the map ${}^tA: \mathscr{E}'(X) \to \mathscr{E}'(X)$;
   b) as a restriction to $C^\infty(X)$ of the map $A: \mathcal{D}'(X) \to \mathcal{D}'(X)$, constructed by duality from the map ${}^tA: C_0^\infty(X) \to C_0^\infty(X)$,
   c) by formula (3.7).

Exercise 3.3. Verify that the operator A is properly supported if and only if ^tA is.

The importance of properly supported ΨDO lies in the fact that they form an algebra where multiplication is the ordinary multiplication (composition) of operators. This statement will be proved below, but presently it is clear from Proposition 3.1, that the composition $A_1 \circ A_2$ of two properly supported ΨDO is defined as a linear continuous operator on the spaces $C_0^\infty(X)$, $\mathscr{E}'(X)$, $C^\infty(X)$ or $\mathscr{D}'(X)$.

Note that it is not immediately clear that (2.18) which determines A via $a(x, y, \xi)$ applies for $u \in \mathscr{D}'(X)$ (or even for $u \in C^\infty(X)$). This is not surprising, since the function $a(x, y, \xi)$ is not uniquely defined by the operator A. However, utilizing this arbitrariness, we can make a better choice of $a(x, y, \xi)$.

Definition 3.2. We say that a function $a(x, y, \xi)$ is *properly supported* if both the projections

$$\Pi_1, \Pi_2: \operatorname{supp}_{x, y} a(x, y, \xi) \to X$$

are proper maps (by $\operatorname{supp}_{x, y} a(x, y, \xi)$ we denote the closure of the projection of $\operatorname{supp} a(x, y, \xi)$ in $X \times X$).

In this case the corresponding ΨDO A is obviously properly supported.

Proposition 3.2. *If $A \in L^m_{\varrho, \delta}(X)$ is properly supported, then A can be put in the form (2.18) with $a(x, y, \xi) \in S^m_{\varrho, \delta}(X \times X \times \mathbb{R}^n)$ being properly supported.*

Proof. Let $\chi(x, y) \in C^\infty(X \times X)$, $\chi(x, y) = 1$ in a neighborhood of $\operatorname{supp} K_A$ and let both projections Π_1, $\Pi_2: \operatorname{supp} \chi \to X$ be proper (verify that such a function exists!). Then, substituting in (2.18) $\chi(x, y) a(x, y, \xi)$ for $a(x, y, \xi)$, the kernel K_A and hence the operator are unchanged while $\chi(x, y) a(x, y, \xi)$ is properly supported. $\square$

Note that if $a(x, y, \xi)$ is properly supported then (2.18), viewed as an iterated integral, is defined for $u \in C^\infty(X)$.

Proposition 3.3. *Any ΨDO A can be written in the form $A = A_0 + A_1$ where A_0 is a properly supported ΨDO and A_1 has kernel $K_{A_1} \in C^\infty(X \times X)$.*

Proof. Given A in the form (2.18) with function $a(x, y, \xi)$, we obtain A_0 and A_1 by substituting $a_0(x, y, \xi) = \chi(x, y) a(x, y, \xi)$ and $a_1(x, y, \xi) = (1 - \chi(x, y)) a(x, y, \xi)$ respectively instead of $a(x, y, \xi)$ where $\chi(x, y)$ equals 1 in a neighborhood of the diagonal $\Delta \subset X \times X$ and is such that both projections Π_1, $\Pi_2: \operatorname{supp} \chi \to X$ are proper. $\square$

Proposition 3.4. *Let A be a ΨDO. Then A is properly supported if and only if the following two conditions are fulfilled;*

a) *for any compact set $K \subset X$ we can find a compact set $K_1 \subset X$ such that $\operatorname{supp} u \subset K$ implies $\operatorname{supp}(Au) \subset K_1$;*

b) *the same condition with ^tA instead of A.*

Proof. The necessity of a) and b) is clear from (3.5). So we are left with proving the sufficiency. For instance let us verify that the projection Π_2: $\operatorname{supp} K_A \to X$ is a proper map. Let K be an arbitrary compact set in X. Find a compact set K_1 as in a) and verify that $\Pi_2^{-1}(K) \cap \operatorname{supp} K_A \subset K_1 \times K$. If $(x_0, y_0) \in (X \backslash K_1) \times K$ and if a smooth function $w(x, y) = u(y)v(x)$ is supported in a neighbourhood of this point, then by a) $\langle K_A, w \rangle = 0$. By linearity and continuity this then also holds for an arbitrary function $w \in C_0^\infty(X \times X)$, supported in a neighbourhood of (x_0, y_0) from which the desired inclusion follows. $\square$

3.2 The symbol of a properly supported pseudodifferential operator. We would like to define the symbol of an arbitrary properly supported ΨDO A by analogy with example 2 of §2.

Let us point out that in this example the following holds

$$\sigma_A(x, \xi) = e_{-\xi}(x)\, A\, e_\xi(x), \tag{3.8}$$

where $e_\xi(x) = e^{ix \cdot \xi}$, and that the right hand side of this expression also makes sense for a properly supported ΨDO.

Definition 3.3. Let A be a properly supported ΨDO. Its *symbol* (or *complete symbol*) is the function $\sigma_A(x, \xi)$ on $X \times \mathbb{R}^n$, defined by (3.8).

Since $e_\xi(x)$ is an infinitely differentiable function of ξ with values in $C^\infty(X)$ and A is a continuous linear operator on $C^\infty(X)$, it is clear that $\sigma_A(x, \xi)$ is also an infinitely differentiable function of ξ taking values in $C^\infty(X)$, therefore $\sigma_A(x, \xi) \in C^\infty(X \times \mathbb{R}^n)$.

Writing $u(x) \in C_0^\infty(X)$ as the inverse Fourier transform

$$u(x) = \int e_\xi(x)\, \hat{u}(\xi)\, d\xi$$

where the integral converges in the topology of $C^\infty(X)$, we see that

$$Au(x) = \int e^{ix \cdot \xi} \sigma_A(x, \xi)\, \hat{u}(\xi)\, d\xi \tag{3.9}$$

or

$$Au(x) = \int e^{i(x-y) \cdot \xi} \sigma_A(x, \xi)\, u(y)\, dy\, d\xi \tag{3.10}$$

(where the integral is viewed as an iterated one), which coincides with the corresponding formulas for differential operators.

As is demonstrated by the formulas (3.8) and (3.9), the symbol $\sigma_A(x, \xi)$ defines A and is also defined by A.

Below it will be shown that if $A \in L^m_{\varrho, \delta}(X)$ and $\delta < \varrho$, then we will have $\sigma_A(x, \xi) \in S^m_{\varrho, \delta}(X \times \mathbb{R}^n)$ implying that (3.10) can also be viewed as an oscillatory integral ((3.9) is absolutely convergent).

Remark. If A is an arbitrary ΨDO on X, then its symbol is frequently defined as the symbol $\sigma_{A_1}(x, \xi)$ of a properly supported ΨDO A_1 on X such that $A - A_1 \in L^{-\infty}$. In this case the symbol is not uniquely defined although, as will be seen later, any two such symbols differ by a function $r(x, \xi) \in S^{-\infty}$.

3.3 Asymptotic expansions in $S^m_{\varrho, \delta}$

Definition 3.4. Let $a_j(x, \theta) \in S^{m_j}_{\varrho, \delta}(X \times \mathbb{R}^N)$, $j = 1, 2, \ldots$, $m_j \to -\infty$ as $j \to +\infty$, and let $a(x, \theta) \in C^\infty(X \times \mathbb{R}^N)$. We will write

$$a(x, \theta) \sim \sum_{j=1}^\infty a_j(x, \theta),$$

if for any integer $r \geq 2$ we have

$$a(x, \theta) - \sum_{j=1}^{r-1} a_j(x, \theta) \in S^{\bar{m}_r}_{\varrho, \delta}(X \times \mathbb{R}^N), \tag{3.11}$$

where $\bar{m}_r = \max_{j \geq r} m_j$.

From this it follows, in particular, that $a \in S^{\bar{m}_1}_{\varrho, \delta}(X \times \mathbb{R}^N)$.

Proposition 3.5. *For a given sequence $a_j \in S^{m_j}_{\varrho, \delta}(X \times \mathbb{R}^N)$, $j = 1, 2, \ldots$, with $m_j \to -\infty$ as $j \to +\infty$, we can always find a function $a(x, \theta)$ such that*

$$a \sim \sum_{j=1}^\infty a_j.$$

If, furthermore, another function a' has the same property $a' \sim \sum_{j=1}^\infty a_j$, then $a - a' \in S^{-\infty}(X \times \mathbb{R}^N)$.

Proof. The second assertion is obvious so let us prove the first one. We can assume that $m_1 > m_2 > m_3 > \ldots$. In fact, if this is not the case, we can always achieve this situation by a simple rearrangement and gathering the terms of the same order. Let $X = \bigcup_{j=1}^\infty X_j$, where X_j are open subsets of X and such that $K_j = \overline{X}_j \subset\subset X$ (i.e. K_j is compact in X). Let $\varphi(\theta) \in C^\infty(\mathbb{R}^N)$ and

$$\varphi(\theta) = \begin{cases} 0 & \text{for } |\theta| \leq 1/2 \\ 1 & \text{for } |\theta| \geq 1 . \end{cases}$$

Put

$$a(x, \theta) = \sum_{j=1}^\infty \varphi\left(\frac{\theta}{t_j}\right) a_j(x, \theta), \tag{3.12}$$

where t_j approaches $+\infty$ so quickly as $j \to +\infty$ that

$$\left| \partial_\theta^\alpha \partial_x^\beta \left[\varphi\left(\frac{\theta}{t_j}\right) a_j(x, \theta) \right] \right| \leq 2^{-j} \langle \theta \rangle^{m_{j-1} - \varrho |\alpha| + \delta |\beta|} \tag{3.13}$$

for $x \in K_l$ and $|\alpha| + |\beta| + l \leq j$. Let us show that this is always possible.

First observe that

$$\left| \partial_\theta^\alpha \varphi \left(\frac{\theta}{t} \right) \right| \le C_\alpha \langle \theta \rangle^{-|\alpha|}, \qquad t \ge 1 , \tag{3.14}$$

where C_α does not depend on t, that is for $\varphi(\theta/t)$ and $t \ge 1$ we have uniform in t estimates of class $S^0_{1,0}$. In fact

$$\partial_\theta^\alpha \varphi (\theta/t) = (\partial_\theta^\alpha \varphi) (\theta/t) \cdot t^{-|\alpha|} \quad \text{and} \quad |\theta| \le t \le 2|\theta|$$

for $\theta \in \operatorname{supp} \partial_\theta^\alpha \varphi (\theta/t)$, from which we obtain (3.14).

Further from (3.14) follows that

$$|\partial_\theta^\alpha \partial_x^\beta [\varphi (\theta/t) \, a_j(x, \theta)]| \le C_j \langle \theta \rangle^{m_j - \varrho |\alpha| + \delta |\beta|},$$

if $x \in K_l$, $t \ge 1$ and $|\alpha| + |\beta| + l \le j$. Let us observe now that

$$\langle \theta \rangle^{m_j - \varrho |\alpha| + \delta |\beta|} \le \varepsilon \langle \theta \rangle^{m_{j-1} - \varrho |\alpha| + \delta |\beta|}$$

for $\langle \theta \rangle^{m_{j-1} - m_j} \ge 1/\varepsilon$. Thus, by the choice of t_j we can achieve (3.13) which implies the convergence of (3.12), together with all its derivatives, uniformly on any compact $K \subset X$, and for arbitrary fixed α, β and l we obtain

$$\left| \partial_\theta^\alpha \partial_x^\beta \left[\sum_{j=r+1}^{\infty} \varphi (\theta/t_j) a_j(x, \theta) \right] \right| \le 2^{-r} \langle \theta \rangle^{m_r - \varrho |\alpha| + \delta |\beta|}, \qquad x \in K_l , \, |\alpha| + |\beta| + l \le r .$$

Thus, $a - \sum_{j=1}^{r} a_j \in S^{m_r}_{\varrho, \delta}(X \times \mathbb{R}^N)$ and since $a_r \in S^{m_r}_{\varrho, \delta}(X \times \mathbb{R}^N)$, we obtain from this that $a - \sum_{j=1}^{r-1} a_j \in S^{m_r}_{\varrho, \delta}(X \times \mathbb{R}^N)$, as required. $\quad \square$

The following proposition facilitates the verification of $a \sim \sum_{j=1}^{\infty} a_j$.

Proposition 3.6. *Let* $a_j \in S^{m_j}_{\varrho, \delta}(X \times \mathbb{R}^N)$, $m_j \to -\infty$ *as* $j \to +\infty$ *and let* $a \in C^\infty(X \times \mathbb{R}^N)$ *so that for each compact* $K \subset X$ *and arbitrary multi-indices* α, β *there exist constants* $\mu = \mu(\alpha, \beta, K)$ *and* $C = C(\alpha, \beta, K)$ *with*

$$|\partial_\theta^\alpha \partial_x^\beta a(x, \theta)| \le C \langle \theta \rangle^\mu, \qquad x \in K . \tag{3.15}$$

Furthermore assume that for any compact $K \subset X$, *there exist numbers* $\mu_l = \mu_l(K)$, $l = 1, 2, \ldots,$ *and constants* $C_l = C_l(K)$, *such that* $\mu_l \to -\infty$ *as* $l \to +\infty$ *and the following estimate holds*

$$\left| a(x, \theta) - \sum_{j=1}^{l-1} a_j(x, \theta) \right| \le C_l \langle \theta \rangle^{\mu_l}, \qquad x \in K . \tag{3.16}$$

Then $a \sim \sum\limits_{j=1}^{\infty} a_j$.

The point of this proposition is that instead of (3.11) one has to verify the remainder estimates only for the functions themselves (and not for all the derivatives) provided the fairly weak estimates (3.15) are guaranteed.

The proof is based on the following well-known lemma.

Lemma 3.1. *Let a function $f(t)$ have continuous derivatives $f'(t)$ and $f''(t)$ for $t \in [-1, 1]$. Put $A_j = \sup\limits_{-1 \le t \le 1} |f^{(j)}(t)|, j = 0, 2$. Then*

$$|f'(0)|^2 \le 4A_0(A_0 + A_2). \tag{3.17}$$

Proof. By the intermediate value theorem we have

$$|f'(t) - f'(0)| \le A_2|t|.$$

Therefore $|f'(t)| \ge 1/2|f'(0)|$ for $A_2|t| \le 1/2|f'(0)|$, $|t| \le 1$. Denoting $\Delta = \min\left\{\dfrac{|f'(0)|}{2A_2}, 1\right\}$, we have $|f'(t)| \ge 1/2|f'(0)|$ for $t \in [-\Delta, \Delta]$. We have

$$2A_0 \ge |f(\Delta) - f(-\Delta)| \ge 2\Delta \frac{|f'(0)|}{2},$$

and consequently,

$$|f'(0)| \le \frac{2A_0}{\Delta} = 2A_0 \max\left\{\frac{2A_2}{|f'(0)|}, 1\right\}.$$

This implies that either $|f'(0)| \le \dfrac{4A_0 A_2}{|f'(0)|}$ or $|f'(0)| \le 2A_0$, i.e. either $|f'(0)|^2 \le 4A_0 A_2$ or $|f'(0)|^2 \le 4A_0^2$ and thus (3.17). $\square$

Lemma 3.2. *Let K_1 and K_2 be two compact sets in $\mathbb{R}^p$ so that $K_1 \subset \operatorname{Int} K_2$ (the set of interior points in K_2). Then there exists a constant $C > 0$, such that for any smooth function f on a neighborhood of K_2, the following estimate holds*

$$\left(\sup_{x \in K_1} \sum_{|\alpha|=1} |D^\alpha f(x)|\right)^2$$

$$\le C \sup_{x \in K_2} |f(x)| \left(\sup_{x \in K_2} |f(x)| + \sup_{x \in K_2} \sum_{|\alpha|=2} |D^\alpha f(x)|\right). \tag{3.18}$$

Proof. Immediate from Lemma 3.1. $\square$

Proof of Proposition 3.6. Let $b \sim \sum\limits_{j=1}^{\infty} a_j$ (such a function exists by Proposition 3.5). Putting $d(x, \theta) = a(x, \theta) - b(x, \theta)$, we have for every compact set $K \subset X$ the estimate

$$|\partial_\theta^\alpha \partial_x^\beta d(x, \theta)| \leqq C \langle \theta \rangle^\mu, \qquad x \in K, \tag{3.19}$$

where C and μ depend on α, β, K, and additionally

$$|d(x, \theta)| \leqq C_r \langle \theta \rangle^{-r}, \qquad x \in K, \tag{3.20}$$

where $C_r = C_r(K)$.

Set $d_\theta(x, \xi) = d(x, \theta + \xi)$. Then

$$\partial_\xi^\alpha \partial_x^\beta d_\theta(x, \xi)|_{\xi=0} = \partial_\theta^\alpha \partial_x^\beta d(x, \theta),$$

and applying Lemma 3.2 with $K_1 = K \times 0$, $K_2 = \hat{K} \times \{|\xi| \leqq 1\}$, where $\hat{K}$ is a compact set in X such that Int $\hat{K} \supset K$, we obtain from (3.20)

$$\left(\sup_{x \in K} \sum_{|\alpha|+|\beta| \leqq 1} |\partial_\theta^\alpha \partial_x^\beta d(x, \theta)| \right)^2 \leqq C \langle \theta \rangle^{-r} (\langle \theta \rangle^{-r} + \langle \theta \rangle^\mu).$$

Here r can be choosen arbitrarily, μ depends on α, β, K and C depends on α, β, K and also on r. Thus it follows that for $x \in K$ and $|\alpha| + |\beta| \leq 1$ the function $\partial_\theta^\alpha \partial_x^\beta d(x, \theta) \to 0$ faster than any power of $\langle \theta \rangle$ as $|\theta| \to +\infty$. By induction we obtain the same for arbitrary α, β which gives $d \in S^{-\infty}(X \times \mathbb{R}^n)$ as required. $\square$

3.4 An expression for the symbol of a properly supported ΨDO in terms of $a(x, y, \xi)$. In this and the following subsections we assume that $\delta < \varrho$.

Theorem 3.1. *Let A be a properly supported ΨDO given by (2.18), and $\sigma_A(x, \xi)$ its symbol. Then*

$$\sigma_A(x, \xi) \sim \sum_\alpha \frac{1}{\alpha!} \partial_\xi^\alpha D_y^\alpha a(x, y, \xi)|_{y=x}, \tag{3.21}$$

where the asymptotic sum runs over all multi-indices α.

Remark. Obviously $\partial_\xi^\alpha D_y^\alpha a(x, y, \xi)|_{y=x} \in S_{\varrho, \delta}^{m-(\varrho-\delta)|\alpha|}$ and the asymptotic sum (3.21) is therefore meaningful.

Proof of Theorem 3.1. Observing that by Proposition 3.2 we can assume that $a(x, y, \xi)$ is properly supported, then (3.8), defining the symbol $\sigma_A(x, \xi)$, can be rewritten in the form

$$\sigma_A(x, \xi) = \iint a(x, y, \theta) \, e^{i(x-y) \cdot \theta} \, e^{i(y-x) \cdot \xi} \, dy \, d\theta, \tag{3.22}$$

Here the integral is regarded as an iterated one and the integral makes sense since for each fixed x, the y-integration is over a compact set. In this way, if K is a compact set in X, the oscillatory integral given by (3.22) depends on a parameter $x \in K$. Making the change of variables $z = y - x$, $\eta = \theta - \xi$ to simplify the exponent we obtain

$$\sigma_A(x, \xi) = \iint a(x, x + z, \xi + \eta)\, e^{-iz \cdot \eta}\, dz\, d\eta\,. \tag{3.23}$$

Expanding $a(x, x + z, \xi + \eta)$ in η near $\eta = 0$, using the Taylor formula, we have:

$$a(x, x + z, \xi + \eta) = \sum_{|\alpha| \leq N-1} \partial_\xi^\alpha a(x, x + z, \xi)\, \eta^\alpha / \alpha! + r_N(x, x + z, \xi, \eta)\,, \tag{3.24}$$

where

$$r_N(x, x + z, \xi, \eta) = \sum_{|\alpha| = N} \frac{N\eta^\alpha}{\alpha!} \int_0^1 (1-t)^{N-1}\, \partial_\xi^\alpha a(x, x + z, \xi + t\eta)\, dt\,. \tag{3.25}$$

Now observe that

$$\iint \partial_\xi^\alpha a(x, x + z, \xi)\, \eta^\alpha\, e^{-iz \cdot \eta}\, dz\, d\eta = \partial_\xi^\alpha D_z^\alpha a(x, x + z, \xi)\big|_{z=0} \tag{3.26}$$

by Fourier inversion formula, this gives the finite terms in formula (3.21).

We would now like to use Proposition 3.6. Let us first get a rough estimate for $\sigma_A(x, \xi)$ of the type (3.15). For this we rewrite (3.23) by integrating by parts

$$\sigma_A(x, \xi) = \iint e^{-iz \cdot \eta} \langle D_z \rangle^\nu a(x, x + z, \xi + \eta) \cdot \langle \eta \rangle^{-\nu}\, dz\, d\eta\,, \tag{3.27}$$

where ν is even and non-negative.

Taking into account the inequality $\langle \xi + \eta \rangle \leq 2 \langle \xi \rangle \cdot \langle \eta \rangle$, we obtain from (3.27) that $|\partial_\xi^\alpha \partial_x^\beta \sigma_A(x, \xi)| \leq C \langle \xi \rangle^{p + \delta\nu} \int \langle \eta \rangle^{p - (1-\delta)\nu}\, d\eta$, where $p = \max(m - \varrho|\alpha| + \delta|\beta|, 0)$, $x \in K$ and ν is sufficiently large. This gives the desired estimates for the derivatives of $\sigma_A(x, \xi)$ of the type (3.15). It remains to estimate the remainder term.

Inserting in (3.23) the expression r_N (formula (3.25)) for $a(x, x + z, \xi + \eta)$ and interchanging the orders of integration over t and over z, η, we see that it is necessary to have a uniform in $t \in (0, 1]$ and $x \in K$ estimate of the integral

$$R_{\alpha, t}(x, \xi) = \iint e^{-iz \cdot \eta} \eta^\alpha\, \partial_\xi^\alpha a(x, x + z, \xi + t\eta)\, dz\, d\eta\,,$$

where $|\alpha| = N$. Integrating by parts, we obtain

$$R_{\alpha, t}(x, \xi) = \iint e^{-iz \cdot \eta} \partial_\xi^\alpha D_z^\alpha a(x, x + z, \xi + t\eta)\, dz\, d\eta\,. \tag{3.28}$$

Let us decompose the integral in (3.28) into two parts:

$$R_{\alpha, t} = R'_{\alpha, t} + R''_{\alpha, t}\,, \tag{3.29}$$

where in $R'_{\alpha,t}$ the integration is over the set $\{(z,\eta)\colon |\eta|\leqq|\xi|/2\}$ and in $R''_{\alpha,t}$ it is over the complement to this set. Note that if $|\eta|\leqq|\xi|/2$ then $|\xi|/2\leqq|\xi+t\eta|$ $\leqq\frac{3}{2}|\xi|$ and moreover in $R'_{\alpha,t}$ the volume of the domain of integration for η doesn't exceed $C|\xi|^n$, hence

$$|R'_{\alpha,t}(x,\xi)|\leqq C\langle\xi\rangle^{m-(\varrho-\delta)N+n},\tag{3.30}$$

where C doesn't depend on ξ and t.

Let us next estimate $R''_{\alpha,t}$. Integrating by parts and using the formula

$$\langle\eta\rangle^{-\nu}\langle D_z\rangle^{\nu}e^{-iz\cdot\eta}=e^{-iz\cdot\eta},$$

where ν is an even and non-negative number, we see that $R''_{\alpha,t}$ can be written as a finite sum of terms of the form

$$R_{\alpha,\beta,t}(x,\xi)=\iint\limits_{|\eta|>|\xi|/2}e^{-iz\cdot\eta}\langle\eta\rangle^{-\nu}\partial_\xi^\alpha D_z^{\alpha+\beta}a(x,x+z,\xi+t\eta)\,dz\,d\eta,\tag{3.31}$$

where $|\beta|\leqq\nu$. For $|\eta|\geqq|\xi|/2$ the expression $\partial_\xi^\alpha D_z^{\alpha+\beta}a(x,x+z,\xi+t\eta)$ is estimated in absolute value by $C\langle\eta\rangle^{m-(\varrho-\delta)N+\delta\nu}$ for $m-(\varrho-\delta)N+\delta\nu\geqq0$ and by C for $m-(\varrho-\delta)N+\delta\nu<0$ (in both cases C is independent of ξ, η and t). Taking into account the factor $\langle\eta\rangle^{-\nu}$ we obtain from (3.31) that for sufficiently large ν

$$|R_{\alpha,\beta,t}(x,\xi)|\leqq C\int\limits_{|\eta|>|\xi|/2}\langle\eta\rangle^{p-(1-\delta)\nu}\,d\eta,$$

where $p=\max\{m-(\varrho-\delta)N,0\}$. If $p-(1-\delta)\nu+n+1<0$, it follows that

$$|R_{\alpha,\beta,t}(x,\xi)|\leqq C\langle\xi\rangle^{p-(1-\delta)\nu+n+1}\int\langle\eta\rangle^{-n-1}\,d\eta\leqq C\langle\xi\rangle^{p-(1-\delta)\nu+n+1},\tag{3.32}$$

where C doesn't depend on x, ξ and t (for $x\in K$, $t\in(0,1]$). Selecting a large enough ν we can make the exponent in (3.32) $p-(1-\delta)\nu+n+1$ negative and as large as we like in absolute value.

Taking (3.29) and (3.30) into account, we obtain for $R_{\alpha,t}$ the estimate

$$|R_{\alpha,t}(x,\xi)|\leqq C\langle\xi\rangle^{m-(\varrho-\delta)N+n},\qquad x\in K,\qquad t\in(0,1],$$

which ensures the applicability of Proposition 3.6 and so finishes the proof. $\square$

Remark. The method of proof of Theorem 3.1 is very typical for the theory of ΨDO and the corresponding arguments are to be found in all versions of this theory independently of the mode of presentation. We therefore strongly urge the reader to carefully study the proof of this key theorem.

3.5 The symbol of the transposed operator and the dual symbol. The transposed operator ${}^tA$ is defined by

$$\langle Au, v \rangle = \langle u, {}^tAv \rangle \tag{3.33}$$

for any $u, v \in C_0^\infty(X)$, where

$$\langle u, v \rangle = \int u(x)\, v(x)\, dx \,.$$

Therefore, if $A \in L_{\varrho,\delta}^m(X)$ is given by (2.18), where $a(x, y, \xi) \in S_{\varrho,\delta}^m(X \times \mathbb{R}^n)$, the transpose ${}^tA$ is given by

$$ {}^tAv(y) = \int\!\!\int e^{i(x-y)\cdot\xi}\, a(x, y, \xi)\, v(x)\, dx\, d\xi \,, $$

which with the change of variable $\eta = -\xi$ gives

$$ {}^tAv(y) = \int\!\!\int e^{i(y-x)\cdot\xi}\, a(x, y, -\eta)\, v(x)\, dx\, d\eta \,. \tag{3.34}$$

It is therefore obvious that ${}^tA \in L_{\varrho,\delta}^m(X)$.

Theorem 3.2. *Let A be a properly supported ΨDO with symbol $\sigma_A(x, \xi)$ and $\sigma_A'(x, \xi)$ the symbol of ${}^tA$, then*

$$\sigma_A'(x, \xi) \sim \sum_\alpha \partial_\xi^\alpha D_x^\alpha\, \sigma_A(x, -\xi)/\alpha! \,. \tag{3.35}$$

Proof. Note that ${}^tA$ is also properly supported (cf. Exercise 3.3). Also, instead of $a(x, y, \xi)$ in the formula (2.18), giving the action of A, we can substitute $\sigma_A(x, \xi)$ (cf. (3.10)). Then (3.34) can be written

$$ {}^tAv(x) = \int\!\!\int e^{i(x-y)\cdot\xi}\, \sigma_A(y, -\xi)\, v(y)\, dy\, d\xi \tag{3.36}$$

This is the standard form for a ΨDO (cf. (2.18)) where the role of $a(x, y, \xi)$ is played by $\sigma_A(y, -\xi)$. It remains only to apply Theorem 3.1. $\square$

Exercise 3.4. Let A be a properly supported ΨDO with symbol $\sigma_A(x, \xi)$ and let A^* be the "adjoint" operator, defined by

$$(Au, v) = (u, A^*v), \quad u, v \in C_0^\infty(X)$$

where $(u, v) = \int u(x)\overline{v(x)}\, dx$. Prove that A^* is a properly supported ΨDO whose symbol satisfies

$$\sigma_{A^*}(x, \xi) \sim \sum_\alpha \partial_\xi^\alpha D_x^\alpha\, \overline{\sigma_A(x, \xi)}/\alpha! \tag{3.37}$$

where the bar denotes the complex conjugation.

We now introduce the *dual symbol* by setting

$$\tilde{\sigma}_A(x, \xi) = \sigma_A'(x, -\xi) \tag{3.38}$$

Taking into account that $'('A) = A$, we obtain from (3.36) that A can be expressed via the dual symbol $\tilde{\sigma}_A(x, \xi)$ in the form

$$Au(x) = \iint e^{i(x-y)\cdot\xi}\, \tilde{\sigma}_A(y, \xi)\, u(y)\, dy\, d\xi$$

or

$$(\widehat{Au})(\xi) = \int e^{-iy\cdot\xi}\, \tilde{\sigma}_A(y, \xi)\, u(y)\, dy \tag{3.39}$$

Theorem 3.3. *The dual symbol $\tilde{\sigma}_A(x, \xi)$ is connected with the symbol $\sigma_A(x, \xi)$ via the asymptotic formula*

$$\tilde{\sigma}_A(x, \xi) \sim \sum_{\alpha}(-\partial_\xi)^\alpha\, D_x^\alpha\, \sigma_A(x, \xi)/\alpha! \tag{3.40}$$

Proof. Obvious from (3.38) and (3.35). $\square$

3.6 The composition formula

Theorem 3.4. *Let A and B be two properly supported ΨDO in a domain $X \subset \mathbb{R}^n$ and let their symbols be $\sigma_A(x, \xi)$ and $\sigma_B(x, \xi)$ respectively. The composition $C = B \cdot A$ is then a properly supported ΨDO, whose symbol satisfies the relation*

$$\sigma_{BA}(x, \xi) \sim \sum_{\alpha}\partial_\xi^\alpha\,\sigma_B(x, \xi)\, D_x^\alpha\sigma_A(x, \xi)/\alpha!. \tag{3.41}$$

Proof. Using formula (3.39) for A and applying formula (3.9) to B, we obtain

$$Cu(x) = \iint e^{i(x-y)\cdot\xi}\, \sigma_B(x, \xi)\, \tilde{\sigma}_A(y, \xi)\, u(y)\, dy\, d\xi.$$

It follows that if $A \in L_{\varrho,\delta}^{m_1}(X)$ and $B \in L_{\varrho,\delta}^{m_2}(X)$ then $C \in L_{\varrho,\delta}^{m_1+m_2}(X)$. Analogously we obtain $'C - {}'A \cdot {}'B \in L_{\varrho,\delta}^{m_1+m_2}(X)$. The fact that the ΨDO C is properly supported now follows from Proposition 3.4 and it remains to compute $\sigma_{BA}(x, \xi)$ using Theorems 3.1 and 3.3.

We have

$$\sigma_{BA}(x, \xi) \sim \sum_{\alpha}\partial_\xi^\alpha D_y^\alpha\,[\sigma_B(x, \xi)\, \tilde{\sigma}_A(y, \xi)]/\alpha!\big|_{y=x}$$

$$= \sum_{\alpha}\partial_\xi^\alpha\,[\sigma_B(x, \xi)\, D_x^\alpha\tilde{\sigma}_A(x, \xi)]/\alpha! \tag{3.42}$$

$$\sim \sum_{\alpha,\beta}\partial_\xi^\alpha\,[\sigma_B(x, \xi)\,(-\partial_\xi)^\beta\, D_x^{\alpha+\beta}\sigma_A(x, \xi)]/\alpha!\,\beta!.$$

Next we state two well-known algebraic lemmas.

Lemma 3.3 (Leibniz' rule). *Let $f(x)$ and $g(x)$ be two smooth functions in an open set $X \subset \mathbb{R}^n$ and α a multi-index. Then*

$$\partial^\alpha (f(x)\, g(x)) = \sum_{\gamma + \delta = \alpha} \frac{\alpha!}{\gamma!\,\delta!} \, [\partial^\gamma f(x)]\,[\partial^\delta g(x)]. \tag{3.43}$$

Lemma 3.4 (Newtons binomial formula). *Let $x, y \in \mathbb{R}^n$ and α a multi-index, then*

$$(x+y)^\alpha = \sum_{\gamma + \delta = \alpha} \frac{\alpha!}{\gamma!\,\delta!} \, x^\gamma y^\delta. \tag{3.44}$$

Exercise 3.5. Prove Lemmas 3.3 and 3.4.

Hint. (3.44) can be shown by induction or by using the Taylor formula for polynomials. (3.43) is obtained from (3.44) by noting that

$$\partial^\alpha [f(x)\, g(x)] = \{(\partial_x + \partial_y)^\alpha\, [f(x)\, g(y)]\}\,|_{y=x}\,.$$

Conclusion of the proof of Theorem 3.4. Rewrite (3.42) using Lemma 3.3.

$$\sigma_{BA}(x,\xi) \sim \sum_{\substack{\alpha,\beta,\gamma,\delta \\ \gamma+\delta=\alpha}} [\partial_\xi^\gamma \sigma_B(x,\xi)]\,[(-\partial_\xi)^\beta\,(\partial_\xi^\delta D_x^{\alpha+\beta}\sigma_A(x,\xi))]/\gamma!\,\beta!\,\delta!$$

$$= \sum_{\beta,\gamma,\delta} (-1)^{|\beta|}\,[\partial_\xi^\gamma \sigma_B(x,\xi)]\,[\partial_\xi^{\beta+\delta} D_x^{\beta+\gamma+\delta}\sigma_A(x,\xi)]/\beta!\,\gamma!\,\delta! \tag{3.45}$$

$$= \sum_\gamma \sum_\varkappa \left(\sum_{\beta+\delta=\varkappa} \frac{(-1)^{|\beta|}}{\beta!\,\delta!} \right) [\partial_\xi^\gamma \sigma_B(x,\xi)]\,[\partial_\xi^\varkappa D_x^{\varkappa+\gamma}\sigma_A(x,\xi)]/\gamma!\,.$$

We then obtain from (3.44) with $x = -y = e$, where $e = (1, 1, \ldots, 1)$ and with $\alpha = \varkappa$;

$$\delta_{|\varkappa|}^0 = (e-e)^\varkappa = \sum_{\beta+\delta=\varkappa} \frac{\varkappa!}{\beta!\,\delta!}\, e^\delta (-e)^\beta = \varkappa! \sum_{\beta+\delta=\varkappa} \frac{(-1)^{|\beta|}}{\beta!\,\delta!}\,,$$

here $\delta_{|\varkappa|}^0$ is the Kronecker symbol, equal to 1 for $\varkappa = 0$ and 0 for $|\varkappa| > 0$. Because of this relation, in (3.45) there only remain terms with $\varkappa = 0$, which proves (3.41). $\square$

Corollary 3.1. *Let $A \in L_{\varrho,\delta}^{m_1}(X)$, $B \in L_{\varrho,\delta}^{m_2}(X)$, $0 \leqq \delta < \varrho \leqq 1$, and assume that B is properly supported. Then the operators AB and BA viewed as operators from $C_0^\infty(X)$ to $C^\infty(X)$ belong to $L_{\varrho,\delta}^{m_1+m_2}(X)$.*

Proof. Decompose A into sum $A = A_1 + R$, where A_1 is properly supported and R has a kernel $R(x,y) \in C^\infty(X \times X)$. It is easily verified that the operators

BR and RB have smooth kernels equal to $B_x R(x, y)$ and ${}^t B_y R(x, y)$ respectively, where B_x operates on x keeping y fixed and analogously for ${}^t B_y$. The assertion of the corollary now follows from Theorem 3.4. $\quad\square$

3.7 Classical symbols and pseudodifferential operators. It is sometimes convenient to consider narrower classes of ΨDO. Here we describe one of this classes, closed under the majority of the necessary conditions.

Definition 3.5. By a *classical symbol* we mean a function $a(x, \theta) \in C^\infty$ $(X \times \mathbb{R}^N)$ such that for some complex m these is an asymptotic expansion

$$a(x, \theta) \sim \sum_{j=0}^{\infty} \psi(\theta)\, a_{m-j}(x, \theta),$$

where $\psi \in C^\infty(\mathbb{R}^N)$, $\psi(\theta) = 0$ for $|\theta| \leq 1/2$, $\psi(\theta) = 1$ for $|\theta| \geq 1$, and $a_{m-j}(x, \theta)$ is positive homogeneous of degree $m - j$ in θ, i.e. $a_{m-j}(x, t\theta) = t^{m-j} a_{m-j}(x, \theta)$ for all $t > 0$ and $(x, \theta) \in X \times (\mathbb{R}^N \setminus 0)$. Denote the class of all symbols fulfilling these requirements by $CS^m(X \times \mathbb{R}^N)$. Furthermore, denote by $CL^m(X)$ the class of ΨDO which can be written in the form (2.18) with $a(x, y, \xi) \in CS^m(X \times X \times \mathbb{R}^n)$. These operators will be called *classical* ΨDO.

If $a_k(x, \theta)$ is positive homogeneous of degree k in 0, then $\partial_\theta^\alpha \partial_x^\beta a_k(x, \theta)$ is positive homogeneous of degree $k - |\alpha|$ in θ. Therefore it is clear that $CS^m(X \times \mathbb{R}^N) \subset S^{\operatorname{Re} m}(X \times \mathbb{R}^N)$.

Proposition 3.7.

a) *If $A \in CL^m(X)$ and is properly supported, then $\sigma_A(x, \xi) \in CS^m(X \times \mathbb{R}^n)$.*

b) *If $A \in CL^{m_1}(X)$ and $B \in CL^{m_2}(X)$ and both are properly supported then $BA \in CL^{m_1 + m_2}(X)$.*

c) *If $A \in CL^m(X)$, then ${}^t A \in CL^m(X)$ and $A^* \in CL^m(X)$.*

Proof. Follows immediately from Theorems 3.1–3.4. $\quad\square$

Thus the class of all classical ΨDO is closed under composition, taking the adjoint, and the transpose. In what follows we will show that it is also closed under changing variables, taking the parametrixes (cf. §5) and complex powers of an clliptic opcrator.

3.8 Exercises and problems

Exercise 3.6. Show the following generalization of Leibniz's rule (3.43): if $p(x, \xi) = \sum_{|\alpha| \leq m} a_\alpha(x)\,\xi^\alpha$, and $p(x, D)$ is the corresponding differential operator, then

$$p(x, D)\,(f(x)\,g(x)) = \sum_{\alpha} [p^{(\alpha)}(x, D)\,f(x)]\,[D^\alpha g(x)]/\alpha!, \tag{3.46}$$

where $p^{(\alpha)}(x, \xi) = \partial_\xi^\alpha p(x, \xi)$.

Exercise 3.7. Derive theorem 3.4 for differential operators from the result of Exercise 3.6.

Exercise 3.8. Let $x_1, x_2, \ldots, x_k$ be n-vectors and α an n-dimensional multi-index. Prove that

$$(x_1 + x_2 + \ldots + x_k)^\alpha = \sum_{\alpha_1 + \ldots + \alpha_k = \alpha} \frac{\alpha!}{\alpha_1! \ldots \alpha_k!} \, x_1^{\alpha_1} \ldots x_k^{\alpha_k}. \tag{3.47}$$

Deduce from this that for any smooth functions $f_1, \ldots, f_k$

$$\partial^\alpha [f_1(x) \ldots f_k(x)] = \sum_{\alpha_1 + \ldots + \alpha_k = \alpha} \frac{\alpha!}{\alpha_1! \ldots \alpha_k!} (\partial^{\alpha_1} f_1)(x) \ldots (\partial^{\alpha_k} f_k)(x). \tag{3.48}$$

Exercise 3.9. Given a function $a(x, \xi) \in S^m_{\varrho, \delta}(X \times \mathbb{R}^n)$, X an open set in $\mathbb{R}^n$, show that there exists a properly supported ΨDO A in X, such that $a - \sigma_A \in S^{-\infty}(X \times \mathbb{R}^n)$.

Hint. Consider the operator given by (2.18) with $a(x, y, \xi) = \chi(x, y) a(x, \xi)$, where $\chi(x, y)$ is the same as in the proof of Proposition 3.3.

Exercise 3.10. Derive from Exercises 2.4 and 3.9 that the operation of taking the symbol defines (for $\delta < \varrho$) an isomorphism

$$L^m_{\varrho, \delta}(X)/L^{-\infty}(X) \simeq S^m_{\varrho, \delta}(X \times \mathbb{R}^n)/S^{-\infty}(X \times \mathbb{R}^n).$$

Problem 3.1. Consider the following operator in $\mathbb{R}^n$

$$Au(x) = \iint e^{i(x-y) \cdot \xi} a(x, \xi) u(y) \, dy \, d\xi, \tag{3.49}$$

where $a(x, \xi)$ satisfies

$$|\partial^\alpha_\xi \partial^\beta_x a(x, \xi)| \leq C_{\alpha\beta} \langle \xi \rangle^{m - \varrho |\alpha| + \delta |\beta|}. \tag{3.50}$$

Assume that $\varrho > 0$ and $\delta < 1$. Attach a meaning to the integral (3.49) in the following two situations; a) $u(x) \in S(\mathbb{R}^n)$; b) $u(x) \in C_b^\infty(\mathbb{R}^n)$ i.e. $|\partial^\alpha_x u(x)| \leq C_\alpha$ for an arbitrary multi-index α. Show that A defines a continuous transformation of the spaces $S(\mathbb{R}^n)$ and $C_b^\infty(\mathbb{R}^n)$ into themselves. Show that the symbol $a(x, \xi)$ is uniquely defined by the action of A on $S(\mathbb{R}^n)$ or $C_b^\infty(\mathbb{R}^n)$.

Problem 3.2. Show that the operators of the form described in Problem 3.1 form an algebra with involution and obtain an asymptotic formula for the symbols of the composition of two operators, of the transpose and of the adjoint operator.

Problem 3.3. Let $K(x, z) \in C^\infty(X \times (\mathbb{R}^n \setminus 0))$ be positive homogeneous in z of degree $-n$ and let

$$\int_{|z| = 1} K(x, z) \, dS_z = 0 \tag{3.51}$$

(integral over the sphere $|z| = 1$). Show that for $u \in C_0^\infty(X)$, the following limit exists

$$Au(x) = \lim_{\varepsilon \to 0} \int_{|x-y| \geq \varepsilon} K(x, x-y)\, u(y)\, dy, \qquad (3.52)$$

defining a ΨDO $A \in CL^0(X)$.

The operator A, defined by (3.52) (under condition (3.51)), is called a *singular integral operator*. We see that such operators are just special cases of ΨDO.

Remark. The solution of Problem 3.2 can be found in one of the works of Kumano-go [1]–[3], and the solution of Problem 3.3 can be extracted from the book of Mihlin [1]. The solutions of these problems are rather laborious but very useful for understanding ΨDO theory.

§4. Change of Variables and Pseudodifferential Operators on Manifolds

4.1 The action of change of variables on a ΨDO. Given a diffeomorphism $\varkappa \colon X \to X_1$ from one open set $X \subset \mathbb{R}^n$ onto another open set $X_1 \subset \mathbb{R}^n$, the induced transformation $\varkappa^* \colon C^\infty(X_1) \to C^\infty(X)$, taking a function u to the function $u \circ \varkappa$, is an isomorphism and transforms $C_0^\infty(X_1)$ into $C_0^\infty(X)$. Let A be a ΨDO on X and define $A_1 \colon C_0^\infty(X_1) \to C^\infty(X_1)$ with the help of the commutative diagram

$$
\begin{array}{ccc}
C_0^\infty(X) & \xrightarrow{\ A\ } & C^\infty(X) \\[2mm]
\varkappa^* \uparrow & & \uparrow \varkappa^* \\[2mm]
C_0^\infty(X_1) & \xrightarrow{\ A_1\ } & C^\infty(X_1)
\end{array}
$$

If $\varkappa_1 = \varkappa^{-1}$, then

$$A_1 u = [A(u \circ \varkappa)] \circ \varkappa_1 . \qquad (4.1)$$

Let A be given by (2.18), then

$$A_1 u(x) = \iint e^{i(\varkappa_1(x) - y)\cdot\xi}\, a(\varkappa_1(x), y, \xi)\, u(\varkappa(y))\, dy\, d\xi$$

and, setting $y = \varkappa_1(z)$, we obtain

$$A_1 u(x) = \iint e^{i(\varkappa_1(x) - \varkappa_1(z))\cdot\xi}\, a(\varkappa_1(x), \varkappa_1(z), \xi)\, |\det \varkappa_1'(z)|\, u(z)\, dz\, d\xi, \qquad (4.2)$$

where $\varkappa_1'$ is the Jacobi matrix of the transformations $\varkappa_1$. It follows from this that A_1 is a FIO with phase function $\Phi(x, y, \xi) = (\varkappa_1(x) - \varkappa_1(y)) \cdot \theta$. We will show, that for $1 - \varrho \leq \delta < \varrho$ the operator A_1 is a ΨDO. This fact can be obviously derived from the following more general theorem.

Theorem 4.1. *Let Φ be a phase function in $X \times X \times \mathbb{R}^n$, such that*

1) $\Phi(x, y, \theta)$ is linear in θ;

2) $\Phi'_\theta(x, y, \theta) = 0 \Leftrightarrow x = y$.

Let A_1 be a FIO with phase function $\Phi(x, y, \theta)$ and $a(x, y, \theta) \in S^m_{\varrho, \delta}$ ($X \times X \times \mathbb{R}^n$) (cf. formula (2.1)), where

$$1 - \varrho \leqq \delta < \varrho \tag{4.3}$$

Then $A_1 \in L^m_{\varrho, \delta}(X)$.

For the proof we need

Lemma 4.1. *Let the phase function Φ satisfy conditions 1) and 2) of theorem 4.1. Then there exists a neighbourhood Ω of the diagonal $\Delta \subset X \times X$ and a C^∞-map $\psi : \Omega \to GL(n, \mathbb{R})$ (non-degenerate matrix-function $\psi(x, y)$), such that*

$$\Phi(x, y, \psi(x, y)\xi) = (x - y) \cdot \xi, \qquad (x, y) \in \Omega, \tag{4.4}$$

where

$$\det \psi(x, x) \cdot \det \Phi''_{x\theta}(x, y, \theta)|_{y = x} = 1. \tag{4.5}$$

Proof. We have

$$\Phi(x, y, \theta) = \sum_{i=1}^{n} \Phi_j(x, y)\theta_j, \tag{4.6}$$

where $\Phi_j(x, x) = 0$ and if $\Phi_j(x, y) = 0$, $j = 1, 2, \ldots, n$, then $x = y$. Further

$$\Phi'_{x, \theta} = \left(\sum_{j=1}^{n} \frac{\partial \Phi_j}{\partial x_1} \theta_j, \ldots, \sum_{j=1}^{n} \frac{\partial \Phi_j}{\partial x_n} \theta_j, \Phi_1, \ldots, \Phi_n \right).$$

Note that differentiation of the relation $\Phi(x, x, \theta) = 0$ with respect to x shows that $\Phi'_y|_{x=y} = -\Phi'_x|_{x=y}$. Now, by definition of the phase function $\Phi'_{x, y, \theta} \neq 0$ for $\theta \neq 0$, so in order that $\Phi'_\theta(x, x, \theta) = 0$, it is necessary that $\Phi'_x(x, x, \theta) \neq 0$, i.e. for arbitrary $\theta \neq 0$ there exists k, $1 \leqq k \leqq n$, such that

$$\sum_{j=1}^{n} \frac{\partial \Phi_j}{\partial x_k} \bigg|_{x=y} \cdot \theta_j \neq 0.$$

It follows that

$$\det \left(\frac{\partial \Phi_j(x, y)}{\partial x_k} \right) \bigg|_{x=y} \neq 0. \tag{4.7}$$

By the Hadamard Lemma we have

$$\Phi_j(x, y) = \sum_{k=1}^{n} \Phi_{kj}(x, y)(x_k - y_k),$$

for close x and y, $\Phi_{kj} \in C^\infty(\Omega')$, Ω' some neighbourhood of the diagonal in $X \times X$ and

$$\Phi_{kj}(x, x) = \frac{\partial \Phi_j(x, y)}{\partial x_k}\bigg|_{x=y} \tag{4.8}$$

Denoting by $\Phi(x, y)$ the matrix $(\Phi_{kj}(x, y))^n_{k,j=1}$, we see from (4.7) and (4.8) that there exists a neighbourhood Ω of the diagonal in $X \times X$ such that $\det \Phi(x, y) \neq 0$ for $(x, y) \in \Omega$. Put

$$\psi(x, y) = \Phi(x, y)^{-1}. \tag{4.9}$$

Since

$$\Phi(x, y, \theta) = \sum_{j,k=1}^{n} \Phi_{kj}(x, y)\, \theta_j(x_k - y_k) = (x - y) \cdot (\Phi(x, y)\theta),$$

and putting $\Phi(x, y)\theta = \xi$, we clearly obtain (4.4). The formula (4.5) follows from (4.8) and (4.9). $\quad\square$

Proof. of Theorem 4.1. In view of Proposition 2.1 and Exercise 2.4 we may assume that A is given by (2.1), where $a(x, y, \theta) = 0$ for $(x, y) \notin \Omega'$ with Ω' any neighbourhood of the diagonal. Making the change of variables $\theta = \psi(x, y)\xi$ in the integral (2.1), we obtain

$$A_1 u(x) = \iint e^{i(x-y)\cdot\xi}\, a(x, y, \psi(x, y)\xi)\, |\det \psi(x, y)|\, u(y)\, dy\, d\xi. \tag{4.10}$$

It remains only to remark that for $a_1(x, y, \xi) = a(x, y, \psi(x, y)\xi)$ condition (4.3) guarantees that $a_1(x, y, \xi) \in S^m_{\varrho,\delta}(X \times X \times \mathbb{R}^n)$, in view of Lemma 1.2. $\quad\square$

4.2 Formulae for transformations of symbols

Theorem 4.2. *Given a diffeomorphism* $\varkappa\colon X \to X_1$ *and a properly supported* ΨDO $A \in L^m_{\varrho,\delta}(X)$ *with* $1 - \varrho \leq \delta < \varrho$, *let* A_1 *be determined by (4.1). Then*

$$\sigma_{A_1}(y, \eta)\big|_{y=\varkappa(x)} \sim \sum_\alpha \frac{1}{\alpha!}\, \sigma_A^{(\alpha)}(x, {}^t\varkappa'(x)\eta) \cdot D_z^\alpha e^{i\varkappa_x''(z)\cdot\eta}\big|_{z=x}, \tag{4.11}$$

where $\sigma_A^{(\alpha)}(x, \xi) = \partial_\xi^\alpha \sigma_A(x, \xi)$ *and* $\varkappa_x''(z)$ *is given by*

$$\varkappa_x''(z) = \varkappa(z) - \varkappa(x) - \varkappa'(x)(z - x). \tag{4.12}$$

Proof. Note first of all that the function $\varkappa_x''(z)$ has a zero of second order for $z = x$. Therefore, denoting

$$\Phi_\alpha(x, \eta) = D_z^\alpha e^{i\varkappa_x''(z)\cdot\eta}\big|_{z=x}, \tag{4.13}$$

we have that $\Phi_\alpha(x, \eta)$ is a polynomial in η of degree no higher than $|\alpha|/2$. Taking Lemma 1.2 into consideration, we see that

$$\sigma_A^{(\alpha)}(x, {}^t\varkappa'(x)\eta)\, D_z^\alpha\, e^{i\varkappa_x''(z)\cdot\eta}\big|_{z=x} \in S_{\varrho,\delta}^{m-(\varrho-1/2)|\alpha|}(X\times\mathbb{R}^n)\,.$$

But from the condition $1 - \varrho \le \delta < \varrho$ it follows that $\varrho > 1/2$, so the asymptotic sum (4.11) is well defined.

To prove formula (4.11), we utilize formula (4.2) with $a(x, y, \theta) = \sigma_A(x, \theta)$. Using the transformation described in the proof of Theorem 4.1, we get

$$A_1 u(x) = \iint e^{i(x-y)\cdot\eta}\, \sigma_A(\varkappa_1(x), \psi(x, y)\eta)\, D(x, y)\, u(y)\, dy\, d\eta, \qquad (4.14)$$

where $D(x, y) = |\det \varkappa_1'(x)|\, |\det \psi(x, y)|$. By Theorem 3.1, we have

$$\sigma_{A_1}(x, \eta) \sim \sum_\alpha \partial_\eta^\alpha D_y^\alpha [\sigma_A(\varkappa_1(x), \psi(x, y)\eta)\, D(x, y)]/\alpha!\,\big|_{y=x}\,. \qquad (4.15)$$

From the terms with multi-index α, we obtain (before substituting $x = y$) a sum of terms of the form

$$c(x, y)\, \eta^\gamma \sigma_A^{(\beta)}(\varkappa_1(x), \psi(x, y)\eta)\,, \qquad (4.16)$$

where $c(x, y)$ depends only on the diffeomorphism (but not on A). For the multi-indices γ and β in (4.16) we have the estimates

$$|\beta| \le 2|\alpha|\,, \qquad (4.17)$$

$$|\gamma| + |\alpha| \le |\beta|\,. \qquad (4.18)$$

Here, (4.17) is obvious and (4.18) follows from the fact that applying D_y to expressions of the type (4.16) does not change $|\beta| - |\gamma|$ and ∂_η increases $|\beta| - |\gamma|$ by 1.

From (4.17) and (4.18) we have

$$|\gamma| \le |\beta| - |\alpha| \le |\beta| - |\beta|/2 = |\beta|/2\,. \qquad (4.19)$$

Note now, that applying (4.5) to $\Phi(x, y, \theta) = (\varkappa_1(x) - \varkappa_1(y))\cdot\theta$ gives $\psi(x, x) = ({}^t\varkappa_1'(x))^{-1}$. Also, rearranging in (4.15) the terms of the form (4.16), collecting together all the terms with the same β, we obtain

$$\sigma_{A_1}(x, \eta) \sim \sum_\beta \sigma_A^{(\beta)}(\varkappa_1(x), ({}^t\varkappa_1'(x))^{-1}\eta)\, \Psi_\beta(x, \eta)/\beta!\,, \qquad (4.20)$$

where $\Psi_\beta(x, \eta)$ is a polynomial in η of degree no higher than $|\beta|/2$ (with $C^\infty(X_1)$-coefficients) and independent of A. Here $\Psi_0 \equiv 1$.

Replacing x by $\varkappa(x)$ in (4.20) we easily obtain the equivalent formula

$$\sigma_{A_1}(\varkappa(x),\eta) \sim \sum_\beta \sigma_A^{(\beta)}(x,\,{}^t\varkappa'(x)\eta)\,\Phi_\beta(x,\eta)/\beta!\,,\tag{4.21}$$

where $\Phi_\beta(x,\eta)$ is a polynomial in η of degree no higher than $|\beta|/2$ (with $C^\infty(X)$-coefficients) independent of A and where $\Phi_0 \equiv 1$. It remains to show that these polynomials are given by (4.13).

We will compute the polynomials $\Phi_\beta(x,\eta)$ with the help of differential operators. For the differential operator A we have

$$\begin{aligned}
\sigma_{A_1}(y,\eta)\,|_{y=\varkappa(x)} &= e^{-iy\cdot\eta}\,A_1\,e^{iy\cdot\eta}\,|_{y=\varkappa(x)}\\
&= e^{-i\varkappa(z)\cdot\eta}\,\sigma_A(z,D_z)\,e^{i\varkappa(z)\cdot\eta}\,|_{z=x}
\end{aligned}\tag{4.22}$$

(here $\sigma_A(z,D_z)$ denotes the operator A, acting on the variable z). We write now

$$\varkappa(z) = \varkappa(x) + \varkappa'(x)\,(z-x) + \varkappa''_x(z),$$

from which

$$\varkappa(z)\cdot\eta = \varkappa(x)\cdot\eta + z\cdot{}^t\varkappa'(x)\eta + \varkappa''_x(z)\cdot\eta - x\cdot{}^t\varkappa'(x)\eta.$$

Putting this into formula (4.22), we obtain

$$\sigma_{A_1}(y,\eta)\,|_{y=\varkappa(x)} = e^{-x\cdot{}^t\varkappa'(x)\eta}\,\{\sigma_A(z,D_z)\,[e^{iz\cdot{}^t\varkappa'(x)\eta}\,e^{i\varkappa''_x(z)\cdot\eta}]\}\,|_{z=x}\,.\tag{4.23}$$

Now use the Leibniz rule (3.46) (Exercise 3.6) to differentiate the product of two exponents in (4.23). We then obtain clearly

$$\sigma_{A_1}(\varkappa(x),\eta) = \sum_\alpha \frac{1}{\alpha!}\,\sigma_A^{(\alpha)}(x,\,{}^t\varkappa'(x)\eta)\cdot D_z^\alpha\,e^{i\varkappa''_x(z)\cdot\eta}\,|_{z=x}\tag{4.24}$$

(we have used here yet another obvious formula for differentiating a linear exponent: $\sigma_A(z,D_z)\,e^{iz\cdot\xi} = e^{iz\cdot\xi}\,\sigma_A(z,\xi)$).

Formula (4.24) signifies the validity for differential operators of (4.13) for the polynomials $\Phi_\alpha(x,\eta)$ in (4.21). But in view of the universality of the polynomials $\Phi_\alpha(x,\eta)$, then (4.13) is valid also in the general case.

Examples. $\Phi_0 \equiv 1$, $\Phi_\beta = 0$ for $|\beta| = 1$, $\Phi_\beta(x,\eta) = D_x^\beta\,(i\varkappa(x)\cdot\eta)$ for $|\beta| = 2$.

Corollary 4.1.

$$\sigma_{A_1}(y,\eta) - \sigma_A(\varkappa_1(y),\,({}^t\varkappa'_1(y))^{-1}\eta) \in S_{\varrho,\delta}^{m-2(\varrho-1/2)}(X_1\times\mathbb{R}^n)\,.\tag{4.25}$$

This statement shows, that modulo symbols of lower order, the symbols of all operators obtained from A by a change of variables form a well-defined function on the cotangent bundle T^*X.

Corollary 4.2. *If $A \in CL^m(X)$, then $A_1 \in CL^m(X_1)$.*

Proof. Obvious from formula (4.11). $\square$

4.3 Pseudodifferential operators on a manifold. Let M be a smooth n-dimensional manifold (of class C^∞). We will denote by $C^\infty(M)$ *and* $C_0^\infty(M)$ the space of all smooth complex-valued functions on M and the subspace of all functions with compact support respectively. Assume that we are given a linear operator

$$A\colon C_0^\infty(M) \to C^\infty(M).$$

If X is some chart in M (not necessarily connected) and $\varkappa\colon X \to X_1$ its diffeomorphism onto an open set $X_1 \subset \mathbb{R}^n$, then let A_1 be defined by the diagram

$$
\begin{array}{ccc}
C_0^\infty(X) & \xrightarrow{\ \ A\ \ } & C^\infty(X) \\[2pt]
\Big\uparrow{\scriptstyle \varkappa^*} & & \Big\uparrow{\scriptstyle \varkappa^*} \\[2pt]
C_0^\infty(X_1) & \xrightarrow{\ \ A_1\ \ } & C^\infty(X_1)
\end{array}
$$

(note, in the upper row is the operator $r_X \circ A \circ i_X$, where i_X is the natural embedding $i_X\colon C_0^\infty(X) \to C_0^\infty(M)$ and r_X is the natural restriction $r_X\colon C^\infty(M) \to C^\infty(X)$; for brevity we denote this operator by the same letter A as the original operator).

Definition 4.1. An operator $A\colon C_0^\infty(M) \to C^\infty(M)$ is called a *pseudo-differential operator on* M if for any chart diffeomorphism $\varkappa\colon X \to X_1$ the operator A_1 defined above is a ΨDO on X_1.

Theorem 4.1 shows that the ΨDO on an open set $X \subset \mathbb{R}^n$ for $1-\varrho \leqq \delta < \varrho$ are ΨDO on the manifold X.

Furthermore, from Lemma 1.2, we see that the class of symbols $S_{\varrho,\delta}^m(T^*M)$, as well as the class of operators $L_{\varrho,\delta}^m(M)$, are well-defined for $1 - \varrho \leqq \delta < \varrho$, and Lemma 4.1 shows that the *principal symbol* is well-defined as an element of the quotient space $S_{\varrho,\delta}^m(T^*M)/S_{\varrho,\delta}^{m-2(\varrho-1/2)}(T^*M)$.

Also, in view of Theorem 4.2 the class of classical ΨDO $CL^m(M)$ is well-defined on M. If $A \in CL^m(M)$, then the principal symbol of A can be considered as a homogenous function $\sigma_A(x,\xi)$ on T^*M with degree of homogeneity equal to m, since two functions $a_1(x,\xi)$ and $a_2(x,\xi)$, positively homogeneous in ξ for $|\xi| \geqq 1$ of degree m, which define the same equivalence class in $S^m(X \times \mathbb{R}^n)$ modulo $S^{m-1}(X \times \mathbb{R}^n)$, must coincide for $|\xi| \geqq 1$.

In conclusion note, that it is essential to allow the use of non-connected charts in definition 4.1, since otherwise we would have to consider the reflection $f(x) \to f(-x)$ in $C^\infty(\mathbb{R}\setminus\{0\})$ as a pseudo-differential operator.

Exercise 4.1. Show that a ΨDO A on a manifold M can be extended by continuity to a mapping

$$A:\ \mathscr{E}'(M) \to \mathscr{D}'(M)$$

where $\mathscr{E}'(M)$ and $\mathscr{D}'(M)$ denote the spaces dual to the spaces of smooth sections and smooth sections with compact support respectively of the line bundle of densities $|\Lambda^n(T^*M)|$. (This bundle can be defined for instance, by choosing a covering of M by charts, regarding the bundle as trivial on each chart and setting the transition functions equal to the absolute values of the Jacobians of the coordinate transformations. The sections of the density bundle can be integrated on the manifold, which cannot be said of exterior n-forms, where one needs an orientation, i.e. essentially an isomorphism between $\Lambda^n(T^*M)$ and $|\Lambda^n(T^*M)|$. If we fixed a smooth positive density on M, then this gives us an isomorphism of the bundle $|\Lambda^n(T^*M)|$ and $M \times \mathbb{R}^1$, which allows us to consider functions as densities and therefore to view the elements in $\mathscr{E}'(M)$ and $\mathscr{D}'(M)$ as functionals on functions.)

The inclusion $C^\infty(M) \hookrightarrow \mathscr{D}'(M)$, inducing the inclusion $C_0^\infty(M) \hookrightarrow \mathscr{E}'(M)$, is defined in a natural way by the formula

$$\langle u, \varphi \rangle = \int_M u \cdot \varphi, \tag{4.26}$$

where $u \in C^\infty(M)$ and φ is a smooth density with compact support (so that $u \cdot \varphi$ is also a smooth density with compact support). Verify the property of pseudolocality for the operator A.

Exercise 4.2. Let E and F be smooth vector bundles on the manifold M; let $\pi: T^*M \to M$ be the natural projection; π^*E and π^*F the induced vector bundles over T^*M. Define a ΨDO $A\ :\ C_0^\infty(M, E) \to C^\infty(M, F)$ ($C_0^\infty(M, E)$ the space of smooth compactly supported sections of E and $C^\infty(M, F)$ that of the smooth sections of F) and show that its principal symbol is a well-defined element in the space

$$S_{\rho,\delta}^m(\mathrm{Hom}\,(\pi^*E,\ \pi^*F))/S_{\varrho,\delta}^{m-2(\varrho-1/2)}(\mathrm{Hom}\,(\pi^*E, \pi^*F)).$$

Problem 4.1. Let A be a differential operator of order m on a manifold M (an operator $A:\ C_0^\infty(M) \to C_0^\infty(M)$, such that any operator $A_1:\ C_0^\infty(X_1) \to C_0^\infty(X_1)$, as defined before, is a differential operator of order m). Give an invariant definition of the principal symbol $a_m(x, \xi)$ as a function on T^*M which is a homogeneous polynomial of order m in ξ (i.e. along the fibres).

Hint. Use the formula

$$a_m(x, \varphi_x') = \lim_{\lambda \to +\infty} \lambda^{-m} e^{-i\lambda\varphi} A(e^{i\lambda\varphi}), \qquad \varphi \in C^\infty(M). \tag{4.27}$$

Problem 4.2. Compute the principal symbol of the (de Rham) exterior differentiation operator:

$$d: \Lambda^p(M) \to \Lambda^{p+1}(M),$$

where $\Lambda^k(M)$ denotes the space of smooth exterior k-forms on M ($k = 0, 1, \ldots, n$).

Problem 4.3. Prove that the one-dimensional singular integral operator on a smooth closed curve $\Gamma \subset \mathbb{C}$

$$Au(t) = a(t)\, u(t) + \lim_{\varepsilon \to 0} \int_{|t-\tau| \geq \varepsilon} \frac{K(t, \tau)}{t - \tau}\, u(\tau)\, d\tau, \qquad t, \tau \in \Gamma,$$

for $a(t) \in C^\infty(\Gamma)$, $K(t, \tau) \in C_0^\infty(\Gamma \times \Gamma)$, is a classical ΨDO and belongs to the class $CL^0(\Gamma)$.

§5. Hypoellipticity and Ellipticity

5.1 Definition of hypoelliptic symbols, operators and examples

Definition 5.1. A function $\sigma(x, \xi) \in C^\infty(X \times \mathbb{R}^n)$, where X is an open set in $\mathbb{R}^n$, is called a *hypoelliptic symbol* if the following conditions are fulfilled:

a) there exist real numbers m_0 and m, such that for an arbitrary compact set $K \subset X$ one can find positive constants R, C_1 and C_2 such that

$$C_1 |\xi|^{m_0} \leq |\sigma(x, \xi)| \leq C_2 |\xi|^m, \qquad |\xi| \geq R, \quad x \in K; \tag{5.1}$$

b) there exist numbers ϱ and δ, with $0 \leq \delta < \varrho \leq 1$, and for each compact set $K \subset X$ a constant R such that for any multi-indices α and β

$$|[\partial_\xi^\alpha \partial_x^\beta \sigma(x, \xi)]\, \sigma^{-1}(x, \xi)| \leq C_{\alpha, \beta, K} |\xi|^{-\varrho|\alpha| + \delta|\beta|}, \qquad |\xi| \geq R, \quad x \in K, \tag{5.2}$$

with some constant $C_{\alpha, \beta, K}$.

Denote by $HS_{\varrho, \delta}^{m, m_0}(X \times \mathbb{R}^n)$ the class of symbols satisfying (5.1) and (5.2) for fixed m, m_0, ϱ and δ. Sometimes we will denote this space simply by $HS_{\varrho, \delta}^{m, m_0}$, if the domain X is obvious (or irrelevant). From (5.1) and (5.2) it obviously follows that

$$HS_{\varrho, \delta}^{m, m_0}(X \times \mathbb{R}^n) \subset S_{\varrho, \delta}^m(X \times \mathbb{R}^n).$$

We will denote by $HL_{\varrho, \delta}^{m, m_0}(X)$, X open, the class of properly supported ΨDO A for which $\sigma_A(x, \xi) \in HS_{\varrho, \delta}^{m, m_0}(X \times \mathbb{R}^n)$.

Definition 5.2. A ΨDO A is called *hypoelliptic* if there exists a properly supported ΨDO $A_1 \in HL^{m,m_0}_{\varrho,\delta}(X)$ such that $A = A_1 + R_1$, where $R_1 \in L^{-\infty}(X)$, i.e. R_1 is an operator with infinitely differentiable kernel.

Note, for any representation of the hypoelliptic operator A in the form $A = A_1 + R_1$, where A_1 is a properly supported ΨDO and R_1 is an operator with smooth kernel, it is true that $A_1 \in HL^{m;m_0}_{\varrho,\delta}(X)$.

Example 5.1. Let A be a differential operator, i.e. $A = \sum\limits_{|\alpha| \leq m} a_\alpha(X)D^\alpha$, with $a_\alpha \in C^\infty(X)$. Denote by $a_m(x,\xi)$ the principal symbol

$$a_m(x,\xi) = \sum_{|\alpha|=m} a_\alpha(x)\xi^\alpha. \tag{5.3}$$

Definition 5.3. A differential operator A is called *elliptic*, if

$$a_m(x,\xi) \neq 0 \quad \text{for} \quad (x,\xi) \in X \times (\mathbb{R}^n \setminus 0). \tag{5.4}$$

Proposition 5.1. *The following conditions are equivalent for a differential operator A:*

a) *A is elliptic;*

b) *$A \in HL^{m;m}_{1,0}(X)$.*

Proof. The implication b) $\Rightarrow$ a) is obvious. To show the converse implication, we introduce the complete symbol of the operator A

$$a(x,\xi) = \sum_{|\alpha| \leq m} a_\alpha(x)\xi^\alpha \tag{5.5}$$

and notice that, if a) is fulfilled, then

$$\frac{a(x,\xi)}{a_m(x,\xi)} = 1 + b_{-1}(x,\xi) + \ldots + b_{-m}(x,\xi),$$

where the functions $b_{-j}(x,\xi) \in C^\infty(X \times (\mathbb{R}^n \setminus 0))$ are homogeneous in ξ of degree $-j$. From this (5.1) follows with $m = m_0$ and (5.2) is obtained similarly. $\quad\square$

Examples of elliptic operators:

the Laplace operator $\quad \Delta = \dfrac{\partial^2}{\partial x_1^2} + \ldots + \dfrac{\partial^2}{\partial x_n^2} \quad$ in $\mathbb{R}^n$;

the Cauchy-Riemann operator $\quad \dfrac{\partial}{\partial \bar{z}} = \dfrac{\partial}{\partial x_1} + i\,\dfrac{\partial}{\partial x_2} \quad$ in $\mathbb{R}^2$.

Example 5.2. The heat operator

$$\frac{\partial}{\partial t} - \Delta = \frac{\partial}{\partial t} - \frac{\partial^2}{\partial x_1^2} - \ldots - \frac{\partial^2}{\partial x_n^2}$$

is hypoelliptic in $\mathbb{R}^{n+1}$, although it is not elliptic.

Exercise 5.1. Verify the hypoellipticity of the heat operator and find the corresponding m, m_0, ϱ, δ.

Example 5.3. Let A be a classical ΨDO with principal symbol $a_m(x, \xi)$. Then the following definition makes sense.

Definition 5.3′. An operator $A \in CL^m(X)$ is called *elliptic*, if its principal symbol $a_m(x, \xi)$ satisfies condition (5.4).

As in the proof of Proposition 5.1 it is easy to verify that if $A \in CL^m(X)$ is properly supported then its ellipticity is equivalent to the inclusion $A \in HL_{1,0}^{m,m}(X)$. Generally, $A \in CL^m(X)$ is elliptic if and only if $A = A_1 + R$ where $A_1 \in HL_{1,0}^{m,m}(X)$ and $R \in L^{-\infty}(X)$.

Examples 5.1 and 5.3 motivate the following

Definition 5.3″. An operator $A \in L_{\varrho,\delta}^m(X)$ is called *elliptic* if $A = A_1 + R$ where $A_1 \in HL_{\varrho,\delta}^{m,m}(X)$ and $R \in L^{-\infty}(X)$.

Proposition 5.1′. *For a properly supported $A \in L_{\varrho,\delta}^m(X)$ to be elliptic it is necessary and sufficient that the condition a) in Definition 5.1 is satisfied for its symbol with $m_0 = m$. Generally $A \in L_{\varrho,\delta}^m(X)$ is elliptic if and only if $A = A_1 + R$ where A_1 is properly supported and the condition a) in Definition 5.1 is satisfied for the symbol of A_1 with $m_0 = m$. In this case this is true for any presentation $A = A_1 + R$ as above.*

Proof. The proof is left to the reader as an excercise. $\square$

5.2 Basic properties of hypoelliptic symbols. First of all, say that $\sigma(x, \xi) \in S_{\varrho,\delta}^m$ *for large* ξ, if for any compact set $K \subset X$ there is an $R = R(K)$, such that $\sigma(x, \xi)$ is defined for $x \in K$, $|\xi| \geq R(K)$ and for these (x, ξ) all the necessary estimates of type (1.10) are fulfilled. If, in addition, the estimates (5.1) and (5.2) are fulfilled, we say that $\sigma(x, \xi) \in HS_{\varrho,\delta}^{m,m_0}$ *for large* ξ.

Note that if $\sigma(x, \xi)$ belongs to $S_{\varrho,\delta}^m$ or $HS_{\varrho,\delta}^{m,m_0}$ for large ξ then, multiplying by a smooth cut-off function $\psi(x, \xi)$, equal to 1 "for large ξ" (e.g. for $x \in K$, $|\xi| \geq R(K) + 2$ for any compact set K) and equal to 0 in a neighbourhood of the set where the symbol σ is not defined (e.g. for $x \in K$, $|\xi| \leq R(K) + 1$), we obtain a symbol $\sigma_1(x, \xi) \in S_{\varrho,\delta}^m(X \times \mathbb{R}^n)$ or $HS_{\varrho,\delta}^{m,m_0}(X \times \mathbb{R}^n)$ respectively, which coincides with $\sigma(x, \xi)$ "for large ξ".

Lemma 5.1. *If $\sigma(x, \xi) \in HS_{\varrho,\delta}^{m,m_0}$ for large ξ, then $\sigma^{-1}(x, \xi) \in HS_{\varrho,\delta}^{-m_0, -m}$ for large ξ. Furthermore, for arbitrary multi-indices α, β we have for large ξ, that*

$$\partial_\xi^\alpha \partial_x^\beta \sigma(x, \xi) / \sigma(x, \xi) \in S_{\varrho,\delta}^{-\varrho|\alpha| + \delta|\beta|}$$

Proof. Let $y = (x, \xi)$ and γ, δ be $2n$-dimensional multi-indices. By induction in $|\delta|$ one verifies that

$$\partial^{\delta}\left[\frac{\partial^{\gamma}\sigma(y)}{\sigma(y)}\right] = \sum_{k=0}^{|\delta|}\ \sum_{\delta_0 + \ldots + \delta_k = \delta} c_{\delta_0 \ldots \delta_k} \frac{\partial^{\delta_0 + \gamma}\sigma(y)}{\sigma(y)} \prod_{l=1}^{k} \frac{\partial^{\delta_l}\sigma(y)}{\sigma(y)}. \qquad (5.6)$$

Obviously, setting $\gamma = (\beta, \alpha)$ we obtain from the definitions all the necessary estimates for the proof of the lemma. $\square$

Lemma 5.2. *If* $\sigma' \in HS_{\varrho,\delta}^{m';m'_0}$ *and* $\sigma'' \in HS_{\varrho,\delta}^{m'';m''_0}$ *then* $\sigma' \circ \sigma'' \in HS_{\varrho,\delta}^{m'+m'',m'_0+m''_0}$.

Proof. Direct from the Leibniz rule. $\square$

Lemma 5.3. *If* $\sigma(x,\xi) \in HS_{\varrho,\delta}^{m,m_0}$ *and* $r(x,\xi) \in S_{\varrho,\delta}^{m_1}$, *where* $m_1 < m_0$, *then* $\sigma + r \in HS_{\varrho,\delta}^{m,m_0}$.

Proof. Writting $\sigma + r = \sigma(1 + r/\sigma)$ and using Lemmas 5.1 and 5.2, we see that it suffices to consider the case $\sigma \equiv 1$ and $m_0 = m = 0$, i.e. $m_1 < 0$. But then the assertion of the lemma is trivial. $\square$

Lemma 5.4. *Let* $\sigma(x,\xi) \in HS_{\varrho,\delta}^{m,m_0}$ *for large* ξ *and let* $\sigma_1(y,\eta) = \sigma(x(y), \xi(y,\eta))$, *where the map* $(y,\eta) \to (x(y), \xi(x,\eta))$ *is a* C^∞ *map from* $X_1 \times (\mathbb{R}^n \setminus 0)$ *into* $X \times (\mathbb{R}^n \setminus 0)$ *and where* $\xi(y,\eta)$ *is positive homogeneous of degree 1 in* η. *Assume that* $1 - \varrho \leqq \delta < \varrho$. *Then* $\sigma_1(y,\eta) \in HS_{\varrho,\delta}^{m,m_0}$ *for large* η.

Proof. Completely analogous to the proof of Lemma 1.2 and is left to the reader. $\square$

5.3 Basic properties of hypoelliptic operators

Proposition 5.2.   *If* $A' \in HL_{\varrho,\delta}^{m';m'_0}(X)$ *and* $A'' \in HL_{\varrho,\delta}^{m'';m''_0}(X)$,   *then* $A' \circ A'' \in HL_{\varrho,\delta}^{m'+m'', m'_0+m''_0}(X)$.

Proof. By theorem 3.4

$$\sigma_{A'A''}(x,\xi) \sim \sigma_{A'}(x,\xi)\,\sigma_{A''}(x,\xi)\left[1 + \sum_{|\alpha| \geqq 1} \frac{\partial_\xi^\alpha \sigma_{A'}}{\sigma_{A'}} \frac{D_x^\alpha \sigma_{A''}}{\sigma_{A''}}\right].$$

and from Lemmas 5.1 and 5.3, we see that the series in square brackets is an asymptotic sum which (in the sense of Proposition 3.4) belongs to $HS_{\varrho,\delta}^{0,0}$. It remains to use Lemma 5.2. $\square$

Proposition 5.3. *If* $A \in HL_{\varrho,\delta}^{m;m_0}(X)$, *then both* ${}^tA$ *and* A^* *belong to* $HL_{\varrho,\delta}^{m;m_0}(X)$.

Proof. Similar to the proof of the preceeding proposition. $\square$

Proposition 5.4. *If* $A \in HL_{\varrho,\delta}^{m;m_0}(X)$ *and* $R \in L_{\varrho,\delta}^{m_1}(X)$ *with* $m_1 < m_0$, *where* R *is properly supported, then* $A + R \in HL_{\varrho,\delta}^{m,m_0}(X)$.

Proof. The statement follows immediately from Lemma 5.3. $\square$

Proposition 5.5. *If* $1 - \varrho \leqq \delta < \varrho$, *then* $HL_{\varrho,\delta}^{m,m_0}(X)$ *is invariant with respect to changes of variables i.e. if we are given a diffeomorphism* $\varkappa : X \to X_1$ *and an operator* A_1 *is defined as in §4 (formula (4.1)), then* $A_1 \in HL^{m,m_0}(X_1)$.

Proof. By Lemma 5.4

$$\sigma_A(\varkappa_1(y),\, ({}^t\varkappa_1'(y))^{-1}\eta)\in HS^{m,\,m_0}_{\varrho,\,\delta}(X)$$

(here $\varkappa_1 = \varkappa^{-1}$). But then, by Theorem 4.2 and Lemma 5.1, it is obvious that

$$\sigma_{A_1}(y,\eta) = \sigma_A(\varkappa_1(y),\, ({}^t\varkappa_1'(y))^{-1}\eta)\,(1+r(y,\eta)),$$

where $r(y,\eta)\in S^{-2(\varrho-1/2)}_{\varrho,\,\delta}$ for large η. The assertion of the proposition now follows from Lemmas 5.2 and 5.3. $\square$

Proposition 5.5 allows us to define the class $HL^{m,\,m_0}_{\varrho,\,\delta}(M)$ of hypo-elliptic ΨDO to an arbitrary manifold M provided $1 - \varrho \leqq \delta < \varrho$.

5.4 The parametrix and the rough regularity theorem

Theorem 5.1. *Let $A \in HL^{m,\,m_0}_{\varrho,\,\delta}(M)$, with either $1 - \varrho \leqq \delta < \varrho$ or $\delta < \varrho$ and M a domain in $\mathbb{R}^n$. Then there exists an operator $B\in HL^{-m_0,\,-m}_{\varrho,\,\delta}(M)$, such that*

$$BA = I + R_1, \qquad AB = I + R_2, \tag{5.7}$$

where $R_j\in L^{-\infty}(M)$, $j=1,2$, and I is the identity operator. If, furthermore, B' is another ΨDO for which either $B'A = I + R_1'$ or $AB' = I + R_2'$ (where $R_j'\in L^{-\infty}(M)$), then $B' - B\in L^{-\infty}(M)$.

Corollary 5.1. *If A is a hypoelliptic ΨDO on M (not necessarily properly supported), then there exists a properly supported ΨDO B, such that (5.7) holds.*

Proof of Theorem 5.1. First let M be a domain X in $\mathbb{R}^n$ and σ_A the symbol of A. Consider a function $b_0(x,\xi)\in HS^{-m_0,\,-m}_{\varrho,\,\delta}(X\times\mathbb{R}^n)$ such that $b_0(x,\xi) = \sigma_A^{-1}(x,\xi)$ for large ξ. Next choose a properly supported operator $B_0\in HL^{-m_0,\,-m}_{\varrho,\,\delta}(X)$ such that $\sigma_{B_0} - b_0\in S^{-\infty}(X\times\mathbb{R}^n)$. Let us verify that

$$B_0A = I + R_0, \tag{5.8}$$

with $R_0\in L^{-(\varrho-\delta)}_{\varrho,\,\delta}(X)$.

In fact, by Theorem 3.4 we have for large ξ

$$\sigma_{B_0A}(x,\xi) \sim 1 + \sum_{|\alpha|\geqq 1}\frac{1}{\alpha!}\,\partial_\xi^\alpha\sigma_A^{-1}D_x^\alpha\sigma_A = 1 + \sum_{|\alpha|\geqq 1}\frac{1}{\alpha!}\,\frac{\partial_\xi^\alpha\sigma_A^{-1}}{\sigma_A^{-1}}\,\frac{D_x^\alpha\sigma_A}{\sigma_A}$$

and it remains to use Lemma 5.1. Now let C_0 be a properly supported ΨDO such that

$$C_0 \sim \sum_{j=0}^{\infty}(-1)^j R_0^j, \tag{5.9}$$

i.e.

$$\sigma_{C_0} \sim \sum_{j=0}^{\infty} (-1)^j \sigma_{R_0^j}.$$ (5.9′)

From (5.9) we clearly have

$$C_0 (I + R_0) - I \in L^{-\infty},$$

so that putting $B_1 = C_0 B_0$, we have

$$B_1 A = I + R_1$$ (5.10)

where $R_1 \in L^{-\infty}(X)$. From the construction it is clear that $B_1 \in HL_{\varrho,\delta}^{-m_0, -m}(X)$. Further, we can similarly construct an operator $B_2 \in HL_{\varrho,\delta}^{-m_0, -m}(X)$, such that

$$A B_2 = I + R_2$$ (5.11)

where $R_2 \in L^{-\infty}(X)$.

Let us now verify, that if B_1 and B_2 are two arbitrary ΨDO, for which (5.10) and (5.11) hold, then $B_1 - B_2 \in L^{-\infty}(X)$. This will then also demonstrate the existence of the required B (for which we may take either of the operators B_1 and B_2) and its uniqueness (modulo $L^{-\infty}(X)$). Note, that B_1 and B_2 can be taken to be properly supported. Multiplying (5.10) on the right by B_2 and using (5.11), we obtain $B_1 - B_2 = R_1 B_2 - B_1 R_2$ and it only remians to note that $B_1 R_2$ and $R_1 B_2$ both belong to $L^{-\infty}(X)$.

Now let $M = \bigcup_\gamma X^\gamma$ be an arbitrary manifold with a covering by charts X^γ. Then, (by the results just shown) there is a properly supported operator B^γ in X^γ, such that

$$B^\gamma \cdot A = I + R_1^\gamma, \qquad A \cdot B^\gamma = I + R_2^\gamma,$$

where R_1^γ and R_2^γ are operators with smooth kernels.

Now let φ_j, $j = 1, 2, \ldots$, be a partition of unity subordinate to the covering of M by the X^γ. This means that the following conditions are fulfilled:

1) $\varphi_j \in C_0^\infty(M)$, $\varphi_j \geq 0$, supp $\varphi_j \subset X^\gamma$ for some $\gamma = \gamma(j)$;

2) for any $x \in M$, there exists a neighbourhood $\mathcal{U}_x$ of x in M, such that $\mathcal{U}_x$ intersects only a finite number of sets supp φ_j;

3) $\sum_j \varphi_j = 1$.

(See e.g. Theorem 6.20 in Rudin [1].) Now let us construct functions $\psi_j \in C_0^\infty(M)$ such that they still satisfy the conditions 1) and 2) above (with the same $\gamma(j)$) and in addition $\psi_j = 1$ in a neighbourhood of supp φ_j.

Denote by Φ_j and Ψ_j the multiplication operators by φ_j and ψ_j respectively. We set then

$$B = \sum_j \Phi_j B^{\gamma(j)} \Psi_j,$$

where necessary operations of restriction and extension by zero are understood. We claim that then B satisfies all the required conditions.

Clearly B is properly supported. Note also that on the intersection $X^\gamma \cap X^{\gamma'}$ the operators B^γ and $B^{\gamma'}$ differ by an operator with a smooth kernel, so modulo operators from $L^{-\infty}$ they may be given by the same symbol. This allows us to calculate the compositions BA and AB modulo $L^{-\infty}$ using the composition formula (3.41). For example, the symbol of BA will locally have the form

$$\sigma_{BA}(x,\xi) \sim \sum_{j,\alpha} \frac{1}{\alpha!} \Phi_j(x)(\partial_\xi^\alpha \sigma_B(x,\xi)) D_x^\alpha(\sigma_A(x,\xi)\Psi_j(x)).$$

If we apply a derivative in x to Ψ_j, then the resulting term will vanish because $\Phi_j D_x^\alpha \Psi_j \equiv 0$ for any $\alpha \neq 0$. Therefore we conclude that

$$\sigma_{BA}(x,\xi) \sim \left(\sum_j \Phi_j(x)\Psi_j(x)\right)\left(\sum_\alpha \frac{1}{\alpha!}(\partial_\xi^\alpha \sigma_B(x,\xi))(D_x^\alpha(\sigma_A(x,\xi))\right) \sim 1,$$

which is equivalent to the first relation in (5.7). The proof of the second relation is not different. $\square$

Remark. The formula for the parametrix B above can be simplified if we add small neighbourhoods of $\mathrm{supp}\,\varphi_j$ to the set of all X^γ. In this case we can simply write

$$B = \sum_\gamma \Phi_\gamma B^\gamma \Psi_\gamma,$$

where φ_γ satisfy the same properties as φ_j above and it is understood that some of the functions φ_γ can be identically 0.

If M is a closed manifold then we can also assume the covering $\{X^\gamma\}$ to be finite, so the formula for B will be a finite sum.

Definition 5.4. An operator B satisfying the condition (5.7) is called a *parametrix* of the operator A.

Corollary 5.2. *Any elliptic operator* $A \in L^m_{\varrho,\delta}(M)$ *has a parametrix* $B \in HL^{-m,\,-m}_{\varrho,\delta}(M)$.

Theorem 5.2. *If A is a hypoelliptic* ΨDO, *then*

$$\mathrm{sing\ supp}\,Au = \mathrm{sing\ supp}\,u, \qquad u \in \mathscr{E}'(M). \tag{5.12}$$

In other words, if Ω is an open submanifold of M, then $Au\,|_\Omega \in C^\infty(\Omega)$ if and only if $u\,|_\Omega \in C^\infty(\Omega)$.

If A is a properly supported hypoelliptic ΨDO, *then (5.12) is true for an arbitrary distribution $u \in \mathscr{D}'(M)$.*

Proof. It obviously suffices to prove the first part of the theorem. For this, the inclusion sing supp $Au \subset$ sing supp u follows from the pseudolocality of A (cf. Proposition 2.4) and it only remains to show the inclusion

$$\text{sing supp } u \subset \text{sing supp } Au \tag{5.13}$$

Let B be a properly supported parametrix of A. Then, from the formula $u = B(Au) - R_1 u$ and the pseudolocality of B, it follows that

$$\text{sing supp } u \subset \text{sing supp } (Au) \cup \text{sing supp } R_1 u,$$

and since $R_1 u \in C^\infty(M)$, we have that sing supp $R_1 u = \emptyset$ proving (5.13). $\square$

Theorem 5.2 is a rough regularity theorem for solutions of hypoelliptic equations of the form $Au = f$. More precise theorems will be proved after we have introduced exact regularity classes of functions, i.e. the scale of Sobolev spaces.

5.5 A parametrix for classical elliptic pseudo-differential operators. In this case a parametrix can be constructed in a much more explicit way.

Let A be a classical ΨDO in an open set $X \subset \mathbb{R}^n$, whose symbol for large ξ admits the asymptotic expansion

$$a(x, \xi) \sim \sum_{j=0}^{\infty} a_{m-j}(x, \xi), \tag{5.14}$$

where $a_{m-j}(x, \xi) \in C^\infty(X \times (\mathbb{R}^n \setminus 0))$, a_{m-j} is positive homogeneous of degree $m - j$ in ξ and also satisfies the ellipticity condition (5.4).

Let B be a parametrix of A. We will show that B is a classical ΨDO, whose symbol $b(x, \xi)$ for large ξ admits an asymptotic expansion

$$b(x, \xi) \sim \sum_{j=0}^{\infty} b_{-m-j}(x, \xi), \tag{5.15}$$

where $b_{-m-j}(x, \xi) \in C^\infty(X \times (\mathbb{R}^n \setminus 0))$ and $b_{-m-j}(x, \xi)$ is positive homogeneous of degree $-m-j$ in ξ.

The composition formula shows that the symbol $b(x, \xi)$ must satisfy the condition

$$\sum_{\alpha} \partial^\alpha a(x, \xi) D_x^\alpha b(x, \xi)/\alpha! \sim 1,$$

or

$$\sum_{\alpha, k, j} \partial_\xi^\alpha a_{m-k}(x, \xi) D_x^\alpha b_{-m-j}(x, \xi)/\alpha! \sim 1. \tag{5.16}$$

Clearly, regrouping the terms in (5.16) by their homogeneity degree we make (5.16) into just an equality of the corresponding homogeneous components, i.e.

$$\sum_{k+j+|\alpha|=p} \partial_\xi^\alpha a_{m-k}(x,\xi)\, D_x^\alpha b_{-m-j}(x,\xi)/\alpha! = \delta_0^p, \qquad p=0,1,\ldots, \qquad (5.17)$$

or, more explicitly

$$a_m \cdot b_{-m} = 1, \qquad (5.17')$$

$$a_m \cdot b_{-m-j} + \sum_{\substack{k+l+|\alpha|=j \\ l<j}} (\partial_\xi^\alpha a_{m-k})\,(D_x^\alpha b_{-m-l})/\alpha! = 0, \qquad j=1,2,\ldots \quad (5.17'')$$

Clearly, from (5.17) the functions $b_{-m-j}(x,\xi)$, positive homogeneous of degrees $-m-j$ $(j=0,1,\ldots)$, are uniquely defined. If we now define $b(x,\xi)$ by (5.15) and find a properly supported ΨDO B such that $\sigma_B(x,\xi)-b(x,\xi) \in S^{-\infty}(X \times \mathbb{R}^n)$, then this operator B is a parametrix of A.

Formula (5.17) defines a parametrix of A also in the case when A is a matrix PDO: in this case $a_{-m-j}(x,\xi)$ are square matrix functions and the ellipticity condition takes the form

$$\det a_m(x,\xi) \neq 0, \qquad (x,\xi) \in (X \times (\mathbb{R}^n \setminus 0) \qquad (5.18)$$

Problem 5.1. Show that the terms $b_{-m-j}(x,\xi)$ $(j>0)$ in the parametrix of the classical elliptic operator A in the scalar case can be expressed via $a_{m-k}(x,\xi)$ by

$$b_{-m-j}(x,\xi) = \sum_{l=2}^{2j+1} c_l(x,\xi)\,(a_m(x,\xi))^{-l}, \qquad (5.19)$$

where $c_l(x,\xi)$ is a function positive homogenous of degree $m(l-1)-j$ in ξ, polynomial in the functions $a_m, a_{m-1}, \ldots, a_{m-j}$ and their derivatives of order $\leqq j$.

The analogous formula in the matrix case is of the form

$$b_{-m-j}(x,\xi) = a_m^{-1}(x,\xi) \sum_{l=2}^{2j+1} \prod_{k=1}^{l} [c_{k,l}(x,\xi)\,a_m^{-1}(x,\xi)]. \qquad (5.20)$$

§6. Theorems on Boundedness and Compactness of Pseudodifferential Operators

6.1 Formulation of the basic boundedness theorem. Let A be a ΨDO in $\mathbb{R}^n$. Consider A as a map

$$A\colon C_0^\infty(\mathbb{R}^n) \to C^\infty(\mathbb{R}^n).$$

Let K_A be the kernel of A in the sense of L. Schwartz. If supp K_A is compact in $\mathbb{R}^n \times \mathbb{R}^n$, then A defines a map

$$A: C_0^\infty(\mathbb{R}^n) \to C_0^\infty(\mathbb{R}^n).$$

Is it possible to extend the operator A to a continuous linear operator

$$A: L^2(\mathbb{R}^n) \to L^2(\mathbb{R}^n)?$$

Clearly, this is so if and only if the following estimate holds

$$\|Au\| \leqq C\|u\|, \qquad u \in C_0^\infty(\mathbb{R}^n), \tag{6.1}$$

where $C > 0$ does not depend on u and $\|\cdot\|$ denotes the norm in $L^2(\mathbb{R}^n)$

Theorem 6.1. *Let $A \in L_{\varrho,\delta}^0(\mathbb{R}^n)$, $0 \leqq \delta < \varrho \leqq 1$, and let* supp K_A *be compact in $\mathbb{R}^n \times \mathbb{R}^n$. Then (6.1) holds and A can be extended to a linear continuous operator on $L^2(\mathbb{R}^n)$.*

6.2 Auxiliary results and proof of Theorem 6.1. In the sequel we will use the notation

$$\varlimsup_{\substack{|\xi|\to\infty \\ x\in K}} |a(x,\xi)| = \lim_{t\to\infty}\ \sup_{\substack{|\xi|\geqq t \\ x\in K}} |a(x,\xi)|.$$

Theorem 6.2. *Let A be a properly supported ΨDO in $L_{\varrho,\delta}^0(X)$, with $0 \leqq \delta < \varrho \leqq 1$ and X an open set in $\mathbb{R}^n$. Suppose there exists a constant M such that*

$$\varlimsup_{\substack{|\xi|\to\infty \\ x\in K}} |\sigma_A(x,\xi)| < M \tag{6.2}$$

for any compact set $K \subset X$. Then there exists a properly supported integral operator R with hermitean kernel $R \in C^\infty(X \times X)$, such that

$$(Au, Au) \leqq M^2(u, u) + (Ru, u), \qquad u \in C_0^\infty(X). \tag{6.3}$$

If, in addition, supp K_A *is compact in $X \times X$, then* supp R *is also compact in $X \times X$.*

Proof that Theorem 6.2 $\Rightarrow$ Theorem 6.1. It suffices to show the boundedness in $L^2(\mathbb{R}^n)$ of an operator R with smooth compactly supported kernel. This, however is well known (one can show for instance that $\|R\|^2 \leqq \int |K_R(x, y)|^2 \, dx\, dy$, where $\|R\|$ is the operator norm of R in $L^2(\mathbb{R}^n)$ and $K_R(x, y)$ is the kernel of the operator R). $\square$

To prove Theorem 6.2 it suffices, in view of the relation $(Au, Au) = (A^*Au, u)$, to construct a properly supported operator $B \in L_{\varrho,\delta}^0$ such that

$$A^*A + B^*B - M^2 = R \tag{6.4}$$

where R has a smooth kernel (in which case R is properly supported since the left hand side of (6.4) is a properly supported operator). Rewriting (6.4) in the form $B^*B = M^2 - A^*A + R$, we note that the symbol of $M^2 - A^*A$ is equal to $M^2 - |\sigma_A(x,\xi)|^2$ modulo symbols of class $S_{\varrho,\delta}^{-(\varrho-\delta)}(X)$, from which we infer

$$\lim_{\substack{|\xi|\to\infty \\ x\in K}} \operatorname{Re}\left[\sigma_{(M^2-A^*A)}(x,\xi)\right] > 0 \tag{6.5}$$

for an arbitrary compact set $K \subset X$. Therefore we derive Theorem 6.2 from the following proposition.

Proposition 6.1. *Let* $C \in L_{\varrho,\delta}^0(X)$ *and be properly supported,* $0 \leqq \delta < \varrho \leqq 1$ *and let* $C^* = C$ *and assume*

$$\lim_{\substack{|\xi|\to\infty \\ x\in K}} \operatorname{Re}\sigma_C(x,\xi) > 0 \tag{6.6}$$

for arbitrary compact sets $K \subset X$. *Then there exists a properly supported operator* $B \in L_{\varrho,\delta}^0(X)$ *such that* $R = B^*B - C$ *has a* C^∞ *kernel.*

Lemma 6.1. *Let* $a(x,\xi) \in S_{\varrho,\delta}^0(X \times \mathbb{R}^n)$ *and let* $a(x,\xi)$, *for arbitrary* $(x,\xi) \in X \times \mathbb{R}^n$, *take values in a compact set* $K \subset \mathbb{C}$. *Let a complex-valued function* $f(z)$ *be defined on a neighbourhood of* K *and be infinitely differentiable as a function of two real variables* $\operatorname{Re} z$ *and* $\operatorname{Im} z$. *Then*

$$f(a(x,\xi)) \in S_{\varrho,\delta}^0(X \times \mathbb{R}^n) \tag{6.7}$$

Proof. Denote $u = \operatorname{Re}z$ and $v = \operatorname{Im}z$. Then we evidently have

$$\partial_y^\gamma f(a(y)) = \sum_{\substack{\varkappa_1+\ldots+\varkappa_p+ \\ +\omega_1+\ldots+\omega_q=\gamma}} c_{p,q,\varkappa_1,\ldots,\varkappa_p,\omega_1,\ldots,\omega_q}(\partial_{u,v}^{(p,q)}f)(a(y)) \times$$
$$\times \partial_y^{\varkappa_1}(\operatorname{Re}a)\ldots\partial_y^{\varkappa_p}(\operatorname{Re}a)\,\partial_y^{\omega_1}(\operatorname{Im}a)\ldots\partial_y^{\omega_q}(\operatorname{Im}a), \tag{6.8}$$

from which (6.7) follows, since $|(\partial_{u,v}^{(p,q)}f)(a(y))| \leqq C_{pq}$. $\quad\square$

Proof of Proposition 6.1. It follows from Lemma 6.1 that $\sqrt{\operatorname{Re}\sigma_C(x,\xi)}$ belongs to $S_{\varrho,\delta}^0$ for large ξ. Therefore there exists a properly supported ΨDO $B_0 \in L_{\varrho,\delta}^0(X)$, such that if $b_0(x,\xi)$ is its symbol then

$$|b_0(x,\xi)|^2 - \operatorname{Re}\sigma_C(x,\xi) \in S_{\varrho,\delta}^{-\infty}\,.$$

From this it follows that

$$C - B_0^*B_0 \in L_{\varrho,\delta}^{-(\varrho-\delta)}(X)\,. \tag{6.9}$$

The operator B_0 serves as the "zero order approximation" to B. We will seek a first order approximation in the form $B_0 + B_1$, where $B_1 \in L_{\varrho,\delta}^{-(\varrho-\delta)}(X)$.

We have

$$C - (B_0^* + B_1^*)\,(B_0 + B_1) = (C - B_0^* B_0) - (B_0^* B_1 + B_1^* B_0) - B_1^* B_1\,. \tag{6.10}$$

The point is to reduce the order of the operator on the left-hand side, taking B_1 to be properly supported with symbol $b_1(x, \xi)$, such that for large ξ

$$2b_1(x, \xi)\,b_0(x, \xi) = \sigma_{C - B_0^* B_0}(x, \xi)\,, \tag{6.11}$$

which is obviously possible, since by Lemma 6.1 $b_0^{-1}(x, \xi) \in S_{\varrho, \delta}^0$ for large ξ. It follows from (6.10) and (6.11), that

$$C - (B_0 + B_1)^*\,(B_0 + B_1) \in L_{\varrho, \delta}^{-2(\varrho - \delta)}(X)\,. \tag{6.12}$$

Arguing by induction, we may in exactly the same way construct properly supported ΨDO $B_j \in L_{\varrho, \delta}^{-j(\varrho - \delta)}(X)$, $j = 0, 1, 2, \ldots$, such that

$$C - (B_0 + \ldots + B_j)^*\,(B_0 + \ldots + B_j) \in L_{\varrho, \delta}^{-j(\varrho - \delta)}(X)\,. \tag{6.13}$$

Now let $b_j(x, \xi)$ be the symbol of B_j. It only remains to construct a properly supported operator B, such that

$$\sigma_B(x, \xi) \sim \sum_{j=0}^{\infty} b_j(x, \xi)\,.$$

It follows easily from (6.13), that this operator will be the one we are looking for. Thus Proposition 6.1 is proved and together with it Theorems 6.1 and 6.2. $\square$

6.3 The compactness theorem. We will derive the compactness theorem from the following much more general statement.

Theorem 6.3. *Let $A \in L_{\varrho, \delta}^0(\mathbb{R}^n)$, $0 \leq \delta < \varrho \leq 1$, let the kernel K_A have compact support in $\mathbb{R}^n \times \mathbb{R}^n$ and let the symbol $\sigma_A(x, \xi)$ satisfy*

$$\varlimsup_{|\xi| \to \infty} |\sigma_A(x, \xi)| < M\,. \tag{6.14}$$

Then there exists an operator A_1 such that $A - A_1 \in L^{-\infty}(\mathbb{R}^n)$, the kernel K_{A_1} has compact support and

$$\|A_1 u\| \leq M \|u\|\,, \qquad u \in C_0^\infty(\mathbb{R}^n)\,. \tag{6.15}$$

Proof. Let $\chi \in C_0^\infty(\mathbb{R}^n)$ be such that $\chi(x) \geq 0$, $\int \chi(x)\,dx = 1$, $0 \leq \hat{\chi}(\xi) \leq 1$. Such a function can be found. Indeed, to begin with let the function $\chi_0(x)$ be such that $\chi_0(x) \in C_0^\infty(\mathbb{R}^n)$, $\chi_0(x) \geq 0$ and $\int \chi_0(x)\,dx = 1$. Then obviously $|\hat{\chi}_0(\xi)| \leq 1$. Put now $\chi(x) = \int \chi_0(x + y)\,\chi_0(y)\,dy$. In view of the fact that $\hat{\chi}(\xi) = |\hat{\chi}_0(\xi)|^2$ the function $\chi(x)$ fulfills all the requirements.

Now put $\chi_\varepsilon(x) = \varepsilon^{-n} \chi(x/\varepsilon)$ and define the operator A_ε by

$$A_\varepsilon u = Au - A(\chi_\varepsilon * u), \tag{6.16}$$

where $(\chi_\varepsilon * u)(x)$, the convolution of χ_ε and u, is defined by

$$(\chi_\varepsilon * u)(x) = \int \chi_\varepsilon(x-y)\, u(y)\, dy = \int u(x-y)\, \chi_\varepsilon(y)\, dy.$$

Now, in view of Theorem 6.2

$$\|A_\varepsilon u\|^2 \leq M^2 \|u - \chi_\varepsilon * u\|^2 + (R(u - \chi_\varepsilon * u),\, u - \chi_\varepsilon * u), \tag{6.17}$$

where R is an operator with kernel $R(x, y) \in C_0^\infty(\mathbb{R}^n \times \mathbb{R}^n)$.

Note that the Fourier transform of $u - \chi_\varepsilon * u$ is $(1 - \hat{\chi}(\varepsilon\xi))\,\hat{u}(\xi)$ and from the condition $0 \leq \hat{\chi} \leq 1$, it follows that

$$\|u - \chi_\varepsilon * u\| \leq \|u\|. \tag{6.18}$$

Further, denote by R_ε the operator which maps u into $R(u - \chi_\varepsilon * u)$, then its kernel is given by the formula

$$R_\varepsilon(x, y) = R(x, y) - \int R(x, z)\, \varepsilon^{-n} \chi\left(\frac{z-y}{\varepsilon}\right) dz,$$

or

$$R_\varepsilon(x, y) = R(x, y) - \int R(x, y + \varepsilon z)\, \chi(z)\, dz,$$

from which it is obvious, that $\operatorname{supp} R_\varepsilon(x, y)$ lies in some fixed compact set K (independent of ε for $0 < \varepsilon \leq 1$) and, in addition,

$$\sup_{x, y} |R_\varepsilon(x, y)| \to 0 \quad \text{for } \varepsilon \to 0.$$

It follows that $\|R_\varepsilon\| \to 0$ for $\varepsilon \to 0$. We now obtain from (6.17) and (6.18) that

$$\|A_\varepsilon u\|^2 \leq M^2 \|u\|^2 + \|R_\varepsilon u\|\, \|u\|. \tag{6.19}$$

From the conditions of the theorem it is evident that we may replace M by $M - \delta$, where δ is sufficiently small. But then it follows from (6.19), that for sufficiently small $\varepsilon > 0$

$$\|A_\varepsilon u\|^2 \leq M^2 \|u\|^2.$$

Put $A_1 = A_\varepsilon$. Since the symbol of the convolution operator with χ_ε is $\hat{\chi}(\varepsilon\xi) \in S^{-\infty}(\mathbb{R}^n \times \mathbb{R}^n)$, it is evident that $A - A_1 \in L^{-\infty}(\mathbb{R}^n)$. It is also easily verified, that the kernel K_{A_1} of A_1 has compact support. $\square$

Theorem 6.4. *Let* $A \in L^0_{\varrho,\delta}(\mathbb{R}^n)$, $0 \leqq \delta < \varrho \leqq 1$, *let the kernel* K_A *have compact support and*

$$\sup_x |\sigma_A(x,\xi)| \to 0 \quad \text{as} \quad |\xi| \to +\infty. \tag{6.20}$$

Then A *extends to a compact operator in* $L^2(\mathbb{R}^n)$.

Proof. By Theorem 6.3 there exists a decomposition $A = A_\varepsilon + R_\varepsilon$ for arbitrary $\varepsilon > 0$, where $\|A_\varepsilon\| < \varepsilon$ and R_ε has a smooth compactly supported kernel (and is thus compact). Therefore $\lim_{\varepsilon \to 0} \|A - R_\varepsilon\| = 0$ and the compactness of A follows. $\square$

Corollary 6.1. *Let* $A \in L^m_{\varrho,\delta}(\mathbb{R}^n)$, $0 \leqq \delta < \varrho \leqq 1$, $m < 0$ *and* K_A *have compact support. Then* A *extends to a compact operator on* $L^2(\mathbb{R}^n)$.

6.4 The case of operators on a manifold. Consider first the case of a closed manifold M (a compact manifold without boundary). Using a partition of unity on M, it is easy to introduce a measure, having a smooth positive density with respect to the Lebesgue measure in any local coordinates. If $d\mu$ is any such measure, the Hilbert space $L^2(M, d\mu)$ is defined. Note, that the elements and the topology in $L^2(M, d\mu)$ do not depend on the choice of $d\mu$. It is therefore meaningful to talk about the space $L^2(M)$ as a topological vector space in which the topology can be defined using some non-uniquely defined Hilbert scalar product. Theorems 6.1 and 6.4 obviously imply

Theorem 6.5. *Let* M *be a closed manifold,* $A \in L^0_{\varrho,\delta}(M)$, $1 - \varrho \leqq \delta < \varrho$. *Then*

1) operator A *extends to a linear continuous operator*

$$A: \ L^2(M) \to L^2(M);$$

2) if the principal symbol $\sigma_A(x,\xi) \in S^0_{\varrho,\delta}(T^*M)/S^{-2\varrho+1}_{\varrho,\delta}(T^*M)$ *satisfies condition (6.20) (it is the same for all representatives of an equivalence class in* $S^0_{\varrho,\delta}(T^*M)/S^{-2\varrho+1}_{\varrho,\delta}(T^*M))$, *then the operator so obtained is a compact operator* $L^2(M) \to L^2(M)$.

Corollary 6.2. *If* $A \in L^m_{\varrho,\delta}(M)$, $1 - \varrho \leqq \delta < \varrho$ *and* $m < 0$, *then* A *extends to a compact operator*

$$A: \ L^2(M) \to L^2(M).$$

We now formulate a version of the boundedness theorem, adequate for non-compact M. For this we introduce the spaces $L^2_{loc}(M)$ and $L^2_{comp}(M)$.

Let f be a complex-valued function on M, defined everywhere except, perhaps on a set of measure 0.

In the sequel, we consider functions f_1 and f_2 as equivalent if they coincide outside some set of measure 0. Indeed, the elements of the spaces $L^2_{comp}(M)$ and

$L^2_{\text{loc}}(M)$ are not functions but equivalence classes, although we will write $f \in L^2_{\text{loc}}(M)$ for a function f, by abuse of language.

We will write that $f \in L^2_{\text{loc}}(M)$ if for an arbitrary diffeomorphism $\varkappa: X \to X_1$ of an open set $X \subset \mathbb{R}^n$ into an open set $X_1 \subset M$ and an arbitrary open subset $X_0 \subset X$, such that $\overline{X}_0$ is compact in X, we have $[\varkappa^*(f|_{X_1})]|_{X_0} \in L^2(X_0, dx)$, where dx is the Lebesgue measure on X_0 induced by the Lebesgue measure on $\mathbb{R}^n$. The topology of $L^2_{\text{loc}}(M)$ is given by a family of seminorms

$$\|f\|_{\varkappa, X_0} = \|[\varkappa^*(f|_{X_1})]|_{X_0}\|_{L^2(X_0, dx)} \,.$$

If M has a countable basis, then $L^2_{\text{loc}}(M)$ is a Fréchet space (a complete metrizable and locally convex space or, what is the same thing, a complete countably normed space).

Further, we will denote by $L^2_{\text{comp}}(M)$ the linear subset of $L^2_{\text{loc}}(M)$, consisting of those elements $f \in L^2_{\text{loc}}(M)$ for which $\operatorname{supp} f$ is compact in M. Given $\varkappa: X \to X_1$ and $X_0 \subset X$ as described above, define the inclusion

$$i_{\varkappa, X_0}: L^2(X_0, dx) \to L^2_{\text{comp}}(M),$$

mapping a function $f^0 \in L^2(X_0, dx)$ into the function $\hat{f}(y)$ on M, equal to $f(x)$ at the point $\varkappa(x)$ of M and to 0 at $y \in M \setminus \varkappa(X_0)$. The topology of $L^2_{\text{comp}}(M)$ is defined as the inductive topology, i.e. the strongest locally convex topology for which all the inclusions $i_{\varkappa, X_0}$ are continuous. From this it follows that the linear operator $A: L^2_{\text{comp}}(M) \to E$, E any locally convex space, is continuous if and only if all the compositions $A \circ i_{\varkappa, X_0}$ are continuous. This circumstance being taken into account, we clearly get the following

Theorem 6.6. *If $A \in L^0_{\varrho, \delta}(M)$, where $1 - \varrho \leq \delta < \varrho$, then A extends to a linear continuous operator*

$$A: \ L^2_{\text{comp}}(M) \to L^2_{\text{loc}}(M)\,.$$

Exercise 6.1. Prove that if $A \in L^0_{\varrho, \delta}(M)$, $1 - \varrho \leq \delta < \varrho$, and if A is properly supported, then it extends to a linear continuous operator

$$A: \ L^2_{\text{comp}}(M) \to L^2_{\text{comp}}(M)$$

and also to a linear continuous operator

$$A: \ L^2_{\text{loc}}(M) \to L^2_{\text{loc}}(M)\,.$$

§7. The Sobolev Spaces

7.1 Definition of the Sobolev spaces

Lemma 7.1. *Let M be an arbitrary manifold. Then for any real s there exists on M a properly supported, classical elliptic ΨDO Λ_s of order s with positive principal symbol (for $\xi \neq 0$).*

Proof. To begin with let $M = X$, a domain in $\mathbb{R}^n$. Then we may take as Λ_s an arbitrary properly supported ΨDO $\Lambda_s \in CL^s(X)$ with principal symbol $|\xi|^s$. Next let M be arbitrary and let there be given a covering of M by charts $M = \bigcup_\gamma X_1^\gamma$.

We will denote by $\varkappa^\gamma$ any coordinate diffeomorphism $\varkappa^\gamma \colon X^\gamma \to X_1^\gamma$, where X^γ is an open set in $\mathbb{R}^n$. We construct on X^γ operators Λ_s^γ, having the required properties and transport these to X_1^γ by the standard procedure (cf. §4) using the diffeomorphisms $\varkappa^\gamma$, producing a properly supported, classical ΨDO $\Lambda_{s,1}^\gamma$ on X_1^γ with positive principal symbol. The operator Λ_s on M can be glued together from the operators $\Lambda_{s,1}^\gamma$ by the process used to construct B from B^γ in the proof of theorem 5.1. $\square$

Definition 7.1. We write $u \in H_{\mathrm{loc}}^s(M)$, if $u \in \mathscr{D}'(M)$ and $\Lambda_s u \in L_{\mathrm{loc}}^2(M)$. Further set $H_{\mathrm{comp}}^s(M) = H_{\mathrm{loc}}^s(M) \cap \mathscr{E}'(M)$. (Concerning $\mathscr{D}'(M)$ and $\mathscr{E}'(M)$, see Exercise 4.1.) If M is a closed manifold, we denote $H_{\mathrm{loc}}^s(M) = H_{\mathrm{comp}}^s(M)$ simply by $H^s(M)$.

If K is a compact in M, we denote by $H^s(K)$ the set of all $u \in H_{\mathrm{comp}}^s(M)$ for which $\operatorname{supp} u \subset K$.

There is a well-defined topology in the spaces $H_{\mathrm{loc}}^s(M)$, $H_{\mathrm{comp}}^s(M)$ and $H^s(K)$, but for the time being only the set of elements in these spaces is essential to us.

Below we will show that these spaces do not depend on the choice of the operator Λ_s.

7.2 The action of ΨDO on Sobolev spaces. The precise regularity theorem.

Theorem 7.1. *If $A \in L_{\varrho,\delta}^m(M)$, $1 - \varrho \leqq \delta < \varrho$, or $\delta < \varrho$ and $M = X$, an open set in $\mathbb{R}^n$, then A defines a map $H_{\mathrm{comp}}^s(M) \to H_{\mathrm{loc}}^{s-m}(M)$. If A, in addition, is properly supported, then A defines maps*

$$A \colon H_{\mathrm{comp}}^s(M) \to H_{\mathrm{comp}}^{s-m}(M),$$

$$A \colon H_{\mathrm{loc}}^s(M) \to H_{\mathrm{loc}}^{s-m}(M).$$

Proof. Without loss of generality, we may assume that the operator Λ_{-s} is a parametrix of Λ_s for arbitrary $s \in \mathbb{R}$, i.e.

$$\Lambda_{-s} \circ \Lambda_s = I + R_s, \tag{7.1}$$

where R_s is a properly supported operator with smooth kernel and therefore transforming $\mathscr{E}'(M)$ into $C_0^\infty(M)$ and $\mathscr{D}'(M)$ into $C^\infty(M)$.

If $u \in H_{\mathrm{comp}}^s(M)$, then, setting $\Lambda_s u = u_0$, we will obtain from (7.1), that $u = \Lambda_{-s} u_0 + v$, where $u_0 \in L_{\mathrm{comp}}^2(M)$, $v \in C_0^\infty(M)$. Therefore

$$\Lambda_{s-m} A u = \Lambda_{s-m} A (\Lambda_{-s} u_0 + v) = \Lambda_{s-m} A \Lambda_{-s} u_0 + \Lambda^s A v,$$

and in view of the fact that $\Lambda_{s-m} A \Lambda_{-s} \in L^0_{\varrho,\delta}(M)$, we obtain from Theorem 6.6, that $\Lambda_{s-m} A \Lambda_{-s} u_0 \in L^2_{\mathrm{loc}}(M)$, $\Lambda_s A v \in C^\infty(M)$, from which $\Lambda_{s-m} A u \in L^2_{\mathrm{loc}}(M)$, i.e. $Au \in H^{s-m}_{\mathrm{loc}}(M)$. Thereby we have proved the first assertion of Theorem 7.1. The remaining ones follow from this one or are shown similarly. $\square$

We are now in a position to give a definition of Sobolev spaces, not depending on the choice of Λ_s. It clearly suffices to define $H^s_{\mathrm{loc}}(M)$.

Definition 7.1′. We will write $u \in H^s_{\mathrm{loc}}(M)$, if $u \in \mathscr{D}'(M)$ and $Au \in L^2_{\mathrm{loc}}(M)$ for any properly supported $A \in L^s_{1,0}(M)$.

The equivalence of Definitions 7.1 and 7.1′ follows in an obvious manner from Theorem 7.1.

Theorem 7.2. *Let $A \in HL^{m,m_0}_{\varrho,\delta}(M)$, where $1 - \varrho \leqq \delta < \varrho$, or $\delta < \varrho$ and $M = X$, an open set in $\mathbb{R}^n$. Then, if $u \in \mathscr{D}'(M)$ and $Au \in H^s_{\mathrm{loc}}(M)$ we have $u \in H^{s+m_0}_{\mathrm{loc}}(M)$.*

Proof. Let B be a parametrix for A, $B \in HL^{-m_0,-m}_{\varrho,\delta}(M)$. Then, by Theorem 7.1 we have $BAu \in H^{s+m_0}_{\mathrm{loc}}(M)$. But $BAu = u + v$, where $v \in C^\infty(M)$ so $u \in H^{s+m_0}_{\mathrm{loc}}(M)$. $\square$

Corollary 7.1. a) *If A is elliptic of order m, $u \in \mathscr{E}'(M)$ and $Au \in H^s_{\mathrm{loc}}(M)$, then $u \in H^{s+m}_{\mathrm{comp}}(M)$.*

b) *If A is properly supported (in particular if A is a differential operator) and elliptic of order m, $u \in \mathscr{D}'(M)$ and $Au \in H^s_{\mathrm{loc}}(M)$, then $u \in H^{s+m}_{\mathrm{loc}}(M)$.*

7.3 Localization. Theorem 7.1 clearly implies that if $u \in H^s_{\mathrm{loc}}(M)$ and $a(x) \in C^\infty(M)$, then $au \in H^s_{\mathrm{loc}}(M)$.

If, in particular, $a(x) \in C^\infty_0(M)$, then $au \in H^s_{\mathrm{comp}}(M)$. The following is a precise version of the converse.

Proposition 7.1. *Let the distribution $u \in \mathscr{D}'(M)$ be such that for any point $x_0 \in M$ one can find a function $\varphi_{x_0} \in C^\infty_0(M)$, such that $\varphi_{x_0}(x_0) \neq 0$ and $\varphi_{x_0} u \in H^s_{\mathrm{comp}}(M)$. Then $u \in H^s_{\mathrm{loc}}(M)$.*

Proof. We may select from the functions φ_{x_0} a system of functions $\{\varphi_\gamma\}$, such that some neighbourhoods of $\operatorname{supp} \varphi_\gamma$ form a locally finite covering of M and for any point $x_0 \in M$ one can find γ such that $\varphi_\gamma(x_0) \neq 0$. Put now

$$\psi_\gamma = \frac{\bar{\varphi}_\gamma \cdot \varphi_\gamma}{\displaystyle\sum_\gamma |\varphi_\gamma|^2}.$$

Then obviously ψ_γ has the same properties as φ_γ and, in addition, $\psi_\gamma \geqq 0$ and $\sum_\gamma \psi_\gamma \equiv 1$. We now have

$$\Lambda_s u = \sum_\gamma \Lambda_s(\psi_\gamma u) \in L^2_{\mathrm{loc}}(M),$$

because, by the fact that Λ_s is properly supported, the sum $\sum_\gamma \Lambda_s(\psi_\gamma u)$ is locally finite. $\square$

Corollary 7.2. *Let $M = \bigcup_{\gamma} X_{\gamma}$ be an open covering of M, $u \in \mathscr{D}'(M)$. Then the condition $u \in H^s_{\mathrm{loc}}(M)$ is equivalent to $u|_{X_{\gamma}} \in H^s_{\mathrm{loc}}(X_{\gamma})$ for any γ.*

Proposition 7.1 shows that it is essentially sufficient to study $H^s(K)$ for K compact in $\mathbb{R}^n$.

7.4 The space $H^s(\mathbb{R}^n)$.

Definition 7.2. Let $s \in \mathbb{R}$, $u \in S'(\mathbb{R}^n)$. We will write $u \in H^s(\mathbb{R}^n)$, if $\hat{u}(\xi) \in L^2_{\mathrm{loc}}(\mathbb{R}^n)$ and

$$\|u\|_s^2 = \int |\hat{u}(\xi)|^2 \langle \xi \rangle^{2s} \, d\xi < +\infty \tag{7.2}$$

(this also serves as a definition of the norm $\| \cdot \|_s$).

Exercise. Show the completeness of $H^s(\mathbb{R}^n)$ (with the norm $\| \cdot \|_s$).
The following Hilbert scalar product can be introduced in $H^s(\mathbb{R}^n)$

$$(u, v)_s = \int \hat{u}(\xi) \, \overline{\hat{v}(\xi)} \, \langle \xi \rangle^{2s} \, d\xi, \tag{7.3}$$

and the map $\langle D \rangle^s$, mapping $u \in S'(\mathbb{R}^n)$ into $F^{-1} \langle \xi \rangle^s \hat{u}(\xi)$ (F is the Fourier transformation), provides an isometric isomorphism

$$\langle D \rangle^s \colon H^s(\mathbb{R}^n) \to L^2(\mathbb{R}^n). \tag{7.4}$$

Lemma 7.2. *Let K be compact in $\mathbb{R}^n$. Then*

$$H^s(K) = \mathscr{E}'(K) \cap H^s(\mathbb{R}^n) \tag{7.5}$$

$\mathscr{E}'(K)$ denotes the set of all $u \in \mathscr{E}'(\mathbb{R}^n)$, such that $\mathrm{supp}\, u \subset K$).

Proof. 1. Since $\langle D \rangle^s \in L^s_{1,0}(\mathbb{R}^n)$ and is elliptic, then by Corollary 7.1 it is clear that $u \in \mathscr{E}'(K)$ and $\langle D \rangle^s u \in L^2(\mathbb{R}^n)$ implies $u \in H^s(K)$, i.e. $\mathscr{E}'(K) \cap H^s(\mathbb{R}^n) \subset H^s(K)$.

2. Now let $u \subset H^s(K)$. We must verify that $\langle D \rangle^s u \subset L^2(\mathbb{R}^n)$ if it is known that $\langle D \rangle^s u \in L^2_{\mathrm{loc}}(\mathbb{R}^n)$. Indeed, a stronger result is valid: if $\varphi(x) \in C_0^{\infty}(\mathbb{R}^n)$, $\varphi = 1$ in a neighbourhood of K, then $(1 - \varphi) \langle D \rangle^s u \in S(\mathbb{R}^n)$. Let us prove this.

Let $\psi \in C_0^{\infty}(\mathbb{R}^n)$ be such that $\varphi = 1$ in a neighbourhood of $\mathrm{supp}\, \psi$ (in particular $(1 - \varphi)\psi \equiv 0$) and moreover $\psi = 1$ in a neighbourhood of K. Then $\psi u = u$ and $(1 - \varphi)\langle D \rangle^s u = (1 - \varphi)\langle D \rangle^s \psi u$. Let us study the operator $(1 - \varphi)\langle D \rangle^s \psi$. Clearly its kernel $K(x, y)$ belongs to $C^{\infty}(\mathbb{R}^n \times \mathbb{R}^n)$. It suffices to verify that $K(x, y) \in S(\mathbb{R}^n \times \mathbb{R}^n)$. But $K(x, y)$ is given by an oscillatory integral:

$$K(x, y) = \int e^{i(x - y) \cdot \xi} (1 - \varphi(x)) \langle \xi \rangle^s \psi(y) \, d\xi,$$

which can be rewritten as

$$K(x, y) = \int e^{i(x-y)\cdot\xi} |x - y|^{-2N} (1 - \varphi(x))\, \psi(y)\, (-\Delta_\xi)^N \langle\xi\rangle^s d\xi,$$

since $|x - y| \geq \varepsilon > 0$ for $x \in \operatorname{supp}(1 - \varphi)$, $y \in \operatorname{supp}\psi$. For $2N > s + n$ we obtain a convergent integral, which can be estimated by $C(1 + |x| + |y|)^{-2N}$. The derivatives of the kernel $K(x, y)$ are estimated in a similar way. $\square$

Noting that $H^s(K)$ is a closed subspace of $H^s(\mathbb{R}^n)$, we see that the scalar product $(u, v)_s$ induces a Hilbert space structure on $H^s(K)$.

Lemma 7.3. *Let $\hat{K}$ be a compact set in $\mathbb{R}^n$ such that $K \subset \operatorname{Int} \hat{K}$. Then for $u \in H^s(K)$, there exists a sequence $u_n \in C_0^\infty(\mathbb{R}^n)$, $\operatorname{supp} u_n \subset \hat{K}$, such that $\|u_n - u\|_s \to 0$ for $n \to +\infty$.*

Proof. Let $\varphi(x) \in C_0^\infty(\mathbb{R}^n)$, $\int \varphi(x)\, dx = 1$ and $\varphi_\varepsilon(x) = \varepsilon^{-n}\varphi(x/\varepsilon)$. We put

$$u_\varepsilon(x) = (\varphi_\varepsilon * u)(x) = \langle u(y), \varphi_\varepsilon(x - y)\rangle, \qquad \varepsilon > 0.$$

It is clear that $u_\varepsilon(x) \in C_0^\infty(\mathbb{R}^n)$ so let us prove that $\lim_{\varepsilon \to 0+} \|u_\varepsilon - u\|_s = 0$. Since $\hat{\varphi}_\varepsilon(\xi) = \hat{\varphi}(\varepsilon\xi)$ and $\hat{\varphi}(0) = 1$, the question reduces to establishing the relation

$$\lim_{\varepsilon \to +0} \int |\hat{\varphi}(\varepsilon\xi) - 1|^2 |\hat{u}(\xi)|^2 \langle\xi\rangle^{2s} d\xi = 0,$$

which is evident from the dominated convergence theorem. $\square$

7.5 Topology in the Sobolev spaces on a manifold. Let $M = \bigcup_\gamma X_1^\gamma$ be a locally finite covering of a manifold M by relatively compact coordinate neighbourhoods X_1^γ, $\varkappa^\gamma : X^\gamma \to X_1^\gamma$ the coordinate diffeomorphisms (X^γ an open set in $\mathbb{R}^n$) and φ^γ a partition of unity on M subordinate to $\{X_1^\gamma\}$. Let K be a compact set in M. Introduce a scalar product in $H^s(K)$, setting

$$(u, v)_s = \sum_\gamma ((\varkappa^\gamma)^*(\varphi^\gamma u), (\varkappa^\gamma)^*(\varphi^\gamma v))_s, \qquad u, v \in H^s(K). \tag{7.6}$$

Proposition 7.2. *The scalar product (7.6) induces a Hilbert space structure on $H^s(K)$.*

Proof. Clearly we only need to verify the completeness of $H^s(K)$ in the norm $\|\cdot\|_s$, defined by the scalar product (7.6). Let the sequence $\{u_m\}_{m=1,2,\ldots}$ of elements in $H^s(K)$ be a Cauchy sequence with respect to $\|\cdot\|_s$. Then

$$(\varkappa^\gamma)^*\varphi^\gamma u_m \to v_1^\gamma \in H^s(\varkappa_1^\gamma(\operatorname{supp}\varphi^\gamma)) \tag{7.7}$$

in the $H^s(\mathbb{R}^n)$ norm (here $\varkappa_1^\gamma = (\varkappa^\gamma)^{-1}$). We may take $v^\gamma = (\varkappa_1^\gamma)^* v_1^\gamma$, thus $v_1^\gamma = (\varkappa^\gamma)^* v^\gamma$. Then $v^\gamma \in H^s(\operatorname{supp}\varphi^\gamma)$ and we put $u = \sum_\gamma v^\gamma$ (which is obviously a finite sum). It is clear that $u \in H^s(K)$ since $v^\gamma \in H^s(K)$. It remains to prove that $\lim_{m \to \infty} \|u_m - u\|_s = 0$. This means in view of (7.6), that

$$\lim_{m \to \infty} \|(\varkappa^\gamma)^* \varphi^\gamma (u_m - u)\|_s = 0$$

for any γ. But it is evident from (7.7) that it suffices to verify $v^\gamma = \varphi^\gamma u$, which is clear since $u_m \to u$ in $\mathscr{D}'(M)$, hence $\varphi^\gamma u_m \to \varphi^\gamma u$ in $\mathscr{D}'(M)$, and (7.7) implies $\varphi^\gamma u_n \to v^\gamma$ in $\mathscr{D}'(M)$ (convergence in $\mathscr{D}'(M)$ in the weak sense). $\square$

We now want to prove that the topology, induced on $H^s(K)$ by the norm $\|\cdot\|_s$, is independent of the choice of arbitrary elements in (7.6) (the covering, the partition of unity and the coordinate diffeomorphisms). For this it is appropriate to give another definition of this topology.

Let Λ_s be as in Lemma 7.1 so that Λ_{-s} is a parametrix for Λ_s (relation (7.1) is satisfied). Then we have

$$u = \Lambda_{-s} \Lambda_s u - R_s u. \tag{7.8}$$

Let $p > s$ be a positive integer and $Q_1, \ldots, Q_N$ differential operators, generating the left $C^\infty(M)$-module of all differential operators of order not greater than p on M. We then set

$$(u, v)'_s = (\Lambda_s u, \Lambda_s v) + \sum_{k=1}^{N} (Q_k R_s u, Q_k R_s v), \tag{7.9}$$

where $u, v \in H^s(K)$, $(\cdot, \cdot)$ the scalar product in $L^2_{\text{comp}}(M)$ induced by any smooth positive density on M. From $(u, u)'_s = 0$ it follows that $\Lambda_s u = 0$ and $R_s u = 0$ and then, in view of (7.8), we have $u = 0$. Therefore the scalar product (7.9) is well defined and we will denote the corresponding Hilbert norm by $\|\cdot\|'_s$.

Proposition 7.2'. *The scalar product (7.9) induces a Hilbert space structure on* $H^s(K)$.

Proof. Once again, we need only verify the completeness. For the beginning note, that from the convergence of a sequence $u_m \in H^s(K)$ with respect to the norm $\|\cdot\|'_s$ the weak convergence in $\mathscr{D}'(M)$ follows. If a sequence $u_m \in H^s(K)$ is a Cauchy sequence with respect to $\|\cdot\|'_s$, then this means, that the following limits exist in the $L^2(\hat{K})$-topology ($\hat{K}$ some compact set in M)

$$\lim_{m \to \infty} \Lambda_s u_m = v, \quad \lim_{m \to \infty} Q R_s u_m = w_Q, \tag{7.10}$$

where Q is any differential operator of order $\leq p$. In particular $\Lambda_s u_m$ and $R_s u_m$ converge in the topology of $L^2_{\text{loc}}(M)$, so that weak convergence of u_m in $\mathscr{D}'(M)$ results from (7.8). Denoting the limit of u_m (in the weak topology of $\mathscr{D}'(M)$) by u, we obviously have

$$u = \Lambda_{-s} v - w_1, \tag{7.11}$$

where $w_1 = \lim_{m \to \infty} R_s u_m$. If as Q in (7.10) we take an elliptic operator of order p, we have

$$Q w_1 = \lim_{m \to \infty} Q R_s u_m = w_Q \in L^2_{\mathrm{loc}}(M),$$

from which $w_1 \in H^p_{\mathrm{loc}}(M)$, therefore $w_1 \in H^s_{\mathrm{loc}}(M)$, but then $w_1 \in H^p_{\mathrm{loc}}(\hat{K})$ for some compact set $\hat{K}$ since the operators Λ_s and R_s are properly supported. Therefore $u \in H^s(K)$.

It remains to verify that

$$\lim_{m \to \infty} \| u_m - u \|'_s = 0.$$

We have

$$u_m - u = u_m - \Lambda_{-s} v + w_1 = \Lambda_{-s} \Lambda_s u_m - R_s u_m - \Lambda_{-s} v + w_1$$
$$= \Lambda_{-s} (\Lambda_s u_m - v) - (R_s u_m - w_1),$$

hence

$$\| u_m - u \|'_s \leq \| \Lambda_s \Lambda_{-s} (\Lambda_s u_m - v) \| + \sum_{k=1}^{N} \| Q_k R_s \Lambda_{-s} (\Lambda_s u_m - v) \|$$
$$+ \| \Lambda_s (R_s u_m - w_1) \| + \sum_{k=1}^{N} \| Q_k R_s (R_s u_m - w_1) \| \tag{7.12}$$

(here the norm $\| \cdot \|$ is induced by the same scalar product $(\cdot, \cdot)$ as in formula (7.9)).

The convergence to 0 of the first, second and last terms in (7.12) follows from the boundedness of the operators $\Lambda_s \Lambda_{-s}$, $Q_k R_s \Lambda_{-s}$ and $Q_k R_s$ as operators from $L^2(K)$ into $L^2(\hat{K})$, and in the case of the third term from the following argument. Take an elliptic differential operator Q of order p and its properly supported parametrix Q_1. Then

$$\Lambda_s (R_s u_m - w_1) = \Lambda_s Q_1 Q (R_s u_m - w_1) + \Lambda_s R (R_s u_m - w_1),$$

where $R \in L^{-\infty}$. The desired result now follows from the fact that $\Lambda_s Q_1$ is a properly supported PDO of order $s - p \leq 0$ continuously mapping $L^2(K)$ into $L^2(\hat{K})$. $\square$

There also exists a third way to introduce a topology on $H^s(K)$, via the seminorms

$$\| u \|_{A,K} = \| A u |_K \|, \qquad A \in L^s_{1,0}(M). \tag{7.13}$$

Proposition 7.3. *The three topologies introduced on $H^s(K)$ above (i.e. via $\| \cdot \|_s$, $\| \cdot \|'_s$ and the seminorms (7.13)) coincide.*

Proof. The equivalence of the topologies determined by the norms $\| \cdot \|_s$ and $\| \cdot \|'_s$ is clear from the closed graph theorem. The topology given by the seminorms (7.13) is obviously stronger than the one given by the norm $\| \cdot \|'_s$, since the latter can be estimated by a sum of $N + 1$ semi norms of the type (7.13), where the corresponding operator A is equal to Λ_s and $Q_k R_s$. To verify the equivalence of the two topologies, it therefore suffices to establish that if $A \in L^s_{1,0}(M)$ and K, $\hat{K}$ are compact in M, then the following estimate holds

$$\| Au|_{\hat{K}} \| \leqq C \|u\|'_s, \qquad u \in H^s(K). \tag{7.14}$$

Now, writing u in the form (7.8) we obtain

$$Au = (A\Lambda_{-s}) \Lambda_s u - AR_s u,$$

and, since $A\Lambda_{-s} \in L^0_{1,0}(M)$, $AR_s \in L^{-\infty}(M)$, (7.14) follows from Theorem 6.6. $\square$

Corollary 7.3. *The topology defined in $H^s(K)$ by the norms $\| \cdot \|_s$ and $\| \cdot \|'_s$, does not depend on arbitrary elements entering the definition of these norms.*

Proposition 7.4. *Let $\hat{K}$ be a compact set in M, such that $K \subset \text{Int } \hat{K}$. Then if $u \in H^s(K)$, there is a sequence $\varphi_n \in C_0^\infty(\hat{K})$, such that $\varphi_n \to u$ as $n \to +\infty$ in the topology of $H^s(\hat{K})$.*

Proof. Follows from Lemma 7.3 taking into account the definition of the norm $\| \cdot \|_s$. $\square$

Proposition 7.5. *Let K be a compact set in M and let $A \in L^m_{\varrho,\delta}(M)$, where either $1 - \varrho \leqq \delta < \varrho$ or $\delta < \varrho$ and $M = X$ an open set in $\mathbb{R}^n$. Then, provided A is properly supported, it defines a linear continuous operator*

$$A\colon H^s(K) \to H^{s-m}(\hat{K}),$$

where $\hat{K}$ is a compact set in M depending on K. Without assuming that A is properly supported, the same holds for the operator φA, $\varphi \in C_0^\infty(M)$.

Proof. It is most convenient to use the norm $\| \cdot \|'_s$. Then, to verify the continuity of A acting from $H^s(K)$ into $H^{s-m}(\hat{K})$, we have to estimate $\|\Lambda_{s-m} Au\|$ and $\|QR_{s-m} Au\|$ by $C\|\Lambda_s u\|$ and $C\|Q'R_s u\|$ (here Q and Q' are differential operators). But from (7.8) we have

$$Au = (A\Lambda_{-s}) (\Lambda_s u) - AR_s u,$$

from which

$$\Lambda_{s-m} Au = (\Lambda_{s-m} A\Lambda_{-s}) (\Lambda_s u) - (\Lambda_{s-m} AR_s) u,$$

$$QR_{s-m} Au = QR_{s-m} \Lambda_{-s} \Lambda_s u - QR_{s-m} R_s u,$$

and the required estimate follows from Theorems 6.1 and 6.6, if we take into consideration that $\Lambda_{s-m}A\Lambda_{-s}\in L^0_{\varrho,\delta}(M)$ and that the operators $\Lambda_{s-m}AR_s$, $QR_{s-m}\Lambda_{-s}$ and QR_{s-m} belong to $L^{-\infty}(M)$. $\square$

We now introduce topologies in $H^s_{\text{loc}}(M)$ and $H^s_{\text{comp}}(M)$.

The topology of $H^s_{\text{loc}}(M)$ is defined as the weakest locally convex topology making all the mappings $M_\varphi\colon H^s_{\text{loc}}(M)\to H^s(\operatorname{supp}\varphi)$, $\varphi\in C^\infty_0(M)$ and $M_\varphi u=\varphi u$, continuous. In other words, this topology is given by the system of seminorms

$$\|u\|_{s,\varphi}=\|\varphi u\|_s, \qquad \varphi\in C^\infty_0(M). \tag{7.15}$$

The topology on $H^s_{\text{comp}}(M)$ is defined as the strongest locally convex topology, making all the embeddings $i_K\colon H^s(K)\to H^s_{\text{comp}}(M)$ continuous. The most important characteristic of this topology (called the *inductive* topology) is that a linear map $f\colon H^s_{\text{comp}}(M)\to E$ to any locally convex space E is continuous if and only if all the compositions $f\circ i_K\colon H^s(K)\to E$ are continuous.

These definitions and Proposition 7.5 imply the following

Theorem 7.3. *Let $A\in L^m_{\varrho,\delta}(M)$ with either $1-\varrho\leqq\delta<\varrho$ or $\delta<\varrho$ and $M=X$, an open set in $\mathbb{R}^n$. Then A is a linear continuous operator for any $s\in\mathbb{R}$*

$$A\colon H^s_{\text{comp}}(M)\to H^{s-m}_{\text{loc}}(M).$$

If A is properly supported it extends to linear continuous operators

$$A\colon H^s_{\text{comp}}(M)\to H^{s-m}_{\text{comp}}(M)$$

and

$$A\colon H^s_{\text{loc}}(M)\to H^{s-m}_{\text{loc}}(M).$$

7.6 Embedding theorems. First, note the completely trivial (and already used) fact, that for $s>s'$, we have the embeddings

$$H^s(K)\subset H^{s'}(K), \qquad H^s_{\text{loc}}(M)\subset H^{s'}_{\text{loc}}(M),$$

$$H^s_{\text{comp}}(M)\subset H^{s'}_{\text{comp}}(M),$$

which are continuous. Less trivial is the following

Theorem 7.4. *Let $s>s'$ and K a compact set in M. Then the embedding operator*

$$i^{s'}_s\colon H^s(K)\to H^{s'}(K)$$

is a compact operator.

Proof. By the equality (7.8), we obtain

$$\Lambda_{s'}u = (\Lambda_{s'}\Lambda_{-s})\,(\Lambda_s u) - (\Lambda_{s'}R_s)\,u$$
$$= (\Lambda_{s'}\Lambda_{-s})\,(\Lambda_s u) - (\Lambda_{s'}\Lambda_{-s})\,(\Lambda_s R_s u) + (\Lambda_{s'}R_s)\,(R_s u).$$

Since $\Lambda_{s'}\Lambda_{-s}\in L_{1,0}^{-(s-s')}(M)$ (hence from Corollary 6.1 for any compact set K_1 one can find a compact set K_2, such that $\Lambda_{s'}\Lambda_{-s}$ is a compact operator from $L^2(K_1)$ to $L^2(K_2)$) it is clear that if u runs through a bounded set in $H^s(K)$ (and, consequently $\Lambda_s u$ and $R_s u$ run through a bounded set in $L^2(K_1)$) then $\Lambda_{s'}u$ runs through a precompact set in $L^2(K_2)$. Similarly one shows that in this case $QR_{s'}u$ runs through a precompact set in $L^2(K_2)$ for any differential operator Q. But this, in view of the equality (7.8) and the definition of the norm, implies the compactness of the corresponding set in $H^{s'}(K_2)$, hence in $H^{s'}(K)$, since, actually, it belongs to $H^{s'}(K)$ and the topology in $H^{s'}(K)$ is induced by the one in $H^{s'}(K_2)$ provided $K \subset K_2$. $\quad\square$

A generalization of Theorem 7.4 is

Theorem 7.5. *Let $A \in L_{\varrho,\delta}^m(M)$, A properly supported, either $1 - \varrho \le \delta < \varrho$ or $\delta < \varrho$ and $M = X$, an open set in $\mathbb{R}^n$. Let the numbers s, $s' \in \mathbb{R}$ be such that $s' < s - m$. Let K be a compact set in M and $\hat{K}$ a compact set in M (depending on K) such that $A\mathscr{E}'(K) \subset \mathscr{E}'(\hat{K})$. Then the operator*

$$A:\ H^s(K) \to H^{s'}(\hat{K})$$

is compact.

Proof. Theorem 7.5 is a consequence of Proposition 7.5 and Theorem 7.4, since the operator $A: H^s(K) \to H^{s'}(\hat{K})$ can be viewed as a composition

$$H^s(K) \xrightarrow{\ \ A\ \ } H^{s-m}(\hat{K}) \xrightarrow[\ i_{s-m}^{s'}\ ]{} H^{s'}(\hat{K}).\quad\square$$

Denote by $C^p(M)$ the space of functions on M having continuous derivatives of order $\le p$ in any local coordinates. The topology in $C^p(M)$ is defined by the seminorms

$$\|u\|_{A,K} = \sup_{x \in K} |Au(x)|, \tag{7.16}$$

where A is any differential operator of order $\le p$. We denote by $C_0^p(K)$ the subspace of the functions $u \in C^p(M)$ with $\operatorname{supp} u \subset K$. It is clear that the topology of $C^p(M)$ induces a topology on $C_0^p(K)$, which can be given by a Banach norm.

Theorem 7.6. *If $s > n/2 + p$, then $H_{\mathrm{loc}}^s(M) \subset C^p(M)$ is a continuous embedding. If K is a compact set in M, then the embedding $H^s(K) \subset C_0^p(K)$ is a compact operator under the same assumption $s > n/2 + p$.*

Proof. Since differential operators of order p are continuous maps $H^s(K) \to H^{s-p}(K)$, it is obvious that it suffices to consider the case $p = 0$. Further, it suffices to verify that for $s > n/2$, we have a continuous embedding

$H^s(K) \subset C_0^0(K)$, since the compactness of this embedding is obtained by writing it as a composition

$$H^s(K) \subset H^{s-\varepsilon}(K) \subset C_0^0(K)$$

($\varepsilon > 0$ such that $s - \varepsilon > n/2$) and using Theorem 7.4. Finally, it is clear that it suffices to consider the case of K lying inside a chart, i.e. the question reduces to the case $M = \mathbb{R}^n$.

Thus, let K be a compact set in $\mathbb{R}^n$ and $s > n/2$. It follows from Lemma 7.3, that it suffices to prove the estimate

$$\sup_{x \in \mathbb{R}^n} |u(x)| \leqq C \|u\|_s, \qquad u \in C_0^\infty(K), \tag{7.17}$$

where C does not depend on u. We will prove this estimate with C even independent of K. We have

$$|u(x)| = |\int e^{ix \cdot \xi} \hat{u}(\xi) \, d\xi| \leqq \int |\hat{u}(\xi)| \, d\xi = \int |\hat{u}(\xi)| \langle \xi \rangle^s \langle \xi \rangle^{-s} d\xi$$

$$\leqq [\int |\hat{u}(\xi)|^2 \langle \xi \rangle^{2s} \, d\xi]^{1/2} [\int \langle \xi \rangle^{-2s} \, d\xi]^{1/2} = C \|u\|_s,$$

where $C = (\int \langle \xi \rangle^{-2s} d\xi)^{1/2} < +\infty$, as required. $\square$

Corollary 7.4. $\bigcap\limits_s H^s_{\mathrm{loc}}(M) = C^\infty(M)$.

This corollary is obvious. Let us also note the dual fact: $\bigcup\limits_s H^s(K) = \mathscr{E}'(K)$ for any compact set $K \subset M$. This fact follows from the well-known statement of distribution theory, that if $u \in \mathscr{E}'(K)$, then u can be written as $u = \sum\limits_{1 \leqq j \leqq N} Q_j v_j$, where $v_j \in L^2(\hat{K})$, $\hat{K}$ is compact and Q_j are differential operators. If m is the greatest order of the Q_j, then $u \in H^{-m}(K)$.

7.7. Duality. Let there be given a smooth positive density $d\mu$ on M. This defines a bilinear form

$$\langle u, v \rangle = \int u(x) \, v(x) \, d\mu(x), \tag{7.18}$$

for instance, if $u \in C_0^\infty(M)$ and $v \in C^\infty(M)$.

Theorem 7.7. *The bilinear form (7.18) extends for any $s \in \mathbb{R}$ to a pairing (separately continuous bilinear mapping)*

$$H^s_{\mathrm{comp}}(M) \times H^{-s}_{\mathrm{loc}}(M) \to \mathbb{C} \tag{7.19}$$

which we will denote as before by $\langle \cdot, \cdot \rangle$. The spaces H^s_{comp} and H^s_{loc} are dual to each other with respect to this pairing, i.e. any continuous linear functional $l(u)$ on $H^s_{\mathrm{comp}}(M)$ can be written in the form $\langle u, v \rangle$ for some $v \in H^{-s}_{\mathrm{loc}}(M)$, and any *continuous linear functional $l(v)$ on $H_{\mathrm{loc}}^{-s}(M)$ can be written as $\langle u, v \rangle$, where $u \in H_{\mathrm{comp}}^{s}(M)$. If the manifold M is closed, then the transformation which attaches to any $v \in H^{-s}(M)$ the linear functional $l_v(u) = \langle u, v \rangle$ is an invertible linear continuous operator from $H^{-s}(M)$ into $(H^s(M))^*$ (where the latter space is endowed with the natural Banach space topology).*

Proof. 1. First let us verify that the form (7.18) extends to the pairing (7.19). Note that the operator Λ_s, appearing in the definition of the Sobolev space can be chosen symmetric with respect to the given density, i.e. such that $\langle \Lambda_s u, v \rangle = \langle u, \Lambda_s v \rangle$ for arbitrary u, $v \in C_0^\infty(M)$. Indeed, we may replace Λ_s by $1/2\,(\Lambda_s + {}^t\Lambda_s)$, without changing the principal symbol. Further, we may suppose that $\Lambda_0 = I$ and Λ_{-s} is a parametrix of Λ_s (this can be achieved, if initially we construct all the Λ_s, $s \geq 0$, as symmetric operators, and then consider their parametrices Λ'_{-s} and take for Λ_{-s} the symmetrization of Λ'_{-s}:

$$\Lambda_{-s} = 1/2\,(\Lambda'_{-s} + {}^t\Lambda'_{-s}))\,.$$

From the definition of the topology on $H_{\mathrm{comp}}^{s}(M)$ it follows, that it suffices to extend (7.18) to a pairing

$$H^s(K) \times H_{\mathrm{loc}}^{-s}(M) \to \mathbb{C} \tag{7.20}$$

where K is any compact set in M. Clearly this is possible for $s = 0$. For $s \neq 0$, we take $u \in C_0^\infty(K)$, $v \in C^\infty(M)$ and write u in the form (7.8). Then

$$\langle u, v \rangle = \langle \Lambda_{-s} \Lambda_s u - R_s u, v \rangle = \langle \Lambda_s u, \Lambda_{-s} v \rangle - \langle R_s u, v \rangle\,, \tag{7.21}$$

from which the extendability of $\langle \cdot\,, \cdot \rangle$ to the pairing (7.20) follows since Λ_s and Λ_{-s} are continuous linear mappings

$$\Lambda_s \colon\ H^s(K) \to L^2(\hat{K})\,, \qquad \Lambda_{-s} \colon\ H_{\mathrm{loc}}^{-s}(M) \to L_{\mathrm{loc}}^2(M)\,.$$

2. Now let $l(\,\cdot\,)$ be a linear continuous functional on $H_{\mathrm{loc}}^{-s}(M)$. We will show that it can be written in the form $l(v) = \langle u, v \rangle$, with $u \in H_{\mathrm{comp}}^{s}(M)$. First of all, since $C^\infty(M) \subset H_{\mathrm{loc}}^{-s}(M)$ is a continuous embedding, the restriction of $l(v)$ to $C^\infty(M)$ can be written in the form $\langle u, v \rangle$, $u \in \mathscr{E}'(K)$ for some compact K in M. The distribution u is thereby uniquely defined and it only remains to verify that $u \in H^s(K)$, i.e. $\Lambda_s u \in L^2(\hat{K})$.

But

$$\langle \Lambda_s u, v \rangle = \langle u, \Lambda_s v \rangle = l(\Lambda_s v)\,,$$

and therefore the desired statement can be derived from the fact that Λ_s is a continuous mapping $L_{\mathrm{loc}}^2(M) \to H_{\mathrm{loc}}^{-s}(M)$ and also from the Riesz theorem, which guarantees the assertion for $s = 0$.

Similarly one shows the representability of the functional $l(\cdot)$ on $H^s_{\mathrm{comp}}(M)$ in the form $l(u) = \langle u, v \rangle$, where $v \in H^{-s}_{\mathrm{loc}}(M)$.

3. Now let M be a closed manifold and let us verify that the map $v \to l_v(\cdot) = \langle \cdot, v \rangle$ is a topological isomorphism between $H^{-s}(M)$ and $(H^s(M))^*$. Obviously this is true for $s = 0$ by the Riesz theorem. Consider the case of an arbitrary $s \in \mathbb{R}$. Since the bijectivity of the map $v \to l_v(\cdot)$ has already been established in 2., it suffices to verify its continuity. But this follows at once from (7.21). $\quad\square$

7.8 Exercises and problems

Exercise 7.1. Verify that $\delta(x) \in H^s(\mathbb{R}^n)$ for $s < -n/2$.

Exercise 7.2. Show that the embedding operator $H^s(\mathbb{R}^n) \subset H^{s'}(\mathbb{R}^n)$ is not compact for any $s, s' (s > s')$.

Exercise 7.3. Let $\mathbb{T}^n = \mathbb{R}^n / 2\pi \mathbb{Z}^n$ be the n-dimensional torus ($\mathbb{Z}^n$ is the lattice of points with integer coordinates in $\mathbb{R}^n$). If $f \in C^\infty(\mathbb{T}^n) (= C_0^\infty(\mathbb{T}^n))$, then f decomposes into a Fourier series

$$f(x) = \sum_{k \in \mathbb{Z}^n} f_k e^{ik \cdot x}, \tag{7.22}$$

where f_k are the Fourier coefficients, given by the formula

$$f_k = (2\pi)^{-n} \int_{\mathbb{T}^n} f(x) \, e^{-ik \cdot x} \, dx. \tag{7.23}$$

The same formula also applies to $f \in \mathscr{D}'(\mathbb{T}^n) (= \mathscr{E}'(\mathbb{T}^n))$, if as the integral in (7.23) we take the value of the functional f at the function $e^{-ik \cdot x}$ (in this case the series (7.22) converges in the weak topology of $\mathscr{D}'(\mathbb{T}^n)$).

Show that if $f \in \mathscr{D}'(\mathbb{T}^n)$, then the condition $f \in H^s(\mathbb{T}^n)$ is equivalent to

$$\sum_{k \in \mathbb{Z}^n} |f_k|^2 (1 + |k|^2)^s < +\infty, \tag{7.24}$$

where the left-hand side of (7.24) defines the square of a norm in $H^s(\mathbb{T}^n)$, equivalent to any of the previous norms.

Exercise 7.4. Show that if A satisfies the conditions of Theorem 6.4, then $A: H^s(K) \to H^s(\hat{K})$ is a compact operator for any $s \in \mathbb{R}$.

Exercise 7.5. Show that the spaces $H^s(\mathbb{R}^n)$ and $H^{-s}(\mathbb{R}^n)$ are dual to each other with respect to the bilinear form $\langle f, g \rangle = \int f(x) g(x) \, dx$.

Problem 7.1. Verify that if N is a submanifold of M of codimension d, then the restriction map $f \to f|_N$ (defined a priori for $f \in C^\infty(M)$) extends for $s > d/2$ to a linear continuous map

$$H^s_{\mathrm{loc}}(M) \to H^{s-d/2}_{\mathrm{loc}}(N).$$

Problem 7.2. Show that the map defined in Problem 7.1 is surjective.

§8. The Fredholm Property, Index and Spectrum

8.1 The basic properties of Fredholm operators

Definition 8.1. Let E_1 and E_2 be Banach spaces and $A: E_1 \to E_2$ a linear continuous operator. It is said to be *Fredholm* if dim Ker $A < +\infty$ and dim Coker $A < +\infty$ (recall that Ker $A = \{x \in E_1: Ax = 0\}$, Coker $A = E_2/\mathrm{Im}\ A$, where Im $A = AE_1$ and the quotient space is meant in the algebraic sense, i.e. regardless of a topology). The *index* of a Fredholm operator is the number

$$\mathrm{index}\ A = \dim \mathrm{Ker}\ A - \dim \mathrm{Coker}\ A. \tag{8.1}$$

We will denote by $\mathscr{L}(E_1, E_2)$ the set of all linear continuous operators $A: E_1 \to E_2$ and the set of all Fredholm operators $A \in \mathscr{L}(E_1, E_2)$ will be denoted by Fred (E_1, E_2).

Lemma 8.1. *Let $A \in \mathscr{L}(E_1, E_2)$ and let* dim Coker $A < +\infty$. *Then* Im A *is a closed subspace in E_2.*

Proof. Clearly, Ker A is a closed subspace of E_1 and therefore the quotient space $E_1/\mathrm{Ker}\ A$ has a natural Banach space structure. The operator A induces a continuous map $A_1: E_1/\mathrm{Ker}\ A \to E_2$ with Im $A_1 = \mathrm{Im}\ A$ and Ker $A_1 = 0$. Now let C denote any finite-dimensional subspace of E_2 for which $E_2 = \mathrm{Im}\ A \oplus C$ (direct sum in the algebraic sense). Define the operator

$$\hat{A}: E_1/\mathrm{Ker}\ A \oplus C \to E_2 \tag{8.2}$$

mapping a pair $\{x, c\}$ into $A_1 x + c \in E_2$. Obviously $\hat{A}$ is bijective and continuous, if the space on the left hand side is considered as a Banach direct sum (e.g. with the norm $\|\{x, c\}\| = \|x\| + \|c\|$, where $\|c\|$ is defined by any norm on C). By the Banach inverse operator theorem, $\hat{A}$ is a topological isomorphism implying that Im A is closed in E_2 since $\hat{A}^{-1}(\mathrm{Im}\ A) = E_1/\mathrm{Ker}\ A \oplus 0$ is closed in $E_1/\mathrm{Ker}\ A \oplus C$. □

Corollary 8.1. *If* dim Coker $A < +\infty$ *and L_1 is a closed subspace in E_1 such that $E_1 = L_1 \oplus \mathrm{Ker}\ A$, then A defines a topological isomorphism $A: L_1 \to \mathrm{Im}\ A$.*

Note that in the case $A \in \mathrm{Fred}\ (E_1, E_2)$ there always exists a subspace L_1 of this type, since by the Hahn-Banach theorem, we may extend the identity map Ker $A \to \mathrm{Ker}\ A$ to a continuous linear operator $P_1: E_1 \to \mathrm{Ker}\ A$ and then put $L_1 = \mathrm{Ker}\ P_1$.

Corollary 8.2. *If $A \in \mathscr{L}(E_1, E_2)$ and* $\dim \operatorname{Coker} A < +\infty$, *then* $\dim \operatorname{Coker} A = \dim \operatorname{Ker} A^*$, *where A^* is the adjoint operator* $A^*: E_2^* \to E_1^*$.

Proof. It is known (and trivial) that

$$\operatorname{Ker} A^* = \{ f \in E_2^* : \langle f, \operatorname{Im} A \rangle = 0 \}.$$

From the closedness of $\operatorname{Im} A$ and the Hahn-Banach theorem it follows that

$$\operatorname{Im} A = \{ x \in E_2 : \langle \operatorname{Ker} A^*, x \rangle = 0 \},$$

implying the desired formula. $\square$

Lemma 8.2. *Let E be a Banach space and let $T \in \mathscr{L}(E, E)$ be of finite rank*, i.e. $\dim \operatorname{Im} T < +\infty$. *Then the operator $I + T$ is Fredholm and* $\operatorname{index}(I + T) = 0$.

Proof. It is easily seen that there exists a decomposition $E = L_0 \oplus L_1$, where L_0 is a closed subspace, $L_0 \subset \operatorname{Ker} T$, $L_1 \supset \operatorname{Im} T$, $\dim L_1 < +\infty$. Then $(I+T)|_{L_0} = I|_{L_0}$, $(I+T) L_1 \subset L_1$, because $T L_1 \subset \operatorname{Im} T \subset L_1$. Therefore L_0 and L_1 are invariant subspaces for $(I+T)$ with $\operatorname{Im}(I+T) \supset L_0$, $\operatorname{Ker}(I+T) \subset L_1$. Therefore $I+T$ is Fredholm and $\operatorname{index}(I+T)$ equals the index of $(I+T)$, viewed as an operator from L_1 into L_1, which means that the whole matter reduces to a trivial statement from linear algebra. $\square$

Lemma 8.3. *If $A \in \operatorname{Fred}(E_1, E_2)$, then there exists an operator $B \in \operatorname{Fred}(E_2, E_1)$ such that*

$$BA = I - P_1, \qquad AB = I - P_2 \tag{8.3}$$

where P_1 is a projection onto $\operatorname{Ker} A$ *and $(I - P_2)$ a projection onto* $\operatorname{Im} A$ *(so that P_1 and P_2 are of finite rank).*

Proof. Let L_1 be a closed complement to $\operatorname{Ker} A$ in E_1 and L_2 any complement to $\operatorname{Im} A$. Define the operator B such that

$$BA|_{L_1} = I|_{L_1}, \qquad B|_{L_2} = 0.$$

From Corollary 8.1 it is clear that $B \in \mathscr{L}(E_2, E_1)$ and $\operatorname{Ker} B = L_2$, $\operatorname{Im} B = L_1$, from which the Fredholm property of B follows. The relation (8.3) is immediately verified. $\square$

Lemma 8.4. *Let $A \in \mathscr{L}(E_1, E_2)$ and let $B_1, B_2 \in \mathscr{L}(E_2, E_1)$ be such that*

$$B_1 A = I + T_1, \qquad AB_2 = I + T_2, \tag{8.4}$$

where T_1, T_2 have finite rank. Then A is Fredholm.

Proof. The statement follows from the obvious inclusions

$$\operatorname{Ker} A \subset \operatorname{Ker}(B_1 A) = \operatorname{Ker}(I + T_1),$$

$$\operatorname{Im} A \supset \operatorname{Im}(A B_2) = \operatorname{Im}(I + T_2)$$

and from Lemma 8.2. $\square$

Lemma 8.5. *Let* $A \in \operatorname{Fred}(E_1, E_2)$, $B \in \operatorname{Fred}(E_2, E_3)$. *Then* $BA \in \operatorname{Fred}(E_1, E_3)$ *and*

$$\operatorname{index} BA = \operatorname{index} A + \operatorname{index} B \tag{8.5}$$

Proof. We show first of all, that there exist closed subspaces $L_j \subset E_j$, $j = 1$, 2, 3, such that $\operatorname{Ker} A|_{L_1} = 0$, $A L_1 = L_2$, $\operatorname{Ker} B|_{L_2} = 0$, $B L_2 = L_3$, where codim $L_j < +\infty$, $j = 1$, 2, 3. (The *codimension* of a subspace L of E is codim $L = \dim(E/L)$. Here, in particular we have codim $L_j = \dim(E_j/L_j)$, $j = 1$, 2, 3.) Indeed, if L_1' is a closed complement of $\operatorname{Ker} A$ in E_1, L_2' a closed complement of $\operatorname{Ker} B$ in E_2, we may put

$$L_2 = L_2' \cap \operatorname{Im} A, \qquad L_1 = (A|_{L_1'})^{-1}(L_2), \qquad L_3 = B L_2.$$

Let us now note the following fact: Let L_1, L_2 be closed subspaces in E_1, E_2 respectively, $A \in \operatorname{Fred}(E_1, E_2)$, $\operatorname{Ker} A|_{L_1} = 0$ and $A L_1 = L_2$. Then, denoting by $\hat{A}: E_1/L_1 \to E_2/L_2$ the map induced by A, we have $\hat{A} \in \operatorname{Fred}(E_1/L_1, E_2/L_2)$ and index $A = \operatorname{index} \hat{A}$. Therefore, using the above constructed subspaces $L_j \subset E_j$ reduces the proof of (8.5) to the case $\dim E_j < +\infty$, $j = 1$, 2, 3, which is evident, since if $A \in \mathscr{L}(E_1, E_2)$ and $\dim E_j < +\infty$, $j = 1$, 2, then index $A = \dim E_1 - \dim E_2$. $\square$

Proposition 8.1. $\operatorname{Fred}(E_1, E_2)$ *is an open subset of* $\mathscr{L}(E_1, E_2)$ *(in the uniform operator topology, i.e. the topology defined by the operator norm) and the function*

$$\operatorname{index}: \operatorname{Fred}(E_1, E_2) \to \mathbb{Z}$$

is continuous (i.e. constant on each connected component of $\operatorname{Fred}(E_1, E_2)$*). In particular, if* A_t *is a continuous (in the norm) operator-valued function, of* $t \in [0, 1]$, *with values in* $\operatorname{Fred}(E_1, E_2)$ *then* index $A_0 = \operatorname{index} A_1$.

Proof. Let $A \in \operatorname{Fred}(E_1, E_2)$. We have to prove the existence of $\varepsilon > 0$ such that if $D \in \mathscr{L}(E_1, E_2)$ and $\|D\| < \varepsilon$ then $A + D \in \operatorname{Fred}(E_1, E_2)$ and index $(A + D) = \operatorname{index} A$.

Let $B \in \mathscr{L}(E_2, E_1)$ be an operator such that

$$BA = I + T_1', \qquad AB = I + T_2' \tag{8.6}$$

with T_1', T_2' of finite rank (which is always possible by Lemma 8.3). We will verify that one may take $\varepsilon = \|B\|^{-1}$. Indeed, let $\|D\| < \varepsilon$. We have $B(A+D) = I + BD + T_1'$ and if we put $B_1 = (I+BD)^{-1}B$, then $B_1(A+D) = I + T_1$, where T_1 is of finite rank. Note that index $B =$ index B_1. Analogously, there is an operator B_2 such that $(A+D)B_2 = I + T_2$, with T_2 of finite rank. By Lemma 8.4, the operator $A + D$ is Fredholm and by Lemmas 8.5 and 8.2 we have

$$\text{index}\,(A+D) \; = \; -\,\text{index}\,B_1 \; = \; -\,\text{index}\,B \; = \; \text{index}\,A. \qquad \square$$

In what follows we will denote by $C(E_1, E_2)$ the set of all compact linear operators from E_1 into E_2.

Lemma 8.6. *Let E be a Banach space and let $R \in C(E, E)$. Then $I + R \in$ Fred (E, E) and* index $(I+R) = 0$.

Proof. Since $I|_{\text{Ker}\,(I+R)} = -R|_{\text{Ker}\,(I+R)}$, the unit ball in $\text{Ker}\,(I+R)$ is compact and therefore $\dim \text{Ker}\,(I+R) < +\infty$. Further, since R^* is also compact, $\dim \text{Ker}\,(I+R)^* < +\infty$ and to show the Fredholm property of $(I+R)$ it only remains to verify the closedness of $\text{Im}\,(I+R)$ (since then $\dim \text{Coker}\,(I+R) = \dim \text{Ker}\,(I+R)^*$).

Let $x_n \in E, n = 1, 2, \ldots,$ and $y_n = (I+R)x_n \to y$ as $n \to +\infty$. We need to verify the existence of an $x \in E$, such that $(I+R)x = y$. Let L be any closed subspace complementary to $\text{Ker}\,(I+R)$ in E. Adding to x_n vectors from $\text{Ker}\,(I+R)$ (which does not change y_n), we may assume that $x_n \in L$ for all n.

Let us show that the sequence x_n is bounded. Indeed, if this is not the case, taking a subsequence of $\{x_n\}$, we may assume that $\|x_n\| \to +\infty$ as $n \to +\infty$. But then, putting $x_n' = x_n/\|x_n\|$, $y_n' = (I+R)x_n'$ we have $y_n' \to 0$ as $n \to +\infty$, and since $\|x_n'\| = 1$ we may assume that $\lim\limits_{n \to +\infty} Rx_n'$ exists. But then also $\lim\limits_{n \to +\infty} x_n' = -\lim\limits_{n} Rx_n' = x$ and clearly $\|x\| = 1$, $x \in L$, $(I+R)x = 0$, contradicting the choice of L.

Thus the sequence $\{x_n\}$ is bounded and we may assume that $\lim\limits_{n \to +\infty} Rx_n$ exists and, consequently so does $\lim\limits_{n \to +\infty} x_n = y - \lim\limits_{n \to +\infty} Rx_n$. Denoting $x = \lim\limits_{n \to +\infty} x_n$, we clearly have $(I+R)x = y$, proving the closedness of $\text{Im}\,(I+R)$, i.e. the Fredholm property of $(I+R)$. By Proposition 8.1, we have index $(I+tR) =$ const for $t \in [0, 1]$ implying index $(I+R) = \text{Index}\,I = 0$. $\quad \square$

Proposition 8.2. *Let $A \in \mathscr{L}(E_1, E_2)$ and let there exist B_1 and B_2 such that*

$$B_1 A = I + R_1, \qquad AB_2 = I + R_2, \tag{8.7}$$

where $R_j \in C(E_j, E_j), j = 1, 2$. Then $A \in$ Fred (E_1, E_2).

Proof. It immediately follows from Lemma 8.6 in a similar way to the proof of Lemma 8.4. $\quad \square$

Proposition 8.3. *Let $A \in$ Fred (E_1, E_2), $R \in C(E_1, E_2)$. Then $A + R \in$ Fred (E_1, E_2) and* index $(A+R) =$ index A.

Proof. Obvious from Proposition 8.2 and Lemmas 8.5 and 8.6. $\square$

8.2 The Fredholm property and the index of elliptic operators on a closed manifold

Theorem 8.1. *Let M be a closed manifold and $A \in HL_{\varrho,\delta}^{m,\,m}(M)$, $1 - \varrho \leq \delta < \varrho$. For any $s \in \mathbb{R}$ construct the operator $A_s \in \mathscr{L}(H^s(M), H^{s-m}(M))$ the extension of A by continuity.*

Then,

a) $A_s \in \mathrm{Fred}(H^s(M), H^{s-m}(M))$;

b) $\mathrm{Ker}\, A_s \subset C^\infty(M)$, therefore $\mathrm{Ker}\, A_s$ does not depend on s and will be denoted simply by $\mathrm{Ker}\, A$;

c) index A_s does not depend on s (so we will denote it simply by index A) and is expressed by the formula

$$\mathrm{index}\, A = \dim \mathrm{Ker}\, A - \dim \mathrm{Ker}\, A^*, \tag{8.8}$$

where A^* is the formal adjoint ΨDO (cf. §3) in the sense of a scalar product determined by any smooth density.

d) if $D \in L_{\varrho,\delta}^{m'}(M)$, where $m' < m$, then index $(A + D)$ $=$ index A.

Proof. By Theorem 5.1 we may construct a parametrix $B \in HL_{\varrho,\delta}^{-m,\,-m}(M)$ of the operator A. In view of Theorem 7.5, the operators $R_j \in L^{-\infty}(M)$ can be extended to operators $R_{j,\,s} \in C(H^s(M), H^s(M))$ for arbitrary $s \in \mathbb{R}$. But now, the Fredholm property for all the operators A_s follows from proposition 8.2.

Further, from Theorem 5.2 statement b) of the theorem follows and since $A^* \in HL_{\varrho,\delta}^{m,\,m}$, we also obtain c).

Finally, d) follows from Proposition 8.3 since if $D \in L_{\varrho,\delta}^{m'}(M)$, $m' < m$, then $D \in C(H^s(M), H^{s-m}(M))$ by Theorem 7.5. $\square$

Remark 8.1. The assertion of this theorem is clearly true not only for scalar operators, but also for operators acting on the sections of vector bundles.

Remark 8.2. For classical elliptic ΨDO, d) says that the index depends only on the principal symbol. It is easy to deduce from Theorem 6.2 that the index does not change with arbitrary continuous deformations of the principal symbol within the class of homogenous elliptic symbols. This is important in the index theory of elliptic operators.

8.3 The spectrum (basic facts). Let M be a closed manifold, $A \in HL_{\varrho,\delta}^{m,\,m}(M)$, $1 - \varrho \leq \delta < \varrho$, $m > 0$. In the space $L^2(M)$ consider the unbounded linear operator defined by A by taking as domain the space $H^m(M)$. We will denote this unbounded operator by A_0, or sometimes just by A if there can be no confusion. So the domain of A_0 is $D_{A_0} = H^m(M)$.

Proposition 8.4. *The operator A_0 is closed, i.e. if for $u_n \in H^m(M)$, $n = 1, 2, \dots$, the limits $u = \lim\limits_{n \to +\infty} u_n$ and $f = \lim\limits_{n \to +\infty} Au_n$ exist in $L^2(M)$ then $u \in H^m(M)$ and $Au = f$.*

Proof. Since convergence in $L^2(M)$ implies convergence in $\mathscr{D}'(M)$ and since A is continuous in $\mathscr{D}'(M)$ (in the sense of, e.g. the weak topology), we obtain $Au = f$, and then $u \in H^m(M)$ in view of Theorem 7.2. $\square$

From the fact that $C^\infty(M)$ is dense in $H^m(M)$ (in the $H^m(M)$-topology), we have

Corollary 8.3. *The operator A_0 is the closure (in $L^2(M)$) of the operator* $A|_{C^\infty(M)}$.

Definition 8.2. The *spectrum* of A is the subset $\sigma(A)$ of the complex plane, defined as follows: for $\lambda \in \mathbb{C}$, $\lambda \notin \sigma(A)$ is equivalent to $(A_0 - \lambda I)$ having a bounded everywhere defined inverse $(A_0 - \lambda I)^{-1}$ in $L^2(M)$.

It is easy to verify that $\sigma(A)$ is a closed subset of $\mathbb{C}$ and that $(A_0 - \lambda I)^{-1}$ is a holomorphic operator-valued function of λ on $\mathbb{C}\backslash\sigma(A)$ with values in $\mathscr{L}(L^2(M), L^2(M))$. The function $R_\lambda = (A_0 - \lambda I)^{-1}$ is called the *resolvent* of A.

Proposition 8.5. *Let a fixed positive smooth density $d\mu(x)$ be fixed on M. Then the conditions $\lambda \notin \sigma(A)$ and $\mathrm{Ker}(A - \lambda I) = \mathrm{Ker}(A^* - \bar{\lambda}I) = 0$ are equivalent.*

Proof. The statement follows from Theorem 8.1, since $(A - \lambda I) \in HL^{m,\,m}_{\varrho,\,\delta}(M)$ because $m > 0$. $\square$

Theorem 8.2. *(the inverse operator theorem) Let $A \in HL^{m,\,m}_{\varrho,\,\delta}(M)$, $1 - \varrho \leq \delta < \varrho$, $m > 0$ and let A_0 be constructed as before. Let also $\lambda \notin \sigma(A)$. Then $(A_0 - \lambda I)^{-1}$ is an extension by continuity (from $C^\infty(M)$) or restriction (from $\mathscr{D}'(M)$) of an operator from $HL^{-m,\,-m}_{\varrho,\,\delta}(M)$ (we denote it by $(A - \lambda I)^{-1}$). In particular, $(A_0 - \lambda I)^{-1}$ is compact in $L^2(M)$.*

Proof. It suffices to consider the case $\lambda = 0$. Let $B \in HL^{-m,\,-m}_{\varrho,\,\delta}(M)$ be a parametrix for A. More precisely

$$AB = I - R \tag{8.9}$$

where R is an operator with smooth kernel $R(x, y)$ (for simplicity we assume that a smooth, positive density on M is fixed, so the kernel $R(x, y)$ is an ordinary function on $M \times M$). It follows from (8.9) that

$$A^{-1} = B + A^{-1}R, \tag{8.10}$$

and it remains to verify that $A^{-1}R$ is an operator with smooth kernel. But by Theorem 7.2, A^{-1} maps $H^s(M)$ into $H^{s+m}(M)$ for any $s \in \mathbb{R}$, and, moreover, is continuous by the closed graph theorem. By the embedding Theorem 7.6, A^{-1} maps $C^\infty(M)$ into $C^\infty(M)$. But then $A^{-1}R$ is given by the smooth kernel $R_1(x, y) = [A^{-1}R(\cdot, y)](x)$. $\square$

Theorem 8.3. *Let a fixed smooth positive density $d\mu(x)$ on a closed manifold M be fixed. Let $A^* = A \in HL_{\varrho,\delta}^{m,m}(M)$, $1 - \varrho \leqq \delta < \varrho$, $m > 0$. Then A_0 is a self-adjoint operator in $L^2(M)$ and there exists in this space a complete orthonormal system $\{\varphi_j\}$, $j = 1, 2, \ldots$ of eigenfunctions of A_0. Here $\varphi_j \in C^\infty(M)$, $A\varphi_j = \lambda_j \varphi_j$ and the eigenvalues λ_j are real, with $|\lambda_j| \to +\infty$ as $j \to +\infty$. The spectrum $\sigma(A)$ coincides with the set of all eigenvalues.*

Proof. Note first of all that $\sigma(A) \subset \mathbb{R}$ in view of Proposition 8.5, since A is symmetric on $C^\infty(M)$ and can thus have no non-real eigenvalues.

Next, we want to show that $\sigma(A) \neq \mathbb{R}$. Assuming $\sigma(A) = \mathbb{R}$ then we could for any $\lambda \in \mathbb{R}$ find a function $\varphi_\lambda \in C^\infty(M)$, such that $A\varphi_\lambda = \lambda\varphi_\lambda$ and $\|\varphi_\lambda\| = 1$. But then $(\varphi_\lambda, \varphi_\mu) = 0$ for $\lambda \neq \mu$ by the symmetry of A, contradicting the separability of $L^2(M)$.

Now take $\lambda_0 \in \mathbb{R} \setminus \sigma(A)$. By Theorem 8.2, $R_{\lambda_0} = (A - \lambda_0 I)^{-1}$ is a compact self-adjoint operator in $L^2(M)$. By a known theorem from functional analysis there is an orthonormal basis $\{\varphi_j\}_{j=1}^\infty$ of eigenfunctions, where the eigenvalues r_j tend to 0 as $j \to +\infty$.

Now note that $r_j \neq 0$ (since $\operatorname{Ker}(A - \lambda_0 I)^{-1} = 0$). The condition $R_{\lambda_0}\varphi_j = r_j\varphi_j$ can therefore be rewritten in the form

$$(A - \lambda_0 I)\,\varphi_j = r_j^{-1}\,\varphi_j$$

or

$$A\varphi_j = (r_j^{-1} + \lambda_0)\,\varphi_j. \tag{8.11}$$

It is obvious from (8.11) that $\varphi_j \in C^\infty(M)$ and the φ_j are eigenfunctions of A with eigenvalues $\lambda_j = r_j^{-1} + \lambda_0$. It is also clear that $|\lambda_j| \to +\infty$ as $j \to +\infty$. The remaining assertions of Theorem 8.3 are obvious. The fact that the spectrum $\sigma(A)$ coincides with the set of all eigenvalues $\{\lambda_j\}$ follows from Proposition 8.5 and the self-adjointness from the representation $A = R_{\lambda_0}^{-1} + \lambda_0 I$. $\quad\square$

The following theorem extends one of the statements of Theorem 8.3 to the non-selfadjoint case.

Theorem 8.4. *Let $A \in HL_{\varrho,\delta}^{m,m}(M)$, $1 - \varrho \leqq \delta < \varrho$ and $m > 0$. Then for the spectrum $\sigma(A)$, there are two possibilities:*

a) *$\sigma(A) = \mathbb{C}$ (which, in particular, is the case if $\operatorname{index} A \neq 0$);*

b) *$\sigma(A)$ is a discrete (maybe empty) subset of $\mathbb{C}$ (subset without limit points).*

If b) *holds and $\lambda_0 \in \sigma(A)$ then there is a decomposition $L^2(M) = E_{\lambda_0} \oplus E'_{\lambda_0}$ such that the following conditions are satisfied:*

1) *$E_{\lambda_0} \subset C^\infty(M)$, $\dim E_{\lambda_0} < +\infty$, and E_{λ_0} is an invariant subspace of A such that there exists a positive integer $N > 0$ with $(A - \lambda_0 I)^N E_{\lambda_0} = 0$ (in other words, the operator $A|_{E_{\lambda_0}}$ has only the eigenvalue λ_0 and is equal to the direct sum of Jordan cells of degree $\leqq N$);*

2) *E'_{λ_0} is a closed subspace of $L^2(M)$, invariant with respect to A_0 (i.e. $A(D_{A_0} \cap E'_{\lambda_0}) \subset E'_{\lambda_0}$) and if we denote by A'_{λ_0} the restriction $A_0|_{E'_{\lambda_0}}$ (understood as*

an unbounded operator in E'_{λ_0} with domain $D_{A_0} \cap E'_{\lambda_0}$), then $A'_{\lambda_0} - \lambda_0 I$ has a bounded inverse (or, in other words, $\lambda_0 \notin \sigma(A|_{E'_{\lambda_0}})$).

Proof. 1. Let $\sigma(A) \neq \mathbb{C}$. Let us prove that $\sigma(A)$ is a discrete subset in $\mathbb{C}$. There is a point $\lambda_0 \in \mathbb{C} \setminus \sigma(A)$ and we may, without loss of generality, assume that $\lambda_0 = 0$, so that by Theorem 8.2 A_0 has a compact inverse A_0^{-1}. Then since $A_0 - \lambda I = (I - \lambda A_0^{-1}) A_0$ the inclusion $\lambda \in \sigma(A)$ is equivalent to $\lambda \neq 0$ and $\lambda^{-1} \in \sigma(A_0^{-1})$. Discreteness of $\sigma(A)$ follows from the fact that $\sigma(A_0^{-1})$ may have only 0 as an accumulation point.

2. Let $\sigma(A) \neq \mathbb{C}$, $\lambda_0 \in \sigma(A)$. Once again, without loss of generality, we may assume that $\lambda_0 = 0$. Let Γ_0 be a contour in the complex plane, encircling 0 and not containing any other points of $\sigma(A)$ (e.g. a circle, sufficiently small and with centre at the origin). Consider the operator

$$P_0 = -\frac{1}{2\pi i} \int_{\Gamma_0} R_\lambda d\lambda. \tag{8.12}$$

Standard arguments (cf. Riesz, Sz.-Nagy [1], Chapter XI) show that P_0 is a projection, of finite rank in view of the compactness of R_λ, commuting with all the operators R_λ (and with A_0 in the sense that $P_0 A_0 \subset A_0 P_0$) and such that if $E_{\lambda_0} = P_0(L^2(M))$, $E'_{\lambda_0} = (I - P_0)(L^2(M))$, then conclusions 1) and 2) of Theorem 8.4 hold.

We leave it as exercise for the reader to take care about the details. We note only that the inclusion $E_{\lambda_0} \subset C^\infty(M)$ follows from $A_0^N E_{\lambda_0} = 0$ if we take into account the ellipticity of A and utilize the regularity Theorem 5.2. $\quad\square$

8.4 Problems

Problem 8.1. Let E be a separable Hilbert space, $\pi_0(\mathrm{Fred}(E, E))$ the set of connected components of $\mathrm{Fred}(E, E)$ provided with the semigroup structure induced by the multiplication. Show that taking the index gives an isomorphism

$$\mathrm{index}\colon \pi_0(\mathrm{Fred}(E, E)) \simeq \mathbb{Z}.$$

Hint. An operator A of index 0 can be written in the form $A = A_0 + T$, where A_0 is invertible and T has finite rank. Show (by use of the polar decomposition) the connectedness of the group of all invertible operators in E.

In all the following problems M is a closed manifold and $1 - \varrho \leq \delta < \varrho$, $m > 0$.

Problem 8.2. Let $A \in HL^{m,\,m_0}_{\varrho,\,\delta}(M)$. Prove that A is a Fredholm operator in $C^\infty(M)$, i.e. that $\dim \mathrm{Ker}\, A < +\infty$, $AC^\infty(M)$ is closed in $C^\infty(M)$ and $\dim \mathrm{Coker}\, A < +\infty$, where $\mathrm{Coker}\, A = C^\infty(M)/AC^\infty(M)$. Show that $AC^\infty(M)$ consists of all $f \in C^\infty(M)$ for which $(f, g) = 0$ for any $g \in \mathrm{Ker}\, A^*$ (here $(\,\cdot\,,\,\cdot\,)$ is a scalar product determined by some smooth positive density and A^* is the adjoint ΨDO with respect to this scalar product).

Problem 8.3. Let $A \in HL^{m;m_0}_{\varrho,\delta}(M)$ and $m_0 > 0$. Let there be given on M a smooth positive density defining the scalar product $(\cdot, \cdot)$ and the formal adjoint ΨDO A^*. Assume $A = A^*$. Let A_0 be the closure of the operator $A|_{C^\infty(M)}$. Then A_0 is self-adjoint in the Hilbert sense in the space $L^2(M)$.

Problem 8.4. Let $A \in HL^{m;m_0}_{\varrho,\delta}(M)$, let A^* be the formal adjoint operator and A_0, A_0^* the closures of $A|_{C^\infty(M)}$ and $A^*|_{C^\infty(M)}$ in $L^2(M)$ respectively. Show that A_0 and A_0^* are adjoint to each other in the sense of the Hilbert space $L^2(M)$.

Hint. Consider the matrix of ΨDO $\mathfrak{A} = \begin{pmatrix} 0 & A^* \\ A & 0 \end{pmatrix}$.

Problem 8.5. Find an example of an operator $A \in HL^{m;m}_{1,0}(M)$ for which $\sigma(A) = \mathbb{C}$.

Problem 8.6. A sequence of Hilbert spaces E_j and linear continuous operators d_j:

$$0 \xrightarrow{d_{-1}} E_0 \xrightarrow{d_0} E_1 \xrightarrow{d_1} \ldots \xrightarrow{d_{N-2}} E_{N-1} \xrightarrow{d_{N-1}} E_N \xrightarrow{d_N} 0 \tag{8.13}$$

is called a *complex* if $d_{j+1}d_j = 0$ for all $j = 0, 1, \ldots, N-2$. Put

$$Z^j = \operatorname{Ker} d_j, \quad B^j = \operatorname{Im} d_{j-1}, \quad H^j = Z^j/B^j, \quad j = 0, 1, \ldots, N.$$

(if (8.13) is a complex, $B^j \subset Z^j$). The spaces H^j are called the *cohomology of the complex* (8.13). The complex is called *Fredholm* if $\dim H^j < \infty$ for all $j = 0, 1, \ldots, N$.

a) show that if the complex (8.13) is Fredholm, then the B^j are closed subspaces of Z^j.

b) Let $\Delta_j = \delta_j d_j + d_{j-1}\delta_{j-1}$, where $\delta_j = d_j^*$. The operators Δ_j are called the *Laplacians* of the complex (8.13) (or the Laplace-Hodge operators). Put $\Gamma^j = \operatorname{Ker} \Delta_j$. Show that for the complex (8.13) to be Fredholm it is necessary and sufficient that all Δ_j are Fredholm operators in E_j, $j = 0, 1, 2, \ldots, N$. In this case

$$\dim H^j = \dim \Gamma^j.$$

More precisely, $\Gamma^j \subset Z^j$ and the map $\Gamma^j \to H^j$ induced by the canonical projection $Z^j \to H^j$ is an isomorphism (in the case of a Fredholm complex).

c) Put now

$$\chi(E) = \sum_{j=0}^{N} (-1)^j \dim H^j$$

(the *Euler characteristic* of the Fredholm complex E). Prove that if $N = 1$, then the Euler characteristic of the complex

$$0 \to E_0 \xrightarrow{\ d_0\ } E_1 \to 0$$

is simply the index of d_0.

Prove that if $\dim E_j < +\infty$, $j = 1, 2, \ldots, N$, then

$$\chi(E) = \sum_{j=0}^{N} (-1)^j \dim E_j .$$

d) Show that $\chi(E)$ does not change under a uniform deformation of all the operators d_j if under this deformation the sequence (8.13) remains a Fredholm complex.

Problem 8.7. Let $V_j (j = 0, 1, \ldots, N)$ be vector bundles on a closed manifold M and $H^s(M, V_j)$ the Sobolev spaces of sections. Let $d_j \colon C^\infty(M, V_j) \to C^\infty (M, V_{j+1})$ be classical ΨDO of the same order m. Let $T_0^*(M)$ be the cotangent bundle over M without the zero section and $\pi_0 \colon T_0^*(M) \to M$ the natural projection. Assume that the operators

$$0 \xrightarrow{\ d_{-1}\ } C^\infty(M, V_0) \xrightarrow{\ d_0\ } C^\infty(M, V_1) \xrightarrow{\ d_1\ } \cdots \xrightarrow{\ d_{N-1}\ } C^\infty(M, V_N) \xrightarrow{\ d_N\ } 0$$

$$(8.14)$$

form a complex. Let $\sigma_{d_j}^0 \colon \pi_0^* V_j \to \pi_0^* V_{j+1}$ be the principal symbols of the operators d_j (homogenous functions in ξ of order m). The complex (8.14) is called *elliptic* if the sequence of vector bundles

$$0 \longrightarrow \pi_0^* V_0 \xrightarrow{\ \sigma_{d_0}^0\ } \pi_0^* V_1 \xrightarrow{\ \sigma_{d_1}^0\ } \cdots \xrightarrow{\ \sigma_{d_{N-1}}^0\ } \pi_0^* V_N \longrightarrow 0$$

is exact (i.e. an exact sequence of vector spaces at every point $(x, \xi) \in T_0^*(M)$).

a) Show that ellipticity of the complex (8.14) is equivalent to ellipticity of all the Laplacians $\Delta_j = \delta_j d_j + d_{j-1}\delta_{j-1}$, where δ_j is the ΨDO adjoint to d_j with respect to some density on M and a Hermitean scalar product on the vector bundles V_j.

b) Show that if (8.14) is an elliptic complex, then for any $s \in \mathbb{R}$, the complex

$$0 \xrightarrow{\ d_{-1}\ } H^s(M, V_0) \xrightarrow{\ d_0\ } H^{s-m}(M, V_1) \xrightarrow{\ d_1\ } \cdots \xrightarrow{\ d_{N-1}\ } H^{s-Nm}(M, V_N) \xrightarrow{\ d_N\ } 0$$

is Fredholm and the dimension of its cohomology (and thus the Euler characteristic) does not depend on s. The cohomology itself can be defined also as the cohomology of the complex (8.14), i.e. putting

$$H^j = \operatorname{Ker}(d_j|_{C^\infty(M, V_j)})/d_{j-1}(C^\infty(M, V_{j-1})) .$$

Problem 8.8. Show that the de Rham complex on a real n-manifold M

$$0 \longrightarrow \Lambda^0(M) \xrightarrow{\ d\ } \Lambda^1(M) \xrightarrow{\ d\ } \Lambda^2(M) \xrightarrow{\ d\ } \cdots \xrightarrow{\ d\ } \Lambda^n(M) \longrightarrow 0$$

($\Lambda^j(M)$ is the space of smooth exterior j-forms on M, d is the exterior differential) and the Dolbeault complex on a complex manifold M, $\dim_{\mathbb{C}} M = n$,

$$0 \longrightarrow \Lambda^{p,0}(M) \xrightarrow{\bar{\partial}} \Lambda^{p,1}(M) \xrightarrow{\bar{\partial}} \cdots \xrightarrow{\bar{\partial}} \Lambda^{p,n}(M) \longrightarrow 0$$

($\Lambda^{p,q}(M)$ is the space of smooth forms of type (p,q) on M and $\bar{\partial}$ is the Cauchy-Riemann-Dolbeault operator) are elliptic complexes.

Derive from this the finite-dimensionality of the de Rham and Dolbeault cohomology in case of a closed M.

Chapter II
Complex Powers of Elliptic Operators

§9. Pseudodifferential Operators with Parameter. The Resolvent

9.1 Preliminaries. Let Λ be a subset of the complex plane (in the applications this will, as a rule, be an angle with the vertex at the origin). In spectral theory it is useful to consider operators depending on a parameter $\lambda \in \Lambda$ (an example of such an operator is the resolvent $(A - \lambda I)^{-1}$).

To begin with, we introduce some symbol classes.

Let X be an open set in $\mathbb{R}^n$ and let $a(x, \theta, \lambda)$ be a function on $X \times \mathbb{R}^N \times \Lambda$, $x \in X$, $\theta \in \mathbb{R}^N$, $\lambda \in \Lambda$.

Definition 9.1. Let m, ϱ, δ, d be real numbers with $0 \leqq \delta < \varrho \leqq 1$, $0 < d < +\infty$. The class $S^m_{\varrho,\delta;d}(X \times \mathbb{R}^N, \Lambda)$ consists of the functions $a(x, \theta, \lambda)$ such that
1) $a(x, \theta, \lambda_0) \in C^\infty(X \times \mathbb{R}^N)$ for every fixed $\lambda_0 \in \Lambda$;

2) For arbitrary multi-indices α and β and for any compact set $K \subset X$ there exist constants $C_{\alpha,\beta,K}$ such that

$$|\partial_\theta^\alpha \partial_x^\beta a(x, \theta, \lambda)| \leqq C_{\alpha,\beta,K}(1 + |\theta| + |\lambda|^{1/d})^{m - \varrho|\alpha| + \delta|\beta|}. \tag{9.1}$$

for $x \in K$, $\theta \in \mathbb{R}^N$, $\lambda \in \Lambda$. As usual we put

$$S^{-\infty}(X \times \mathbb{R}^N, \Lambda) = \bigcap_{m \in \mathbb{R}} S^m_{\varrho,\delta;d}(X \times \mathbb{R}^N, \Lambda)$$

(the right-hand side does not depend on ϱ, δ and d).

If $a(x, y, \xi, \lambda) \in S^m_{\varrho,\delta;d}(X \times X \times \mathbb{R}^n, \Lambda)$, we may construct a ΨDO A_λ, depending on the parameter $\lambda \in \Lambda$:

$$(A_\lambda u)(x) = \int\int e^{i(x-y)\cdot\xi} a(x, y, \xi, \lambda) u(y)\, dy\, d\xi, \tag{9.2}$$

for $u \in C_0^\infty(X)$. In this case we will write

$$A_\lambda \in L^m_{\varrho,\delta;d}(X, \Lambda).$$

Note that $A_\lambda \in L^{-\infty}(X, \Lambda)$ if and only if the operator A_λ has a smooth kernel $K_\lambda(x, y)$ for any fixed $\lambda \in \Lambda$ and there exist constants $C^{(N)}_{\alpha,\beta,K}$ (K a compact set in X,

α and β multi-indices and N a positive integer) such that the following estimate holds

$$|\partial_x^\alpha \partial_y^\beta K_\lambda(x, y)| \leqq C_{\alpha,\beta,K}^{(N)} (1 + |\lambda|)^{-N}, \qquad x, y \in K. \tag{9.3}$$

Many of the statements about ΨDO without a parameter (cf. §§3–7) can also be proved for the case with a parameter λ. We indicate now some of these statements, which are necessary in what follows.

First, note that the whole theory of asymptotic summation (Definition 3.4 and Propositions 3.5 and 3.6) carries over to symbols depending on a parameter. The corresponding formulations are obtained by changing $S_{\varrho,\delta}^m(X \times \mathbb{R}^N)$ to $S_{\varrho,\delta;d}^m(X \times \mathbb{R}^N, \Lambda)$ and the proofs are almost verbatim repetitions of the arguments in 3.3 and are left for the reader as an exercise. We only state that the role of $\langle\theta\rangle$ in these proofs (as in the following) is now played by $(1 + |\theta|^2 + |\lambda|^{2/d})^{1/2}$.

Further we will call an operator $A_\lambda \in L_{\varrho,\delta;d}^m(X, \Lambda)$ *properly supported* if it is uniformly properly supported in λ, i.e. there exists a closed set $L \subset X \times X$, having proper projections on each factor in $X \times X$, such that supp $K_{A_\lambda} \subset L$ for all $\lambda \in \Lambda$.

Note that any operator $A \in L_{\varrho,\delta;d}^m(X, \Lambda)$ can be decomposed into a sum $A = A_1 + R_1$, where A_1 *(depending on a parameter) is* properly supported in the sense described and $R_1 \in L_{\varrho,\delta;d}^{-\infty}(X, \Lambda)$. For properly supported ΨDO A_λ depending on a parameter, the symbol $\sigma_{A_\lambda}(x, \xi) = \sigma_A(x, \xi, \lambda)$ is defined and a theorem of type 3.1 is valid. Naturally, we have to interpret formula (3.21) taking the parameter into account, i.e.

$$\sigma_A(x, \xi, \lambda) - \sum_{|\alpha| \leqq N-1} \frac{1}{\alpha!} \partial_\xi^\alpha D_y^\alpha a(x, y, \xi, \lambda)|_{y=x} \in$$

$$\in S_{\varrho,\delta;d}^{m-(\varrho-\delta)N}(X \times \mathbb{R}^n, \Lambda).$$

In an analogous way Theorems 3.2–3.4 on the transpose and adjoint operators and composition can be generalized.

Exercise 9.1. Prove all the statements in sec. 3 in the case of operators and symbols depending on a parameter.

Further, repeating the arguments of §4, for $1 - \varrho \leqq \delta < \varrho$, we may introduce the classes $L_{\varrho,\delta;d}^m(M, \Lambda)$ on a manifold M.

Let us now pass to considering hypoellipticity and ellipticity.

We introduce the class $HS_{\varrho,\delta;d}^{m;m_0}(X \times \mathbb{R}^n, \Lambda)$ of symbols $\sigma(x, \xi, \lambda)$ (we will call them *hypoelliptic with parameter*), belonging to $S_{\varrho,\delta;d}^m(X \times \mathbb{R}^n, \Lambda)$ and satisfying the estimates

$$C_1 (|\xi| + |\lambda|^{1/d})^{m_0} \leqq |\sigma(x, \xi, \lambda)| \leqq C_2(|\xi| + |\lambda|^{1/d})^m, \tag{9.4}$$

for $x \in K$ (K compact in X), $|\xi| + |\lambda| \geqq R$, $C_1 > 0$ and R, C_1, C_2 may depend on K;

$$|[\partial_\xi^\alpha \partial_x^\beta \sigma(x, \xi, \lambda)]\, \sigma^{-1}(x, \xi, \lambda)| \leq C_{\alpha, \beta, K}(|\xi| + |\lambda|^{1/d})^{-\varrho|\alpha| + \delta|\beta|}, \qquad (9.5)$$

for $x \in K$, $|\xi| + |\lambda| \geq R$ (here, as above, R may depend on K).

We will denote by $HL_{\varrho, \delta; d}^{m, m_0}(X, \Lambda)$ the class of properly supported ΨDO (depending on the parameter $\lambda \in \Lambda$), whose symbols belong to $HS_{\varrho, \delta; d}^{m, m_0}(X \times \mathbb{R}^n, \Lambda)$. We have an analogue of Theorem 5.1:

If $A \in HL_{\varrho, \delta; d}^{m, m_0}(X, \Lambda)$, then there exists an operator $B_\lambda \in HL_{\varrho, \delta; d}^{-m_0; -m}(X, \Lambda)$ called the *parametrix* of the operator A_λ such that

$$B_\lambda A_\lambda = I + R'_\lambda, \qquad A_\lambda B_\lambda = I + R''_\lambda \qquad (9.6)$$

where R'_λ, $R''_\lambda \in L^{-\infty}(X, \Lambda)$. The same statement is also true when X is a manifold.

Exercise 9.2. Prove this analogue of Theorem 5.1.

It is natural also to consider classical ΨDO depending on a parameter. In this case Λ is assumed to be an angle with the vertex at 0. The corresponding symbols $a(x, \xi, \lambda)$ admit asymptotic expansions (for $|\xi| + |\lambda|^{1/d} \geq 1$) of the form

$$a(x, \xi, \lambda) \sim \sum_{j=0}^{+\infty} a_{m-j}(x, \xi, \lambda), \qquad (9.7)$$

where $a_{m-j}(x, \xi, \lambda)$ is positive homogeneous in $(\xi, \lambda^{1/d})$ of degree $m-j$, i.e.

$$a_{m-j}(x, t\xi, t^d\lambda) = t^{m-j} a_{m-j}(x, \xi, \lambda) \qquad (9.8)$$

for $t > 0$, $\lambda \in \Lambda$ and $t^d \lambda \in \Lambda$. Here m can be any complex number. This class of symbols will be denoted by $CS_d^m(X \times \mathbb{R}^n, \Lambda)$ and the corresponding class of operators by $CL_d^m(X, \Lambda)$. This class is stable under composition, taking the transpose and taking the adjoint.

We will say that the operator $A_\lambda \in CL_d^m(X, \Lambda)$ is *elliptic with parameter* if it is properly supported and

$$a_m(x, \xi, \lambda) \neq 0 \quad \text{if} \quad x \in X \quad \text{and} \quad |\xi| + |\lambda|^{1/d} \neq 0. \qquad (9.9)$$

It clearly follows that $A_\lambda \in HL_{1; 0; d}^{m; m}(X, \Lambda)$. There exists a parametrix $B_\lambda \in CL_d^{-m}(X, \Lambda)$ of a classical elliptic operator with parameter, which is also an elliptic operator with parameter.

Example 9.1. Let A be a differential operator in X of degree m and I the unit operator. Then $A - \lambda I \in CL_m^m(X, \mathbb{C})$ and the principal symbol is given by the formula

$$a_m(x, \xi, \lambda) = a_m(x, \xi) - \lambda, \qquad (9.10)$$

where $a_m(x, \xi)$ is the principal symbol of A. If Λ is a closed angle in the complex plane with vertex at the origin such that $a_m(x, \xi)$ for $|\xi| = 1$ does not take values in Λ, then the operator $A - \lambda I$ is elliptic with parameter (and, in particular, belongs to $HL^{m,\,m}_{1,\,0;\,m}(X, \Lambda)$).

9.2 Norms of operators with parameter. In this subsection we will consider operators with parameter of two kinds:

1) operators A_λ in $\mathbb{R}^n$, such that supp K_{A_λ} lies in a fixed compact set $\hat{K} \subset \mathbb{R}^{2n}$ (where $0 \leq \delta < \varrho \leq 1$);

2) operators on a closed manifold M (here, as usual, we assume $1 - \varrho \leq \delta < \varrho$).

In what follows we will write $A_\lambda \in L^m_{\varrho,\,\delta;\,d}(X, \Lambda)$ keeping in mind that $X = \mathbb{R}^n$ or $X = M$ and that 1) or 2) is fulfilled.

We denote by $\|A\|_{s,\,s-l}$ the norm of A viewed as an operator from $H^s(\mathbb{R}^n)$ into $H^{s-l}(\mathbb{R}^n)$ or from $H^s(M)$ into $H^{s-l}(M)$ (here l and s are real numbers). Our aim is to study the dependence of $\|A_\lambda\|_{s,\,s-l}$ on λ for large $|\lambda|$.

Theorem 9.1. *Let $A_\lambda \in L^m_{\varrho,\,\delta;\,d}(X, \Lambda)$, $l \geq m$ and $s, l \in \mathbb{R}$. Let one of the numbers δ, s, $s - l$ be equal to 0. Then*

$$\|A_\lambda\|_{s,\,s-l} \leq C_{s,\,l}(1 + |\lambda|^{1/d})^m, \quad \text{if} \quad l \geq 0, \tag{9.11}$$

$$\|A_\lambda\|_{s,\,s-l} \leq C_{s,\,l}(1 + |\lambda|^{1/d})^{-(l-m)}, \quad \text{if} \quad l \leq 0. \tag{9.12}$$

Corollary 9.1. *If $A_\lambda \in L^m_{\varrho,\,\delta;\,d}(X, \Lambda)$, where $m \leq 0$, then*

$$\|A_\lambda\| \leq C(1 + |\lambda|^{1/d})^m, \tag{9.13}$$

where $\|A_\lambda\|$ denotes the operator norm of A_λ in $L^2(X)$.

We will need the following useful
Lemma 9.1 (Schur Lemma). *If A is an operator with the Schwartz kernel K_A such that*

$$\sup_x \int |K_A(x, y)|dy \leq C \quad \text{and} \quad \sup_y \int |K_A(x, y)|dx \leq C,$$

then
$$\|A : L^2 \to L^2\| \leq C.$$

(This lemma holds for integral operators in L^2 on any measure space, or for integral operators $L^2(Y) \to L^2(X)$ for any measure spaces X, Y.)

Proof. By the Cauchy-Schwarz inequality we have

$$|Au(x)|^2 \leq \left(\int |K_A(x, y)||u(y)|dy \right)^2$$

$$= \left(\int |K_A(x, y)|^{1/2}|K_A(x, y)|^{1/2}|u(y)|dy \right)^2$$

$$\leq \int |K_A(x, y)|dy \int |K_A(x, y)||u(y)|^2 dy$$

$$\leq C \int |K_A(x, y)||u(y)|^2 dy.$$

Now integrating with respect to x and changing the order of integration we obtain the desired norm estimate. $\square$

Proof of Theorem 9.1. 1. Note first of all that a partition of unity reduces the case $X = M$ to the case $X = \mathbb{R}^n$, which we will now study.

2. Consider first the case $s = l = m = 0$. The statement of the theorem reduces to the estimate

$$\|A_\lambda\| \leqq C, \tag{9.14}$$

which is proved by repeating verbatim the argument in §6 (we recommend the reader to work this through as an exercise). Note now that the estimate (9.14) could be proved directly utilizing the results from §6 if the constants

$$C_{\alpha,\beta}(\lambda) = \sup_{x,\xi} |\partial_\xi^\alpha \partial_x^\beta a(x,\xi,\lambda)| \langle\xi\rangle^{\varrho|\alpha| - \delta|\beta|} \tag{9.15}$$

where bounded as $|\lambda| \to +\infty$. This evidently follows from (9.1) for $\delta = 0$, but for $\delta > 0$ some of the constants $C_{\alpha,\beta}(\lambda)$ can grow as $|\lambda| \to +\infty$. Therefore, for $\delta > 0$ it is indeed necessary to repeat the argument from §6.

3. Now consider in $\mathbb{R}^n$ the standard operator-valued function $\Phi_m(\lambda) \in L^m_{1,0;d}(\mathbb{R}^n, \mathbb{C})$ with the symbol $\varphi_m(x,\xi,\lambda) = (1 + |\xi|^2 + |\lambda|^{2/d})^{m/2}$. This ΨDO with parameter will be useful to us, although it does not satisfy condition 1).

Let us estimate the norm of the operator $\Phi_m(\lambda)$. The operator $\Phi_m(0)$ induces an isometric isomorphism of $H^m(\mathbb{R}^n)$ onto $L^2(\mathbb{R}^n)$. Therefore

$$\|\Phi_m(\lambda)\|_{s,\,s-l} = \|\Phi_{s-l}(0)\,\Phi_m(\lambda)\,\Phi_{-s}(0)\|.$$

But $\Phi_{s-l}(0)\,\Phi_m(\lambda)\,\Phi_{-s}(0) = \Phi_m(\lambda)\,\Phi_{-l}(0)$ is simply the multiplication operator by $(1+|\xi|^2+|\lambda|^{2/d})^{m/2}\,(1+|\xi|^2)^{-l/2}$ of the Fourier transform $\hat{u}(\xi)$ in $L^2(\mathbb{R}^n)$ and therefore its norm is equal to

$$\psi_{ml}(\lambda) = \sup_{\xi \in \mathbb{R}^n} (1+|\xi|^2+|\lambda|^{2/d})^{m/2}\,(1+|\xi|^2)^{-l/2}.$$

We obviously have

$$\psi_{ml}(\lambda) \leqq C \sup_{\xi \in \mathbb{R}^n} (1+|\xi|+|\lambda|^{1/d})^m\,(1+|\xi|)^{-l},$$

where C only depends on m and l (but not on λ).

Now, from the easily verified relation

$$\sup_{x \geqq 0} (1+x+t)^m (1+x)^{-l} = \begin{cases} (1+t)^m, & \text{if } l \geqq 0, \\ C_{ml}\,t^{m-l}, & \text{if } l \leqq 0 \quad \text{and} \quad t \geqq R_{ml} \end{cases}$$

(we assume $m \leqq l$ everywhere) it follows that

$$\psi_{ml}(\lambda) \leqq C_{ml}(1+|\lambda|^{1/d})^m, \qquad \text{if } l \geqq 0,$$
$$\psi_{ml}(\lambda) \leqq C_{ml}(1+|\lambda|^{1/d})^{m-l}, \qquad \text{if } l \leqq 0,$$

i.e. for $\Phi_m(\lambda)$ one of the norm-estimates (9.11), (9.12) holds as asserted in the theorem.

4. Consider now the general case $A_\lambda \in L^m_{\varrho,\delta;d}(\mathbb{R}^n, \Lambda)$. We obviously have

$$\begin{aligned}
\|A_\lambda\|_{s,s-l} &= \|\Phi_m(\lambda)\cdot(\Phi_{-m}(\lambda)\cdot A_\lambda)\|_{s,s-l} \\
&\leqq \|\Phi_m(\lambda)\|_{s,s-l}\cdot\|\Phi_{-m}(\lambda)A_\lambda\|_{s,s}
\end{aligned} \tag{9.16}$$

and analogously

$$\begin{aligned}
\|A_\lambda\|_{s,s-l} &= \|(A_\lambda\Phi_{-m}(\lambda))\Phi_m(\lambda)\|_{s,s-l} \\
&\leqq \|A_\lambda\cdot\Phi_{-m}(\lambda)\|_{s-l,s-l}\|\Phi_m(\lambda)\|_{s,s-l}.
\end{aligned} \tag{9.17}$$

Using the already proven norm-estimate for $\|\Phi_m(\lambda)\|_{s,s-l}$ we see that in order to complete the proof of the theorem when $s = 0$ or $s - l = 0$ it suffices to verify that

$$\|\Phi_{-m}(\lambda)A_\lambda\| \leqq C, \qquad \|A_\lambda\cdot\Phi_{-m}(\lambda)\| \leqq C. \tag{9.18}$$

5. Let us define $\tilde{\Phi}_{-m}(\lambda)$ as an operator with the Schwartz kernel which is obtained by multiplying the Schwartz kernel of $\Phi_{-m}(\lambda)$ by $\varphi(x-y)$, where $\varphi \in C_0^\infty(\mathbb{R}^n)$, $\varphi = 1$ in a neighbourhood of $0 \in \mathbb{R}^n$. Then $\tilde{\Phi}_{-m}(\lambda)$ is a uniformly properly supported ΨDO in $L^m_{1,0;d}(\mathbb{R}^n, \mathbb{C})$. Let us write

$$\Phi_{-m}(\lambda) = \tilde{\Phi}_{-m}(\lambda) + R_{-m}(\lambda) \tag{9.19}$$

and investigate the remainder operator $R_{-m}(\lambda)$ which has a Schwartz kernel $K_{R_{-m}}$ vanishing in a ε-neighbourhood of the diagonal. In fact it is a convolution operator, so $K_{R_{-m}}$ depends on $x - y$ and λ only. It is easy to see from the construction of $R_{-m}(\lambda)$ that $K_{R_{-m}}(x, y; \lambda) = r_{-m}(x-y, \lambda)$ with

$$r_{-m}(z, \lambda) = \int_{\mathbb{R}^n} e^{iz\cdot\xi}(1-\varphi(z))(1+|\xi|^2+|\lambda|^{2/d})^{-m/2}\,d\xi,$$

where $\varphi \in C_0^\infty(\mathbb{R}^n)$, $\varphi = 1$ is a neighbourhood of 0.

Now it is easy to prove that $r_{-m}(\cdot, \lambda) \in S(\mathbb{R}^n)$ for every fixed $\lambda \in \mathbb{C}$. Moreover, all seminorms of $r_{-m}(\cdot, \lambda)$ in $S(\mathbb{R}^n)$ decay as $|\lambda| \to \infty$ faster than any power of $|\lambda|$. Indeed, we can apply the standard integration by parts to get

$$r_{-m}(z, \lambda) = \int_{\mathbb{R}^n} e^{iz\cdot\xi}|z|^{-2N}(1-\varphi(z))\left[(-\Delta_\xi)^N(1+|\xi|^2+|\lambda|^{2/d})^{-m/2}\right]d\xi,$$

hence

$$z^{\alpha} D_z^{\beta} r_{-m}(z, \lambda) = \int_{\mathbb{R}^n} z^{\alpha} D_z^{\beta} \left[e^{iz \cdot \xi} |z|^{-2N} (1 - \varphi(z)) \right]$$

$$\times \left[(-\Delta_{\xi})^N (1 + |\xi|^2 + |\lambda|^{2/d})^{-m/2} \right] d\xi,$$

for an arbitrary integer $N \geq 0$ and any fixed multiindices α, β. The integrand can be estimated by

$$C(1 + |z|)^{|\alpha|-2N} (1 + |\xi|)^{|\beta|} (1 + |\xi| + |\lambda|^{1/d})^{-m-2N}$$

with a constant $C > 0$. We can assume that $m + 2N \geq 0$ and use the obvious estimate

$$(1 + |\xi| + |\lambda|^{1/d})^{-m-2N} \leq (1 + |\xi|)^{-m/2-N} (1 + |\lambda|^{1/d})^{-m/2-N}$$

to arrive to the estimate

$$|z^{\alpha} D_z^{\beta} r_{-m}(z, \lambda)| \leq C_{\alpha\beta N} (1 + |\lambda|^{1/d})^{-m/2-N},$$

which holds for sufficiently large N and implies the desired result.

6. Now we will sketch two possible proofs of estimates of type (9.18) for $R_{-m}(\lambda) A_{\lambda}$ and $A_{\lambda} R_{-m}(\lambda)$.

Let us recall that it is assumed that the Schwartz kernels of A_{λ} are supported in a fixed (independent of λ) compact subset of $\mathbb{R}^n \times \mathbb{R}^n$.

(a) Note that the proof of the boundedness result (e.g. Proposition 7.5) implies that

$$\|A_{\lambda}\|_{s,s-m} \leq C(1 + |\lambda|)^M,$$

where $M = M(s, A)$ is independent of λ. This is a rough estimate and it is easy to obtain by following the steps of the proof of Proposition 7.5 and of the necessary results from Sect. 6.

Now note that $R_{-m}(\lambda)$ is infinitely smoothing in the Sobolev scale $H^s(\mathbb{R}^n)$, and, more precisely,

$$\|R_{-m}(\lambda)\|_{s,t} \leq C_{s,t,N} (1 + |\lambda|)^{-N}$$

for all $s, t \in \mathbb{R}$ and $N \geq 0$. This holds because the convolution operator $R_{-m}(\lambda)$ can be presented as the multiplication of the Fourier transform by a function $\tilde{r}_{-m}(\xi, \lambda) = F_{z \to \xi} r(z, \lambda)$ which is in $S(\mathbb{R}^n)$ with respect to ξ with seminorms which decay as $|\lambda| \to \infty$ faster than any power of $|\lambda|$ due to the estimates obtained in the first part of my comments to this question above.

Combining the two inequalities above we immediately obtain the desired estimates of the type (9.18) for $R_{-m}(\lambda) A_{\lambda}$ and $A_{\lambda} R_{-m}(\lambda)$.

(b) Another way to establish these estimates is to study first the structure of the operators $R_{-m}(\lambda) A_{\lambda}$ and $A_{\lambda} R_{-m}(\lambda)$. It suffices to consider the operator $A_{\lambda} R_{-m}(\lambda)$ and then use the relation

$$(R_{-m}(\lambda)A_\lambda)^* = A_\lambda^* R_{-m}(\lambda)$$

to establish the same estimates for the operators of the type $R_{-m}(\lambda)A_\lambda$.

Clearly $A_\lambda R_{-m}(\lambda)$ can be written in the standard form (3.9) with the symbol $\sigma_{\lambda,m}(x,\xi) = \sigma_{A_\lambda}(x,\xi)\tilde{r}_{-m}(\xi,\lambda)$, where the function $\tilde{r}$ was defined above. Note that $\sigma_{A_\lambda}(x,\xi)$ has a compact support with respect to x, uniformly in ξ. Taking into account the behavior of $\tilde{r}$ we see that $\sigma_{\lambda,m}(x,\xi)$ satisfies the estimates

$$|\partial_\xi^\alpha D_x^\beta \sigma_{\lambda,m}(x,\xi)| \leq C_{\alpha,\beta,m,M,N}(1+|\xi|)^{-M}(1+|\lambda|)^{-N}$$

for any $M, N \geq 0$ and any multiindices α, β. The Schwartz kernel of the operator $A_\lambda R_{-m}(\lambda)$ has the form

$$K_\lambda(x,y) = \int\limits_{\mathrm{IR}^n} e^{i(x-y)\cdot\xi}\sigma_\lambda(x,\xi)\,d\xi.$$

The estimates for σ_λ immediately imply that

$$|K_\lambda(x,y)| \leq C_N(1+|x-y|)^{-N},$$

and the required norm estimates follow from the Schur lemma.

7. Now let $\delta = 0$. It is clear from (9.16) that we need to show that

$$\|\Phi_{-m}(\lambda)A_\lambda\|_{s,s} \leq C, \tag{9.20}$$

where C does not depend on λ (but may depend on s). Clearly

$$\|\Phi_{-m}(\lambda)A_\lambda\|_{s,s} = \|\Phi_s(0)(\Phi_{-m}(\lambda)A_\lambda)\Phi_{-s}(0)\|. \tag{9.21}$$

Acting as in the part 5 of this proof, we may replace $\Phi_t(\lambda)$ by a properly supported ΨDO $\tilde{\Phi}_t(\lambda)$ (for any $t \in \mathrm{IR}$) and instead of (9.20) prove the estimate

$$\|\tilde{\Phi}_s(0)\tilde{\Phi}_{-m}(\lambda)A_\lambda\tilde{\Phi}_{-s}(0)\| \leq C. \tag{9.22}$$

Denote the symbol of $\tilde{\Phi}_{-m}(\lambda)A_\lambda$ by $b(x,\xi,\lambda)$. We claim that

$$b(x,\xi,\lambda) \in S^0_{\varrho,0;d}(\mathrm{IR}^n;\Lambda)$$

in the sense of the uniform classes in IR^n (see problems 3.1 and 3.2), and in particular $b(x,\xi,\lambda) \in S^0_{\varrho,0}(\mathrm{IR}^{2n})$ uniformly in λ, i.e.

$$\sup_{x,\xi,\lambda}[|\partial_\xi^\alpha\partial_x^\beta b(x,\xi,\lambda)|\langle\xi\rangle^{\varrho|\alpha|}] < +\infty$$

Using an appropriate composition formula (e.g. in the uniform classes discussed in Problems 3.1 and 3.2) we see that the same estimates hold for the symbol $a(x,\xi,\lambda)$ of the operator $\tilde{\Phi}_s(0)\tilde{\Phi}_{-m}(\lambda)A_\lambda\tilde{\Phi}_{-s}(0)$.

Now applying the boundedness theorem (again extended to uniform operators in IR^n) we get the desired estimate (9.22) for the operator norm.

Another possible way of arguing (to avoid using uniform operators in IR^n): introduce appropriate cut-off functions to reduce everything to functions supported in a fixed compact set, and then investigate remainders, using the Schur Lemma.   □

9.3 The inverse of operators with parameter. In this part we will consider only operators on a closed manifold M.

Theorem 9.2. *Let* $A_\lambda \in HL^{m;\,m_0}_{\varrho,\,\delta;\,d}(M,\Lambda)$. *Then there exists* $R > 0$, *such that for* $|\lambda| \geqq R$, *the operator* A_λ *is invertible with*

$$A_\lambda^{-1} \in HL^{-m_0;\,-m}_{\varrho,\,\delta;\,d}(M,\Lambda_R)\,, \tag{9.23}$$

where $\Lambda_R = \Lambda \cap \{\lambda \mid |\lambda| \geqq R\}$. *More precisely, if* B_λ *is a parametrix for the operator with parameter* A_λ, *i.e. condition* (9.6) *is fulfilled, then*

$$A_\lambda^{-1} - B_\lambda \in L^{-\infty}(M,\Lambda_R)\,. \tag{9.24}$$

Proof. Let B_λ be a parametrix for the operator A_λ. Then it is obvious from (9.6) that it suffices to prove that $I + R_\lambda$ with $R_\lambda \in L^{-\infty}(M,\Lambda_R)$ is invertible for small λ and

$$(I + R_\lambda)^{-1} - I \in L^{-\infty}(M,\Lambda_R)\,. \tag{9.25}$$

Note that for arbitrary $N > 0$ and $s,\,t \in \mathbb{R}$

$$\|R_\lambda\|_{s,\,t} \leqq C^{(N)}_{s,\,t}(1 + |\lambda|)^{-N}\,. \tag{9.26}$$

From this it follows, in particular, that there exists $R > 0$ such that $\|R_\lambda\| < \frac{1}{2}$ for $|\lambda| \geqq R$ and hence $(I + R_\lambda)^{-1}$ exists for $\lambda \in \Lambda_R$ at least in the space $L^2(M)$. Now, $I + R_\lambda$ is Fredholm in each of the spaces $H^s(M)$ (formally this is a consequence of the ellipticity of $I + R_\lambda$ and Theorem 8.1, although it is easy to obtain directly) and has everywhere the same kernel and cokernel, so that the invertiblity of $I + R_\lambda$ for $|\lambda| \geqq R$ is guaranteed in each of the spaces $H^s(M)$.

To prove (9.25) it is convenient to use

$$(I + R_\lambda)^{-1} - I = -R_\lambda(I + R_\lambda)^{-1} \tag{9.27}$$

and (9.26). Denoting the left hand side of (9.27) by Q_λ we see that estimates of the form (9.26) hold for Q_λ.

The kernel $Q_\lambda(x,y)$ of Q_λ can be expressed by the formula

$$Q_\lambda(x,y) = [Q_\lambda\delta(\cdot - y)](x)\,, \tag{9.28}$$

where $\delta(z - y)$ is the δ-function (in z) at a point $y \in M$ on which it depends as a parameter. The operator Q_λ in (9.28) acts on the variable z and the result is taken at the point x. Note that if $s < -n/2$, then $\delta(\cdot - y) \in H^s(M)$. Further, $\delta(\cdot - y)$ is a differentiable function of y with values in $H^{s-1}(M)$ and is more generally a k times differentiable function of y with values in $H^{s-k}(M)$. Therefore from (9.26) it follows that estimates of the form (9.3) hold for the kernel $Q_\lambda(x, y)$, which also demonstrates that $Q_\lambda \in L^{-\infty}(M, \Lambda_R)$. The inclusions (9.23) and (9.24) then also readily follow. $\square$

9.4 The resolvent of an elliptic operator. Returning to example 9.1 in the case of an operator on a manifold and applying the results obtained we get

Theorem 9.3. *Let A be a differential operator on a closed manifold M with principal symbol $a_m(x, \xi)$ and Λ a closed angle in the complex plane $\mathbb{C}$ with vertex $0 \in \mathbb{C}$. Let A be elliptic with parameter relative to Λ, i.e. for $\xi \neq 0$, $a_m(x, \xi)$ does not take values in Λ. Then*

a) *there exists $R > 0$ such that $A - \lambda I$ is invertible for $\lambda \in \Lambda_R$ with*

$$(A - \lambda I)^{-1} \in CL_m^{-m}(M, \Lambda_R);\qquad(9.29)$$

b) *the following norm estimate holds*

$$\|(A - \lambda I)^{-1}\|_{s,\,s+l} \leq C_{s,\,l}/|\lambda|^{1-l/m}, \qquad 0 \leq l \leq m, \qquad \lambda \in \Lambda_R,\qquad(9.30)$$

where s is any real number.

Proof. a) follows from Theorem 9.2 and b) from a) and Theorem 9.1. $\square$

Corollary 9.2. *Under the conditions of Theorem 9.3 we have*

$$\|(A - \lambda I)^{-1}\| \leq C/|\lambda|, \qquad \lambda \in \Lambda_R.\qquad(9.31)$$

Corollary 9.3. *Let A be an elliptic self-adjoint differential operator on a closed manifold M with principal symbol $a_m(x, \xi)$. Assume that $a_m(x, \xi) > 0$ for all (x, ξ), $\xi \neq 0$. Then A is semi-bounded from below, i.e. there is a constant $C > 0$ such that $A \geq -CI$ or*

$$(Au, u) \geq -C(u, u), \qquad u \in C^\infty(M).\qquad(9.32)$$

We mention here another important fact, namely that under the condition of ellipticity with parameter, the resolvent $(A - \lambda I)^{-1}$ differs from the parametrix only by an operator of $L^{-\infty}(M, \Lambda)$. This is used in the theory of complex powers along with the explicit construction of a parametrix as given in the elliptic theory (cf. section 5.5).

Problem 9.1. Extend the theory of elliptic operators with parameter to the case of matrix operators, i.e. to systems.

Hint. The condition of ellipticity with parameter for a matrix function $a_m(x, \xi)$, means that $\det(a_m(x, \xi) - \lambda) \neq 0$ for $\xi \neq 0$ and $\lambda \in \Lambda$ or, equivalently, that the eigenvalues of $a_m(x, \xi)$ do not belong to Λ for $\xi \neq 0$.

Problem 9.2. Let A be an elliptic differential operator on a closed manifold M, and suppose that for some angle Λ the operator $A - \lambda I$ is elliptic with parameter for $\lambda \in \Lambda$. Show that index $A = 0$.

§10. Definition and Basic Properties of the Complex Powers of an Elliptic Operator

10.1 Definition of the holomorphic semigroup A_z. Let A be an elliptic differential operator of order m on a closed n-dimensional manifold M and $a_m(x, \xi)$ the principal symbol of A. Assume that $a_m(x, \xi)$ does not take values in a closed angle Λ of the complex plane $\mathbb{C}$ for $\xi \neq 0$ (here the vertex of Λ is assumed to be at $0 \in \mathbb{C}$). In other words, in the notation of §9, $A - \lambda I \in CL_m^m(M, \Lambda)$ and satisfies the condition of ellipticity with parameter.

It follows from Theorem 9.3 that the resolvent $R_\lambda = (A - \lambda I)^{-1}$ is defined for $|\lambda| \geq R$, $\lambda \in \Lambda$ i.e. for $\lambda \in \Lambda_R$. Now, in view of Theorem 8.4, we see that the spectrum $\sigma(A)$ of A is a discrete subset of $\mathbb{C}$. Hence, in the angle there can be only a finite number of points of $\sigma(A)$. We may therefore draw a ray L_0, starting at 0 and running inside Λ such that $\sigma(A) \cap L_0$ is either empty or consists of the point 0 only. In what follows, we assume for convenience firstly that $0 \notin \sigma(A)$, i.e. A^{-1} exists as an operator (cf. §8), and secondly that $L_0 = (-\infty, 0]$. Neither of these assumptions is very essential; we may get rid of the first one by replacing A with $A + \varepsilon I$ and the second by studying $e^{i\theta} A$ instead of A.

So, finally, our assumptions are as follows:

$$1) \quad a_m(x, \xi) - \lambda \neq 0 \quad \text{for} \quad \xi \neq 0 \quad \text{and} \quad \lambda \in (-\infty, 0]; \tag{10.1}$$

$$2) \quad \sigma(A) \cap (-\infty, 0] = \emptyset. \tag{10.2}$$

It follows from conditions 1) and 2) that for some angle Λ of the form $\{\pi - \varepsilon \leq \arg \lambda \leq \pi + \varepsilon\}$, with $\varepsilon > 0$, the following hold:

$$1') \; a_m(x, \xi) - \lambda \neq 0 \text{ for } \xi \neq 0 \text{ and } \lambda \in \Lambda; \tag{10.1'}$$

$$2') \; \sigma(A) \cap \Lambda = \emptyset. \tag{10.2'}$$

In what follows we shall assume that Λ has been chosen in this way.

Since $0 \notin \sigma(A)$, we see that $\sigma(A)$ does not intersect a disk $|\lambda| < 2\varrho$ in the complex λ-plane. Now select in this plane a contour $\Gamma = \Gamma_\varrho$ of the form $\Gamma = \Gamma_1 \cup \Gamma_2 \cup \Gamma_3$, where (Fig. 1)

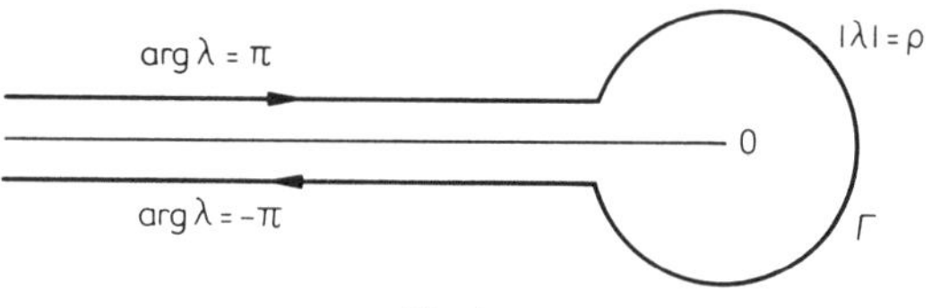

Fig. 1

$$\lambda = re^{i\pi} \ (+\infty > r > \varrho) \qquad \text{on } \Gamma_1,$$
$$\lambda = \varrho e^{i\varphi} \ (\pi > \varphi > -\pi) \qquad \text{on } \Gamma_2,$$
$$\lambda = re^{-i\pi} \ (\varrho < r < +\infty) \qquad \text{on } \Gamma_3.$$

Consider the integral

$$A_z = \frac{i}{2\pi} \int_\Gamma \lambda^z (A - \lambda I)^{-1} \, d\lambda, \tag{10.3}$$

where $z \in \mathbb{C}$, λ^z is defined as a holomorphic function of λ for $\lambda \in \mathbb{C} \setminus (-\infty, 0]$, equal to $e^{z \ln \lambda}$ for $\lambda > 0$ (here it is assumed that $\ln \lambda \in \mathbb{R}$ for $\lambda > 0$). In other words, on Γ we set

$$\lambda^z = e^{z \ln \lambda} = e^{z \ln |\lambda| + iz \arg \lambda} = |\lambda|^z e^{iz \arg \lambda}, \tag{10.4}$$

where $-\pi \leq \arg \lambda \leq \pi$ ($\arg \lambda$ is described more precisely in the definition of the contour Γ).

Note that in view of the estimate (9.31) (Corollary 9.2) the integral (10.3) converges in the operator norm on $L^2(M)$ for $\operatorname{Re} z < 0$ and also A_z is a bounded operator on $L^2(M)$. In the same way, by Theorem 9.3, the integral (10.3) converges in the operator norm on $H^s(M)$ for arbitrary $s \in \mathbb{R}$ and also, for $\operatorname{Re} z < 0$, A_z maps $H^s(M)$ into $H^s(M)$ hence also maps $C^\infty(M)$ into $C^\infty(M)$ as well as $\mathcal{D}'(M)$ into $\mathcal{D}'(M)$, since

$$C^\infty(M) = \bigcap_s H^s(M) \qquad \text{and} \qquad \mathcal{D}'(M) = \bigcup_s H^s(M).$$

Proposition 10.1. a) *For* $\operatorname{Re} z < 0$ *and* $\operatorname{Re} w < 0$ *we have the semigroup property*

$$A_z A_w = A_{z+w} \tag{10.5}$$

b) *If* $k \in \mathbb{Z}$ *and* $k > 0$ *then*

$$A_{-k} = (A^{-1})^k. \tag{10.6}$$

c) *For arbitrary* $s \in \mathbb{R}$, A_z *is a holomorphic operator-function of* z *(for* $\operatorname{Re} z < 0$*) with values in the algebra of bounded operators on the Hilbert space* $H^s(M)$.

Proof. a) Construct a contour Γ' (Fig. 2) of the form $\Gamma' = \Gamma_1' \cup \Gamma_2' \cup \Gamma_3'$, where

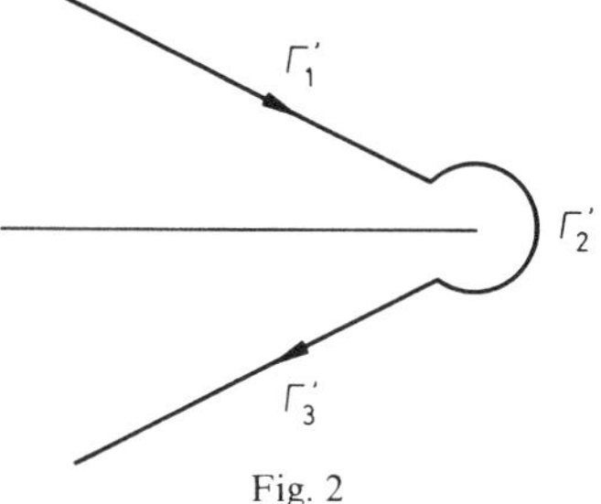

Fig. 2

$$\lambda = re^{i(\pi-\varepsilon)} \ (+\infty > r > \tfrac{3}{2}\varrho) \qquad\qquad \text{on } \Gamma_1'$$
$$\lambda = \tfrac{3}{2}\varrho e^{i\varphi} \ (\pi-\varepsilon > \varphi > -\pi+\varepsilon) \qquad\qquad \text{on } \Gamma_2'$$
$$\lambda = re^{i(-\pi+\varepsilon)} \ (\tfrac{3}{2}\varrho < r < +\infty) \qquad\qquad \text{on } \Gamma_3'.$$

The number $\varepsilon > 0$ is chosen so that (10.1$'$) and (10.2$'$) are satisfied. The contour Γ is contained "within" Γ', and in view of (9.31) and the condition on Γ' it is obvious that the integral (10.3) does not change if we replace Γ by Γ'.

Utilizing this fact, we obtain

$$A_z A_w = -\frac{1}{4\pi^2} \int_{\Gamma'} \int_{\Gamma} (A-\lambda I)^{-1} (A-\mu I)^{-1} \lambda^z \mu^w \, d\mu \, d\lambda$$

$$= -\frac{1}{4\pi^2} \int_{\Gamma'} \int_{\Gamma} \frac{\lambda^z \mu^w}{\lambda-\mu} [(A-\lambda I)^{-1} - (A-\mu I)^{-1}] \, d\mu \, d\lambda$$

$$= \frac{i}{2\pi} \int_{\Gamma'} \lambda^{w+z}(A-\lambda I)^{-1} \, d\lambda + \frac{1}{4\pi^2} \int_{\Gamma} \int_{\Gamma'} (A-\mu I)^{-1} \frac{\lambda^z \mu^w}{\lambda-\mu} \, d\lambda \, d\mu$$

$$= A_{z+w} + 0 = A_{z+w}.$$

In this computation we have used the Cauchy formula and the so-called Hilbert identity

$$(A-\lambda I)^{-1} (A-\mu I)^{-1} = \frac{1}{\lambda-\mu} [(A-\lambda I)^{-1} - (A-\mu I)^{-1}], \qquad (10.7)$$

which is clear if we, for instance, multiply both sides by $(A-\lambda I)(A-\mu I)$.

b) Note that if $z = -1, -2, \ldots$, then $(re^{i\pi})^z = (re^{-i\pi})^z$ and the integrals along the straight line parts of Γ in (10.3) cancel. Therefore

$$A_{-k} = \frac{i}{2\pi} \int_{\Gamma_2} \lambda^{-k}(A-\lambda I)^{-1} \, d\lambda,$$

where $\Gamma_2 = \{|\lambda| = \varrho\}$, traversed clockwise. Now make a change of variables, putting $\lambda = 1/\mu$ which gives

$$A_{-k} = -\frac{i}{2\pi} \int_{\Gamma_2'} \mu^k (A - \mu^{-1} I)^{-1} \mu^{-2} \, d\mu,$$

where $\Gamma_2' = \{|\mu| = 1/\varrho\}$, also traversed clockwise. Taking into account that $(A - \mu^{-1} I)^{-1} = \mu A^{-1} (\mu I - A^{-1})^{-1}$, we may now write

$$A_{-k} = -\frac{iA^{-1}}{2\pi} \int_{\Gamma_2'} \mu^{k-1} (\mu I - A^{-1})^{-1} \, d\mu = A^{-1} (A^{-1})^{k-1} = (A^{-1})^k,$$

since the entire spectrum of the bounded operator A^{-1} is situated inside the contour Γ_2' and we may use the Cauchy formula.

c) Differentiating the integral (10.3) with respect to z we obtain the integral

$$\frac{i}{2\pi} \int_{\Gamma} \lambda^z (\ln \lambda) (A - \lambda I)^{-1} \, d\lambda, \tag{10.8}$$

converging in operator norm (in $H^s(M)$) uniformly for $\operatorname{Re} z \leqq -\varepsilon < 0$. We conclude, that the operator function A_z is holomorphic in z and the derivative $\frac{d}{dz} A_z$ is equal to the integral (10.8). $\square$

10.2 Definition of the complex powers of an operator

Definition 10.1. Let $z \in \mathbb{C}$ and $k \in \mathbb{Z}$ be such that $\operatorname{Re} z < k$. Put, on $C^\infty(M)$ or $\mathscr{D}'(M)$

$$A^z = A^k A_{z-k}. \tag{10.9}$$

We need to verify that this is well-defined and that is the content of the first part of the following theorem.

Theorem 10.1. a) *The operator A^z as defined by (10.9) is independent of the choice of integer k, provided $\operatorname{Re} z < k$.*

b) *If $\operatorname{Re} z < 0$, then $A^z = A_z$.*

c) *The group property holds*

$$A^z A^w = A^{z+w}, \qquad z, w \in \mathbb{C} \tag{10.10}$$

d) *If $k \in \mathbb{Z}$, then (10.9) with $z = k$ gives the usual k-th power of the operator A (in particular $A^0 = I$, $A^1 = A$ and A^{-1} is the inverse to A).*

e) *For arbitrary $k \in \mathbb{Z}$ and $s \in \mathbb{R}$, the function A^z is a holomorphic operator function of z in the half-plane* $\mathrm{Re}\, z < k$ *with values in the Banach space $\mathscr{L}\,(H^s(M), H^{s-mk}(M))$ of bounded linear operators from $H^s(M)$ to $H^{s-mk}(M)$.*

Proof. a) Let $z \in \mathbb{C}$, $k, l \in \mathbb{Z}$ be such that $\mathrm{Re}\, z < k$, $\mathrm{Re}\, z < l$. We need to verify that

$$A^k A_{z-k} = A^l A_{z-l} \tag{10.11}$$

Assume that $k > l$ and put $k - l = p$ and $z - k = w$. Then (10.11) reduces to the equality $A_w = A^{-p} A_{w+p}$ with p a positive integer and $\mathrm{Re}\,(w+p) < 0$. This however, follows at once from Proposition 10.1 since $A^{-p} = A_{-p}$ by (10.6) and it only remains to use the semi-group property (10.5).

b)–d) These properties follow in an obvious manner from a) and Proposition 10.1.

e) This statement follows straightforwardly from a) and c) in Proposition 10.1, if we remember that A^k, for k an integer, maps $H^s(M)$ continuously into $H^{s-mk}(M)$ (this follows from Theorem 7.3 (on boundedness) and the fact that $A^{-1} \in CL^{-m}(M)$, as is clear from Theorem 8.2 (about the inverse operator)). $\square$

10.3 The self-adjoint case. Let a smooth positive density on M be given defining a scalar product on $L^2(M)$ and A a self-adjoint elliptic differential operator of order m on M. Then its principal symbol $a_m(x, \xi)$ is real-valued. In this case, conditions (10.1) and (10.2), which we assume to hold, mean that

$$a_m(x, \xi) > 0, \qquad \xi \neq 0, \tag{10.12}$$

$$A \geq \delta I, \qquad \delta > 0, \tag{10.13}$$

i.e. $(Au, u) \geq \delta\,(u, u)$ for any $u \in C^\infty(M)$. Let $\{\varphi_j\}_{j=1}^\infty$ be a complete orthonormal system of eigenfunctions for A with eigenvalues $\{\lambda_j\}_{j=1}^\infty$. Remember that $\lambda_j \to +\infty$ as $j \to \infty$ (cf. Theorem 8.3). It follows from (10.13) that $\lambda_j \geq \delta > 0$ for all $j = 1, 2, \ldots$.

Now, any distribution $f \in \mathscr{D}'(M)$ may be represented as a Fourier series

$$f \sim \sum_{j=1}^\infty f_j \varphi_j(x), \qquad x \in M, \tag{10.14}$$

where

$$f_j = (f, \varphi_j). \tag{10.15}$$

Here, when $f \in L^2(M)$ we of course have the usual scalar product in $L^2(M)$. If, however $f \in \mathscr{D}'(M)$, then (f, φ_j) denotes $\langle f, \bar{\varphi}_j d\mu \rangle$, where $d\mu$ is the fixed density on M (recall that the distributions are linear continuous functionals on the space of smooth densities). We now describe the properties of the Fourier series of smooth functions and distributions.

Proposition 10.2. *For a series*

$$\sum_{j=1}^{\infty} c_j \varphi_j(x) \tag{10.16}$$

with complex coefficients c_j the following properties are equivalent:
   a) *the series (10.16) converges in the $C^{\infty}(M)$-topology;*
   b) *the series (10.16) is the Fourier series of some function $f \in C^{\infty}(M)$;*
   c) *for any integer N*

$$\sum_{j=1}^{\infty} |c_j|^2 \lambda_j^{2N} < +\infty . \tag{10.17}$$

Furthermore, conditions d), e) *and* f) *are also equivalent:*
   d) *the series (10.16) converges in the weak topology of $\mathscr{D}'(M)$;*

   e) *the series (10.16) is the Fourier series of some distribution $f \in \mathscr{D}'(M)$;*

   f) *the exists an integer N (perhaps negative), such that (10.17) is fulfilled.*

Proof. The basic idea of the proof is to use the relations

$$C^{\infty}(M) = \bigcap_s H^s(M), \qquad \mathscr{D}'(M) = \bigcup_s H^s(M),$$

and the fact that A^N is a topological isomorphism between the spaces $H^{mN}(M)$ and $L^2(M)$. Now, $L^2(M)$ is easily characterized in terms of the coefficients of Fourier series by the Parceval equality. Therefore the topology of $C^{\infty}(M)$ may be determined via the seminorms $\|f\|_{A;N}$ where

$$\|f\|_{A;N}^2 = \sum_{j=1}^{\infty} |f_j|^2 \lambda_j^{2N} ,$$

because $A^N f$ has the Fourier coefficients $\lambda_j^N f_j$. From this the equivalence of conditions a), b) and c) is obvious.

Let us verify the equivalence of d), e) and f). If d) is satisfied denote by f the sum of the series (10.16), so that $f \in \mathscr{D}'(M)$ and

$$(f, \varphi) = \sum_{j=1}^{\infty} c_j (\varphi_j, \varphi)$$

for any $\varphi \in C^{\infty}(M)$. In particular, taking $\varphi = \varphi_j$ we obtain $c_j = f_j$ i.e. precisely e). Further, if e) is satisfied, i.e. $c_j = (f, \varphi_j)$, with $f \in \mathscr{D}'(M)$, then selecting an integer N such that $A^N f \in L^2(M)$ we also see that f) is satisfied. Finally, if f) is satisfied, then the series (10.16) converges in the norm of $H^{mN}(M)$, thus giving d). $\square$

We now introduce a "spectral" characterization of the complex powers of a self-adjoint operator A in terms of the coefficients of the eigenfunction expansion associated with the eigenfunctions φ_j.

Proposition 10.3. *Let $f \in \mathscr{D}'(M)$ and let $f(x) = \sum_{j=1}^{\infty} f_j \varphi_j(x)$ be the Fourier series expansion of f in the eigenfunctions of the operator A. Then*

$$A^z f = \sum_{j=1}^{\infty} \lambda_j^z f_j \varphi_j(x). \tag{10.18}$$

In particular, $\varphi_j(x)$ are the eigenfunctions of the operator A^z with eigenvalues λ_j^z.

Proof. The operator A^z maps $C^\infty(M)$ continuously into itself. In view of the easily verified relation $(A^z)^* = A^{\bar{z}}$ we see that A^z being the adjoint of $A^{\bar{z}}$ continuously maps $\mathscr{D}'(M)$ into $\mathscr{D}'(M)$ provided $\mathscr{D}'(M)$ is endowed with its weak topology. Since the series on the right hand side of (10.18) converges weakly in $\mathscr{D}'(M)$ by Proposition 10.2, in order to verify (10.18) in the general case it suffices to do so for $f = \varphi_j$, $\operatorname{Re} z < 0$. But, $f = \varphi_j$ and $\operatorname{Re} z < 0$ imply

$$A^z \varphi_j = \frac{i}{2\pi} \int_\Gamma \lambda^z (A - \lambda I)^{-1} \varphi_j \, d\lambda = \frac{i \varphi_j}{2\pi} \int_\Gamma \lambda^z (\lambda_j - \lambda)^{-1} \, d\lambda = \lambda_j^z \varphi_j$$

by the Cauchy formula. But this is exactly (10.18) for $f = \varphi_j$. $\quad\square$

Exercise 10.1. Let A satisfy (10.1) but instead of (10.2) require the weaker condition

$$\sigma(A) \cap (-\infty, 0) = \emptyset,$$

so that $\operatorname{Ker} A$ may be a non-empty finite-dimensional subspace in $C^\infty(M)$. Define A_z via the contour integral (10.3) and let E_0, E_0' be the invariant subspaces of the operator A introduced in Theorem 8.4.

a) Show that $A_z E_0 = 0$ and E_0' is an invariant subspace of A_z for $\operatorname{Re} z < 0$.

b) Show that for the operators A_z the semigroup property $A_z A_w = A_{z+w}$, $\operatorname{Re} z < 0$, $\operatorname{Re} w < 0$ holds, allowing A^z to be well-defined for all z by formula (10.9), so that the group property (10.10) holds for the operators A^z.

c) Verify that A^z for a sufficiently large positive integer z is the usual power of A whereas

$$A^0 = I - P_0$$

where P_0 is the projection onto the subspace E_0 parallel to E_0' and that, analogously, A_{-k} for a sufficiently large positive integer k is the inverse of the operator A^k on E_0' and equals 0 on E_0.

Problem 10.1. Let A be an elliptic differential operator with principal symbol $a_m(x, \xi)$ on a closed manifold M. Assume that $\operatorname{Re} a_m(x, \xi) < 0$ for $\xi \neq 0$. Show that the Cauchy problem

$$\frac{\partial u}{\partial t} = Au, \qquad t > 0; \qquad u|_{t=0} = \varphi(x); \tag{10.19}$$

has a unique solution in $C^\infty(M)$ and $\mathscr{D}'(M)$.

Hint. The solution $u(t, x)$ will necessarily be of the form

$$u = e^{tA}\varphi, \tag{10.20}$$

where the operator e^{tA} is determined via the contour integral

$$e^{tA} = \frac{i}{2\pi} \int_\Gamma e^{\lambda t}(A - \lambda I)^{-1}\, d\lambda. \tag{10.21}$$

Assuming that the spectrum $\sigma(A)$ is situated in the half-plane $\operatorname{Re}\lambda < 0$ (which can be achieved by changing A into $A - CI$ or substituting $u = v e^{ct}$ in (10.19)), it suffices to take $\Gamma = \Gamma_1 \cup \Gamma_2$, where Γ_1 and Γ_2 are the following two rays:

$$\lambda = r e^{i\left(\frac{\pi}{2} - \varepsilon\right)} \quad (+\infty > r > 0) \quad \text{on } \Gamma_1,$$

$$\lambda = r e^{i\left(-\frac{\pi}{2} + \varepsilon\right)} \quad (0 < r < +\infty) \quad \text{on } \Gamma_2.$$

The uniqueness of the generalized solution of the problem (10.19) is demonstrated using the Holmgren principle (cf. e.g. Gel'fand I. M., Šilov G. E. [1], vol. 3).

§11. The Structure of the Complex Powers
of an Elliptic Operator

11.1 The symbol of the resolvent. Let A be an elliptic differential operator on a closed manifold M. We shall next construct in local coordinates the symbol of a special parametrix of the operator with parameter $A - \lambda I$ (which we view here as an operator in $CL_m^m(M, \Lambda)$, Λ a closed angle in $\mathbb{C}$ with vertex at 0). We assume that A satisfies the conditions for ellipticity with parameter relative to Λ, where the angle Λ is as described in §10 (i.e. it satisfies (10.1′) and (10.2′), and Λ contains the semi-axis $(-\infty, 0]$).

The parametrix will be constructed in a chart $X \subset M$ and we will identify X with an open set in $\mathbb{R}^n$ using a coordinate system on X. Let the operator A on X be of the form

$$A = \sum_{|\alpha| \leq m} a_\alpha(x) D^\alpha. \tag{11.1}$$

Its total symbol

$$a(x, \xi) = \sum_{|\alpha| \leq m} a_\alpha(x) \xi^\alpha \tag{11.2}$$

may be decomposed into homogeneous components

$$a_j(x, \xi) = \sum_{|\alpha| = j} a_\alpha(x) \xi^\alpha, \qquad j = 0, 1, \ldots, m. \tag{11.3}$$

The total symbol $a(x, \xi, \lambda) = a(x, \xi) - \lambda$ of $A - \lambda I$ may be decomposed into components homogeneous in $(\xi, \lambda^{1/m})$ given by the formulas

$$a_m(x, \xi, \lambda) = a_m(x, \xi) - \lambda, \tag{11.4}$$

$$a_j(x, \xi, \lambda) = a_j(x, \xi), \qquad j = 0, 1, \ldots, m - 1. \tag{11.5}$$

The condition of ellipticity with parameter means that

$$a_m(x, \xi, \lambda) \neq 0 \quad \text{for} \quad x \in X, \quad \lambda \in \Lambda, \quad |\xi| + |\lambda|^{1/m} \neq 0 \tag{11.6}$$

It is natural to look for the symbol of the parametrix of $A - \lambda I$ in the form of an asymptotic sum of functions homogeneous in $(\xi, \lambda^{1/m})$. Denote these functions by $b^0_{-m-j}(x, \xi, \lambda), j = 0, 1, 2, \ldots,$ where the lower index indicates the degree of homogeneity:

$$b^0_{-m-j}(x, t\xi, t^m\lambda) = t^{-m-j} b^0_{-m-j}(x, \xi, \lambda), \quad t > 0, \quad |\xi| + |\lambda|^{1/m} \neq 0. \tag{11.7}$$

These functions are recursively defined by the relations

$$a_m(x, \xi, \lambda)\, b^0_{-m}(x, \xi, \lambda) = 1, \tag{11.8}$$

$$a_m(x, \xi, \lambda) b^0_{-m-j}(x, \xi, \lambda)$$

$$+ \sum_{\substack{k + l + |\alpha| = j \\ l < j}} \partial_\xi^\alpha a_{m-k}(x, \xi, \lambda) D_x^\alpha b^0_{-m-l}(x, \xi, \lambda)/\alpha! = 0, \tag{11.9}$$

$$j = 1, 2, \ldots$$

(compare the construction of a parametrix for the classical elliptic PDO in §5, formulas (5.17′) and (5.17″)).

To obtain a real parametrix from these functions $b^0_{-m-j}(x, \xi, \lambda)$ it is first of all necessary to eliminate their singularities for $|\xi| + |\lambda|^{1/m} = 0$ by multiplication with a cut-off function and, secondly, to glue together the various local parametrices using a partition of unity (cf. §5).

11.2 The Symbols of complex powers. We shall construct the homogeneous components $b^{(z),\,0}_{mz-j}(x,\xi)$ of the symbol of A^z using the homogeneous components of the parametrix constructed earlier, in exactly the same way as the powers A^z were constructed using the resolvent $(A-\lambda I)^{-1}$. Indeed, from the condition of ellipticity with a parameter there follows the existence of $\varrho = \varrho(x,\xi)$ such that $a_m(x,\xi,\lambda) \neq 0$ for $|\lambda| < 2\varrho$, $\xi \neq 0$ and $\lambda \in \mathbb{C}$. From (11.8) and (11.9) it is clear that $b_{-m-j}(x,\xi,\lambda)$ is holomorphic in λ in the disc $|\lambda| < 2\varrho$. Forming the contour Γ as in §10, we may define the functions $b^{(z),\,0}_{mz-j}(x,\xi)$ for $\operatorname{Re} z < 0$ by the formulas

$$b^{(z),\,0}_{mz-j}(x,\xi) = \frac{i}{2\pi} \int_{\Gamma} \lambda^z b^0_{-m-j}(x,\xi,\lambda)\,d\lambda, \qquad j = 0,\,1,\,2,\,\ldots, \tag{11.10}$$

where the branch λ^z is defined as in §10.

In particular, for $j = 0$, we obtain by the Cauchy formula that

$$b^{(z),\,0}_{mz}(x,\xi) = \frac{i}{2\pi} \int_{\Gamma} \lambda^z (a_m(x,\xi)-\lambda)^{-1}\,d\lambda = a_m^z(x,\xi). \tag{11.11}$$

Let us note that for a sufficiently small ϱ the integral (11.10) is independent of the choice of ϱ (in the disc $|\lambda| < 2\varrho$ there are no singularities of the functions $b_{-m-j}(x,\xi,\lambda)$). The function $b^{(z),\,0}_{mz-j}(x,\xi)$ is positively homogeneous in ξ of degree $mz - j$, i.e.

$$b^{(z),\,0}_{mz-j}(x,t\xi) = t^{mz-j} b^{(z),\,0}_{mz-j}(x,\xi), \quad t > 0, \quad \xi \neq 0. \tag{11.12}$$

To prove this, it is necessary to perform a change of variables in the integral (11.10) and use the homogeneity of $b^0_{-m-j}(x,\xi,\lambda)$:

$$
\begin{aligned}
b^{(z),\,0}_{mz-j}(x,t\xi) &= \frac{i}{2\pi} \int_{\Gamma} \lambda^z b^0_{-m-j}(x,t\xi,\lambda)\,d\lambda \\
&= \frac{i}{2\pi} \int_{\Gamma^t} (t^m\mu)^z b_{-m-j}(x,t^m\mu)\cdot t^m\,d\mu \\
&= t^{mz-j}\,\frac{i}{2\pi} \int_{\Gamma^t} \mu^z b^0_{-m-j}(x,\xi,\mu)\,d\mu \\
&= t^{mz-j}\,\frac{i}{2\pi} \int_{\Gamma} \mu^z b^0_{-m-j}(x,\xi,\mu)\,d\mu = t^{mz-j} b^{(z),\,0}_{mz-j}(x,\xi).
\end{aligned}
$$

Here Γ^t is the contour $t^{-m}\Gamma$ (it has the same shape as Γ but the radius of the curved part is $t^{-m}\varrho$ instead of ϱ).

Now it is necessary to extend the definition of $b^{(z),\,0}_{mz-j}(x,\xi)$ to all $z \in \mathbb{C}$. This is done in the same way as the construction of A^z for $z \in \mathbb{C}$ in §10. The following analogue of Proposition 10.1 holds.

Proposition 11.1. a) *For* $\operatorname{Re} z < 0$ *and* $\operatorname{Re} w < 0$ *we have the semigroup property*

$$\sum_{|\alpha| + p + q = j} \partial_\xi^\alpha b_{mz-p}^{(z),\,0}(x, \xi)\, D_x^\alpha b_{mw-q}^{(w),\,0}(x, \xi)/\alpha! = b_{m(z+w)-j}^{(z+w),\,0}(x, \xi),$$

$$j = 0, 1, 2, \ldots \tag{11.13}$$

b) *If* $k \in \mathbb{Z}$ *and* $k > 0$, *then the set* $b_{-mk-j}^{(-k)}(x, \xi)$, $j = 0, 1, 2, \ldots$, *is the set of homogeneous components of the parametrix of* A^k.

c) *For any multi-indices* α, β *the derivative* $\partial_\xi^\alpha \partial_x^\beta b_{mz-j}^{(z),\,0}(x, \xi)$ *is a holomorphic function of* z *for* $\operatorname{Re} z < 0$ *and* $\xi \neq 0$.

Proof. This is achieved by repeating on the symbol level the proof of Proposition 10.1; recommended to the reader as a useful exercise. $\quad\square$

In what follows it is convenient to denote by $a_j^{(k)}(x, \xi)$, $k > 0$ and integer, the homogeneous components (of degree j) of the symbol $a^{(k)}(x, \xi)$ of the operator A^k, so that

$$a^{(k)}(x, \xi) = \sum_{j=0}^{mk} a_j^{(k)}(x, \xi).$$

If k is an integer and $k < 0$, then by $a_j^{(k)}(x, \xi)$ we denote the homogeneous components of the symbol of the parametrix to the operator A^k, or, what is the same thing, the homogeneous components of the symbol of A^{-k}. They are defined recursively by the relations

$$a_{-mk}^{(-k)}(x, \xi) \cdot a_{mk}^{(k)}(x, \xi) = 1, \tag{11.14}$$

$$a_{-mk}^{(-k)}(x, \xi) \cdot a_{mk-j}^{(k)}(x, \xi) +$$
$$+ \sum_{\substack{p+q+|\alpha|=j \\ q<j}} \partial_\xi^\alpha a_{-mk-p}^{(-k)}(x, \xi) \cdot D_x^\alpha a_{mk-q}^{(k)}(x, \xi)/\alpha! = 0, \quad j = 1, 2, \ldots \tag{11.15}$$

Definition 11.1. Let $z \in \mathbb{C}$ and $k \in \mathbb{Z}$ be so chosen that $\operatorname{Re} z < k$. Put

$$b_{mz-j}^{(z),\,0}(x, \xi) = \sum_{p+q+|\alpha|=j} \partial_\xi^\alpha a_{mk-p}^{(k)}(x, \xi) \cdot D_x^\alpha b_{m(z-k)-q}^{(z-k),\,0}(x, \xi)/\alpha!,$$

$$j = 0, 1, \ldots \tag{11.16}$$

Theorem 11.1. a) *The function* $b_{mz-j}^{(z),\,0}(x, \xi)$ *as defined via formula* (11.16) *is independent of the choice of the integer* k *as long as* $\operatorname{Re} z < k$.

b) *If* $\operatorname{Re} z < 0$, *then the functions* $b_{mz-j}^{(z),\,0}(x, \xi)$ *obtained by formula* (11.16) *coincide with the functions, denoted by the same symbol, obtained via the contour integral* (11.10).

c) *The group property* (11.13) *holds for any* $z, w \in \mathbb{C}$.

d) *If* $k \in \mathbb{Z}$, *then* $b_{mk-j}^{(k),\,0}(x, \xi) = a_{mk-j}^{(k)}(x, \xi)$.

e) *For any multi-indices* α, β, *any* x, $\xi\,(\xi \neq 0)$ *and any* $j = 0, 1, \ldots$, $\partial_\xi^\alpha \partial_x^\beta b_{mz-j}^{(z),\,0}(x, \xi)$ *is an entire function in* z.

Proof. Statements a)–d) are proved in exactly the same way as the corresponding statements in Theorem 10.1 (it is only necessary to pass from the operator algebra to the symbol algebra consisting of formal series of homogeneous functions and with multiplication given by the composition formula for symbols). The proof of e) is obtained immediately by looking at the formulae defining the functions $b_{mz-j}^{(z),0}(x, \xi)$. $\square$

11.3 Smoothed resolvent symbols. Let $\omega(\tau) \in C^\infty(\mathbb{R}^1)$, so that $\omega(\tau) = 0$ for $\tau < \frac{1}{2}$, $\omega(\tau) = 1$ for $\tau \geq 1$. Put

$$\theta(\xi, \lambda) = \omega(|\xi|^2 + |\lambda|^{2/m}) \tag{11.17}$$

and let us define

$$b_{-m-j}(x, \xi, \lambda) = \theta(\xi, \lambda)\, b^0_{-m-j}(x, \xi, \lambda). \tag{11.18}$$

With the help of the function $b_{-m-j}(x, \xi, \lambda)$ we construct on the manifold M a parametrix of the operator with parameter $A - \lambda I$ in a way similar to the second part of the proof of Theorem 5.1. Now let $M = \bigcup_\gamma X^\gamma$ be a finite covering of M by charts, φ^γ a subordinated partition of unity and the functions $\psi^\gamma \in C_0^\infty(X^\gamma)$ be such that $\psi^\gamma \equiv 1$ in a neighbourhood of $\operatorname{supp} \varphi^\gamma$. Further let Φ^γ, Ψ^γ be the multiplication operators by φ^γ and ψ^γ and $B^\gamma_{-m-j}(\lambda)$ a pseudo-differential operator on X^γ with the symbol $b^\gamma_{-m-j}(x, \xi, \lambda)$ constructed by formula (11.18) in the coordinate neighbourhood X^γ.

Now put

$$B_{-m-j}(\lambda) = \sum_\gamma \Phi^\gamma B^\gamma_{-m-j}(\lambda)\, \Psi^\gamma \tag{11.19}$$

and

$$B_{(N)}(\lambda) = \sum_{j=0}^{N-1} B_{-m-j}(\lambda). \tag{11.20}$$

Proposition 11.2.

$$(A - \lambda I)^{-1} - B_{(N)}(\lambda) \in CL_m^{-m-N}(M, \Lambda). \tag{11.21}$$

Proof. First we construct an exact parametrix $B(\lambda)$ of the operator $A - \lambda I$, putting

$$B(\lambda) \sim \sum_{j=0}^\infty B_{-m-j}(\lambda). \tag{11.22}$$

The precise meaning of this formula is: construct in each X^γ an asymptotic sum

$$b^\gamma(x, \xi, \lambda) \sim \sum_{j=0}^\infty b^\gamma_{-m-j}(x, \xi, \lambda), \tag{11.23}$$

then take the operator $B^\gamma(\lambda)$ in X^γ with the symbol $b^\gamma(x, \xi, \lambda)$ and finally put

$$B(\lambda) = \sum_{\gamma} \Phi^{\gamma} B^{\gamma} \Psi^{\gamma}, \tag{11.24}$$

where Φ^{γ}, Ψ^{γ} are as in formula (11.19). Then we obviously have

$$B(\lambda) - B_{(N)}(\lambda) \in CL_m^{-m-N}(M, \Lambda). \tag{11.25}$$

But from Theorem 9.2 it follows that

$$(A - \lambda I)^{-1} - B(\lambda) \in L^{-\infty}(M, \Lambda). \tag{11.26}$$

Equating (11.25) and (11.26) we arrive at (11.21). $\square$

11.4 Smoothed symbols of complex powers and the structure theorem. Let $\omega(\tau)$ be the same function on $\mathbb{R}^1$ as at the beginning of 11.3. Put

$$\theta(\xi) = \omega(|\xi|) \tag{11.27}$$

and define

$$b_{mz-j}^{(z)}(x, \xi) = \theta(\xi)\, b_{mz-j}^{(z),\,0}(x, \xi). \tag{11.28}$$

The construction of $b_{mz-j}^{(z),\,0}(x, \xi)$ and $b_{mz-j}^{(z)}(x, \xi)$ can be carried out in any coordinate neighbourhood X^{γ} (the corresponding functions will be denoted by $b_{mz-j}^{(z),\,0,\,\gamma}(x, \xi)$ and $b_{mz-j}^{(z),\,\gamma}(x, \xi)$ if we need to know exactly which coordinate neighbourhood). Denoting by $B_{mz-j}^{(z),\,\gamma}$ the operator in X^{γ} with symbol $b_{mz-j}^{(z),\,\gamma}(x, \xi)$, we once again put

$$B_{mz-j}^{(z)} = \sum_{\gamma} \Phi^{\gamma} B_{mz-j}^{(z),\,\gamma} \Psi^{\gamma} \tag{11.29}$$

and

$$B_{(N)}^{(z)} = \sum_{j=0}^{N-1} B_{mz-j}^{(z)}. \tag{11.30}$$

For the statement of the basic structure theorem, we shall also need the definition of holomorphic families of ΨDO, which constitute the set $\mathcal{O}(G, L_{\varrho,\delta}^m(M))$, where G is a domain in the complex plane $\mathbb{C}$.

Definition 11.2. Let X be an open set in $\mathbb{R}^n$, G a domain in $\mathbb{C}$, $a(x, \xi, z) \in C^{\infty}(X \times \mathbb{R}^n \times G)$ where $a(x, \xi, z)$ is holomorphic in z. We shall write $a(x, \xi, z) \in \mathcal{O}(G, S_{\varrho,\delta}^m(X))$ if for any multi-indices α, β any $k \in \mathbb{Z}_+$ and any compacts $K_1 \subset X$, $K_2 \subset G$ there exists a constant $C = C(\alpha, \beta, k, K_1, K_2)$ such that

$$|\partial_{\xi}^{\alpha} \partial_x^{\beta} \partial_z^k a(x, \xi, z)| \leqq C\langle\xi\rangle^{m-\varrho|\alpha|+\delta|\beta|}, \tag{11.31}$$

for $(x, \xi, z) \in K_1 \times \mathbb{R}^n \times K_2$. If $A(z)$ is a ΨDO in X, depending on the parameter z, then we shall write that $A(z) \in \mathcal{O}(G, L_{\varrho,\delta}^m(X))$ if $A(z) = A_1(z) + R(z)$, where $A_1(z)$ is a properly supported ΨDO on X with symbol $a(x, \xi, z) \in \mathcal{O}(G, S_{\varrho,\delta}^m(X))$

and $R(z)$ has kernel $R(x, y, z) \in C^\infty (X \times X \times G)$, holomorphic in z. Finally, we shall write that $A(z) \in \mathcal{O}(G, L^m_{\varrho, \delta}(M))$, if $A(z)$ is a ΨDO on M, depending on the parameter $z \in G$ such that for any coordinate neighbourhood $X \subset M$ and any local coordinates $\varkappa: X \to X_1$, X_1 an open set in $\mathbb{R}^n$, the family of operators $A^\varkappa(z)$ on X_1 induced by the operators $A(z)$ on M via the diffeomorphism $\varkappa$ is such that

$$A^\varkappa(z) \in \mathcal{O}(G, L^m_{\varrho, \delta}(X_1)).$$

Repeating the arguments of §4 we see that to verify that $A(z) \in \mathcal{O}(G, L^m_{\varrho, \delta}(M))$ for $1 - \varrho \leqq \delta < \varrho$ it suffices to do so for a fixed coordinate covering of M and fixed coordinate diffeomorphisms.

Also, repeating the argument in §3 and §5, we see that composition, taking the adjoint and forming the parametrix of a hypoelliptic operator (we assume that the condition for hypoellipticity is fulfilled uniformly in z) does not take us outside the holomorphic families, specified in Definition 11.2.

We shall now formulate the basic structure theorem.

Theorem 11.2. *For any $z \in \mathbb{C}$*

$$A^z \in CL^{mz}(M), \tag{11.32}$$

moreover for any integer $N \geqq 0$ and $t \in \mathbb{R}$

$$A^z - B^{(z)}_{(N)} \in \mathcal{O}(\operatorname{Re} z < t, \ L^{mt-N}_{1,0}(M)). \tag{11.33}$$

Proof. 1. First let us show that (11.32) follows from (11.33). For fixed $z \in \mathbb{C}$ set

$$B^{(z)} \sim \sum_{j=0}^{\infty} B^{(z)}_{mz-j} \tag{11.34}$$

(this formula can be decoded in the same way as (11.22)). Then it is obvious that we may assume

$$B^{(z)} \in CL^{mz}(M). \tag{11.35}$$

Now, fixing in (11.33) $t \in \mathbb{R}$ such that $t > \operatorname{Re} z$ and then letting N tend to $+\infty$, we obtain

$$A^z - B^{(z)} \in L^{-\infty}(M), \tag{11.36}$$

from which (11.32) follows.

2. Let us prove (11.33). Set

$$R^{(z)}_N = A^z - B^{(z)}_{(N)}. \tag{11.37}$$

First of all note, that the group property of the complex powers A^z and its, for the time being hypothetical, symbols $b_{mz-j}^{(z),0}$ (Theorem 10.1 and 11.1) together with the fact that the composition of holomorphic families yields again a holomorphic family allow us to reduce the proof to the case $t = 0$ (N may be arbitrary). In other words, it suffices to verify that

$$R_{(N)}^{(z)} \in \mathcal{O}\,(\operatorname{Re} z < 0,\, L_{1,0}^{-N}(M))\,. \tag{11.38}$$

In order to make use of Proposition 11.2 it is convenient to consider for $\operatorname{Re} z < 0$ and together with the operator $B_{(N)}^{(z)}$, the operator

$$'B_{(N)}^{(z)} = \frac{i}{2\pi} \int_{\Gamma} \lambda^z\, B_{(N)}(\lambda)\, d\lambda\,. \tag{11.39}$$

It is easily verified that

$$'B_{(N)}^{(z)} - B_{(N)}^{(z)} \in \mathcal{O}\,(\operatorname{Re} z < 0,\, L^{-\infty}(M))\,. \tag{11.40}$$

Indeed, we obviously have

$$'B_{(N)}^{(z)} = \sum_{i=0}^{N-1} {}'B_{mz-j}^{(z)}\,, \tag{11.41}$$

where

$$'B_{mz-j}^{(z)} = \frac{i}{2\pi} \int_{\Gamma} \lambda^z\, B_{-m-j}(\lambda)\, d\lambda\,, \tag{11.42}$$

and this is why the symbol $'b_{mz-j}^{(z)}(x, \xi)$ of the operator $'B_{mz-j}^{(z)}$ is expressed in some local coordinate system by the formula

$$'b_{mz-j}^{(z)}(x, \xi) = \frac{i}{2\pi} \int_{\Gamma} \lambda^z\, b_{-m-j}(x, \xi, \lambda)\, d\lambda\,. \tag{11.43}$$

From this it obviously follows that

$$'b_{mz-j}^{(z)}(x, \xi) = b_{mz-j}^{(z),0}(x, \xi) \qquad \text{for} \quad |\xi| > 1 \tag{11.44}$$

But then the same holds for the symbols $b_{mz-j}^{(z)}(x, \xi)$ of the operators $B_{mz-j}^{(z)}$, the sum of which gives the operator $B_{(N)}^{(z)}$ (cf. the formulas (11.28)–(11.30)). Taking the obvious estimates for the derivatives with respect to z into account, (11.40) is obtained at once.

Now, in view of (11.40), it is clear that it suffices to verify a membership of the type (11.38) for $'R_{(N)}^{(z)} = A^z - 'B_{(N)}^{(z)}$. Denote by $'r_{(N)}^{(z)}(x, \xi)$ the symbol of $'R_{(N)}^{(z)}$ in some local coordinate system and by $r_{(N)}(x, \xi, \lambda)$ the symbol of $R_{(N)}(\lambda) = (A - \lambda I)^{-1} - B_{(N)}(\lambda)$. Then

$$'r_{(N)}^{(z)}(x, \xi) = \frac{i}{2\pi} \int_\Gamma \lambda^z r_{(N)}(x, \xi, \lambda)\, d\lambda \tag{11.45}$$

and we have

$$\partial_\xi^\alpha \partial_x^\beta \partial_z^k\, 'r_{(N)}^{(z)}(x, \xi) = \frac{i}{2\pi} \int_\Gamma \lambda^z (\ln \lambda)^k\, \partial_\xi^\alpha \partial_x^\beta r_{(N)}(x, \xi, \lambda)\, d\lambda. \tag{11.46}$$

But in view of Proposition 11.2, we have the estimate

$$|\partial_\xi^\alpha \partial_x^\beta r_{(N)}(x, \xi, \lambda)| \leqq C_{\alpha, \beta, K}(1 + |\xi| + |\lambda|^{1/m})^{-m - N - |\alpha|}, \qquad x \in K, \tag{11.47}$$

where K is some compact set in the coordinate neighbourhood under consideration. From this it follows that

$$|\partial_\xi^\alpha \partial_x^\beta r_{(N)}(x, \xi, \lambda)| \leqq C_{\alpha, \beta, K}(1 + |\xi|)^{-N - |\alpha|}(1 + |\lambda|)^{-1}, \qquad x \in K, \tag{11.48}$$

and via (11.46) we get (11.38) for $'R_{(N)}^{(z)}$. $\square$

Exercise 11.1. Extend Theorem 11.2 to the situation described in Exercise 10.1.

§12. Analytic Continuation of the Kernels of Complex Powers

12.1 Statement of the problem. Expressing the kernel in terms of the symbol. Let M be a closed manifold, A an elliptic operator on M, satisfying (10.1) and (10.2), which makes it possible to construct the complex powers. For $\operatorname{Re} z < -n/m$ we denote by $A_z(x, y)\, dy$ the kernel of A^z (this then depends on the parameter $x \in M$ and is a density on M and may, in local coordinates defined for $y \in Y$, be expressed as $A_z(x, y)\, dy$, where dy is the Lebesgue measure defined by the local coordinates and $A_z(x, y)$ is a continuous function on $M \times Y$). By a abuse of language, this function $A_z(x, y)$, which depends on z and on the local coordinates in a neighbourhood of y, is called the *kernel*.

Our immediate goal is to construct an analytic continuation (in z) of the kernel $A_z(x, y)$ to the entire complex z-plane $\mathbb{C}$. Note, that if X, Y are open subsets of M, then $A_z(x, y)$ is uniquely defined for $x \in X$, $y \in Y$ by the values $(A^z u, v)$ for $u \in C_0^\infty(X)$ and $v \in C_0^\infty(Y)$.

Now let X be a coordinate neighbourhood (not necessarily connected), which we identify with an open subset of $\mathbb{R}^n$. If we write the ΨDO $B \in L^m(X)$ in the form

$$Bu(x) = \int e^{i(x-y)\cdot\xi} b(x, \xi)\, u(y)\, dy\, d\xi,$$

where $b \in S^l(X \times \mathbb{R}^n)$, then for $l < -n$ the kernel $A(x, y)$ is continuous and is of the form

$$A(x, y) = \int e^{i(x-y)\cdot\xi}\, b(x, \xi)\, d\xi\,.$$

The restriction of the ΨDO A^z to X (cf. 4.3) can be represented in the form $A^z = A_{1,z} + R_{1,z}$, where $A_{1,z}$ is a properly supported ΨDO and $R_{1,z}$ is an operator with kernel $R_1(x, y, z) \in C^\infty(X \times X \times \mathbb{C})$, holomorphic in z and equal to zero for x and y close to each other. Denoting by $a(x, \xi, z)$ the symbol of $A_{1,z}$ we will also by abuse of language, call this symbol the symbol of A^z. The kernel $A_z(x, y)$, for $x, y \in X$ close to each other, may be represented in the form

$$A_z(x, y) = \int e^{i(x-y)\cdot\xi}\, a(x, \xi, z)\, d\xi\,. \tag{12.1}$$

The kernel $A_z(x, y)$, for $\operatorname{Re} z < -n/m$, is continuous and holomorphic in z. For $x = y$ we obtain

$$A_z(x, x) = \int a(x, \xi, z)\, d\xi\,. \tag{12.2}$$

Note that the result of the integration in (12.2) (and in (12.1) for x and y close to each other) does not depend on the choice of the "symbol" $a(x, \xi, z)$.

12.2 Statement of the result. In the statement of the result we will make use of the homogeneous components $b^{(z),\,0}_{mz-j}(x, \xi)$ of the symbol of A^z, which were constructed in §11. Note here that for $\operatorname{Re} z < j/m$, these homogeneous components are given by the formulas

$$b^{(z),\,0}_{mz-j}(x, \xi) = \frac{i}{2\pi}\ \int_\Gamma \lambda^z\, b^0_{-m-j}(x, \xi, \lambda)\, d\lambda\,, \qquad j = 0, 1, 2, \ldots, \tag{12.3}$$

where $b^0_{-m-j}(x, \xi, \lambda)$ are the homogeneous components of the symbol of the parametrix for the operator $A - \lambda I$, also constructed in §11.

Earlier (12.3) was applied only for $\operatorname{Re} z < 0$, but the integral in (12.3) converges for $\operatorname{Re} z < j/m$, hence both parts of (12.3) are holomorphic in z for $\operatorname{Re} z < j/m$ demonstrating their equality for these z.

We shall also need the functions

$$d^{(z),\,0}_{mz-j}(x, \xi) = \int_0^\infty r^z\, b^0_{-m-j}(x, \xi, -r)\, dr\,, \qquad j = 0, 1, 2, \ldots, \tag{12.4}$$

defined for $-1 < \operatorname{Re} z < j/m$ and positively homogeneous in ξ of degree $mz - j$.

Theorem 12.1. *Let X be a fixed arbitrary coordinate neighbourhood on M, $A_z(x, y)$ the kernels of the complex powers A^z of the elliptic operator A, defined for $\operatorname{Re} z < -n/m$ and for $x, y \in X$. Then*

1) for $x \neq y$ the function $A_z(x, y)$ can be extended to an entire function of z, equal to 0 for $z = 0, 1, 2, \ldots$;

2) $A_z(x, x)$ can be extended to a meromorphic function in the whole complex z-plane with at most simple poles, which may be situated only at the points of the arithmetic progression $z_j = \dfrac{j - n}{m}$, $j = 0, 1, \ldots$, and the residue of $A_z(x, x)$ at z_j is equal to

$$\gamma_j(x) = -\frac{1}{m} \int\limits_{|\xi|=1} b_{-n}^{(z_j),\, 0}(x, \xi)\, d\xi', \tag{12.5}$$

where $d\xi' = (2\pi)^{-n} d\xi'$ and $d\xi'$ is the surface element of the sphere $|\xi| = 1$;

3) if $z_i = l$ is a non-negative integer then $\gamma_i(x) = 0$ and the value $\varkappa_l(x) = A_l(x, x)$ of the analytically extended kernel at $z = l$ is given by the formula

$$\varkappa_l(x) = (-1)^l \frac{1}{m} \int\limits_{|\xi|=1} d_{-n}^{(l),\, 0}(x, \xi)\, d\xi'. \tag{12.6}$$

Statement 1) is valid uniformly in $x \in K_1$, $y \in K_2$; K_1 and K_2 being disjoint compact sets in X, i.e. $K_z(x, y)$ can be continued to an entire function of z with values in $C(K_1 \times K_2)$. Similarly, statements 2) and 3) are uniform in $x \in K$, where K is a compact set, i.e. the map $z \to A_z(x, x)$ viewed as a function of z with values in $C(K)$, can be extended meromorphically to the whole complex z-plane with poles at the points $z_j, j = 0, 1, \ldots$, with residues $\gamma_j(x)|_K$ at these poles and values $\varkappa_l(x)|_K$ at $l = 0, 1, 2, \ldots$.

Remarks. 1) In formula (12.5) the function $b_{-n}^{(z_j),\, 0}(x, \xi)$ appears, which may, according to the notation in §11, be written also

$$b_{-n}^{(z_j),\, 0}(x, \xi) = b_{mz_j - j}^{(z_j),\, 0}(x, \xi), \tag{12.7}$$

since $mz_j - j = -n$.

2) Since $z_j < j/m$, then instead of (12.5) we may directly write an expression for $\gamma_j(x)$ in terms of $b^0_{-m-j}(x, \xi, \lambda)$:

$$\gamma_j(x) = -\frac{i}{m \cdot 2\pi} \int\limits_{|\xi|=1} d\xi' \int\limits_{\Gamma} \lambda^{\frac{j-n}{m}} b^0_{-m-j}(x, \xi, \lambda)\, d\lambda. \tag{12.8}$$

A similar expression can be written also for $\varkappa_l(x)$:

$$\varkappa_l(x) = (-1)^l \frac{1}{m} \int\limits_{|\xi|=1} d\xi' \int\limits_0^\infty r^l b^0_{-m(l+1)-n}(x, \xi, -r)\, dr, \tag{12.9}$$

where l is a non-negative integer (the subscript $-m(l+1) - n$ in (12.9) is obtained by expressing j in terms of l by $\dfrac{j-m}{n} = l$).

3) Note the special formula for the residue of the left-most pole $z_0 = -n/m$:

$$\gamma_0(x) = -\frac{1}{m} \int\limits_{|\xi|=1} a_m^{-n/m}(x,\xi)\,d\xi'. \tag{12.10}$$

In the important special case, when $a_m(x,\xi) > 0$ for $\xi \neq 0$, it follows from this formula that $\gamma_0(x) \neq 0$. In what follows in this case we transform (12.10) to a form more convenient for applications.

Proof of Theorem 12.1 1. We will make use of the structure theorem 11.2 and the notation used there, viz. $B_{(N)}^{(z)}$ and $R_{(N)}^{(z)} = A^z - B_{(N)}^{(z)}$. Let us denote by $R_{(N)}^{(z)}(x,y)$ the kernel of $R_{(N)}^{(z)}$ and by $r_{(N)}^{(z)}(x,\xi)$ its symbol in some chart X. Then, if $x, y \in X$ we have

$$R_{(N)}^{(z)}(x,y) = \int r_{(N)}^{(z)}(x,\xi)\, e^{i(x-y)\cdot\xi}\,d\xi. \tag{12.11}$$

This integral converges for $\mathrm{Re}\,z < \dfrac{N-n}{m}$ and defines for these z a holomorphic function of z with values in $C(K_1 \times K_2)$, where K_1, K_2 are arbitrary compact sets in X. Therefore the statements about holomorphy, poles and residues reduce to the corresponding statements about the kernels of the operators $B_{(N)}^{(z)}$.

The kernel of $B_{(N)}^{(z)}$, in its turn, is a sum of terms of the form

$$B_{mz-j}^{(z)}(x,y) = \int e^{i(x-y)\cdot\xi}\, b_{mz-j}^{(z)}(x,\xi)\,d\xi, \tag{12.12}$$

therefore in what follows we will study integrals of the form (12.12).

2. Let K_1 and K_2 be disjoint compact sets in X. We will show that $B_{mz-j}^{(z)}(x,y)$ is an entire function of z with values in $C(K_1 \times K_2)$.

Integrating by parts in (12.12) for $x \neq y$, we obtain

$$B_{mz-j}^{(z)}(x,y) = \int e^{i(x-y)\cdot\xi}\, |x-y|^{-2M}\, \Delta_\xi^M\, b_{mz-j}^{(z)}(x,\xi)\,d\xi, \tag{12.13}$$

where M is a positive integer. This integral converges already for $\mathrm{Re}\,z < \dfrac{2M+j-n}{m}$ and defines a holomorphic function of these z with range in $C(K_1 \times K_2)$, since $|x-y| \geq \varepsilon > 0$ for $x \in K_1$, $y \in K_2$. Since M is arbitrary, it is clear that $B_{mz-j}^{(z)}(x,y)$ is an entire function of z with values in $C(K_1 \times K_2)$.

3. We have for $x = y$

$$B_{mz-j}^{(z)}(x,x) = \int b_{mz-j}^{(z)}(x,\xi)\,d\xi,$$

or

$$B_{mz-j}^{(z)}(x,x) = \int \theta(\xi)\, b_{mz-j}^{(z),\,0}(x,\xi)\,d\xi, \tag{12.14}$$

where $\theta(\xi) = \omega(|\xi|)$, $\omega(\tau) \in C^\infty(\mathbb{R}^1)$, $\omega(\tau) = 0$ for $\tau \leq \frac{1}{2}$, $\omega(\tau) = 1$ for $\tau \geq 1$. Passing to polar coordinates in the integral (12.14), we obtain

$$B^{(z)}_{mz-j}(x, x) = \left(\int_0^\infty \omega(r)\, r^{mz-j+n-1}\, dr \right) \left(\int_{|\xi|=1} b^{(z),\,0}_{mz-j}(x, \xi)\, d\xi' \right). \qquad (12.15)$$

Obviously, the second factor is an entire function of z with values in $C(K)$ (K any compact set in X). The first factor may be decomposed into the sum

$$\int_0^\infty \omega(r)\, r^{mz-j+n-1}\, dr = \int_0^1 \omega(r)\, r^{mz-j+n-1}\, dr + \int_1^\infty r^{mz-j+n-1}\, dr. \qquad (12.16)$$

The first integral in (12.16) is an entire function of z and the second one can be computed

$$\int_1^\infty r^{mz-j+n-1}\, dr = -\frac{1}{mz-j+n} = -\frac{1}{m} \cdot \frac{1}{z-z_j}, \qquad (12.17)$$

where $z_j = \dfrac{j-n}{m}$. Therefore $B^{(z)}_{mz-j}(x, x)$ has one pole with the residue $\gamma_j(x)$, given by the formula (12.5).

Let us verify that $\gamma_j(x) = 0$ if $z_j = l$, a non-negative integer. This is clear from Theorem 11.1, part d) (the functions $b^{(l),\,0}_{ml-j}(x, \xi)$ are the homogenous components of the symbol of the differential operator A^l and therefore vanish for $ml - j < 0$ and, in particular, for $ml - j = -n$).

4. To conclude the proof of 1) and 2) in Theorem 12.1 it remains to show that $A_l(x, y) = 0$ for $x \neq y$ and $l \in \mathbb{Z}_+$. For this, it suffices to show that if u, $v \in C_0^\infty(X)$ and $\operatorname{supp} u \cap \operatorname{supp} v = \emptyset$, then

$$\int A_l(x, y)\, u(y)\, v(x)\, dy\, dx = 0. \qquad (12.18)$$

This however is equivalent to the fact that $\langle A^l u, v \rangle = 0$, since in view of Theorem 10.1 part e), the function $\langle A^z u, v \rangle$ is an entire function of z.

Thus we have shown 1) and 2) of Theorem 12.1 and the absence of poles for $z \in \mathbb{Z}_+$. In what follows we will in fact give a new independent (although also more intricate) proof, allowing us to compute $A_l(x, x)$ for $l \in \mathbb{Z}_+$. The reader who is not interested in this computation may omit the remainder of the proof. without any loss of understanding of the later parts of the book.

5. To investigate the values of $A_z(x, x)$ for $z \in \mathbb{Z}_+$, it is convenient to use another approximation of $A_z(x, y)$ which is obtained if we smooth off the symbols of the parametrix $b^0_{-m-j}(x, \xi, \lambda)$ instead of the symbols $b^{(z),\,0}_{mz-j}(x, \xi)$ of the complex powers. This was essentially done in 11.3, where we introduced the operators $B_{-m-j}(\lambda)$, $B_{(N)}(\lambda)$, $R_{(N)}(\lambda)$, $'B^{(z)}_{mz-j}$, $'B^{(z)}_{(N)}$ and $'R^{(z)}_{(N)}$ and their symbols, defined in any coordinate neighbourhood X, $b_{-m-j}(x, \xi, \lambda)$, $b_{(N)}(x, \xi, \lambda)$, $r_{(N)}(x, \xi, \lambda)$, $'b^{(z)}_{mz-j}(x, \xi)$, $'b^{(z)}_{(N)}(x, \xi)$, $'r^{(z)}_{(N)}(x, \xi)$ (cf. formulae (11.17)–(11.48)). As

an approximation of the kernel $A_z(x, y)$ we will use here the kernels $'B^{(z)}_{(N)}(x, y)$ of the operators $'B^{(z)}_{(N)}$:

$$'B^{(z)}_{(N)}(x, y) = \sum_{j=0}^{N-1} {}'B^{(z)}_{mz-j}(x, y), \qquad (12.19)$$

where $'B^{(z)}_{mz-j}(x, y)$ is the kernel of the operator $'B^{(z)}_{mz-j}$, expressed by the formula

$$'B^{(z)}_{mz-j}(x, y) = \int e^{i(x-y)\cdot\xi} \, 'b^{(z)}_{mz-j}(x, \xi) \, d\xi \qquad (12.20)$$

or

$$'B^{(z)}_{mz-j}(x, y) = \frac{i}{2\pi} \int d\xi \int_\Gamma \lambda^z e^{i(x-y)\cdot\xi} b_{-m-j}(x, \xi, \lambda) \, d\lambda. \qquad (12.21)$$

Here an important role is played by the choice of contour Γ, since, generally speaking, the integral (12.21) depends on the radius of the curved part of the contour (the cut-off function $\theta(\xi, \lambda)$ entering the definition of $b_{-m-j}(x, \xi, \lambda)$ is not holomorphic in λ). We shall denote by ϱ^m, the radius of the curved part of the contour Γ, $\varrho > 0$. In view of the homogeneity of $b^0_{-m-j}(x, \xi, \lambda)$ in $(\xi, \lambda^{1/m})$, it is clear that there is a constant $L > 0$ such that the function $b^0_{-m-j}(x, \xi, \lambda)$ is holomorphic in λ for $|\lambda| < L^m |\xi|^m$. We take the radius of the curved part of Γ equal to ϱ^m, $\varrho > 0$ and such that

$$\varrho < L/(2\sqrt{L^2 + 1}). \qquad (12.22)$$

Then, if $|\xi|^2 + |\lambda|^{2/m} \geq 1/4$ and $\lambda \in \Gamma$, we have either $|\lambda| > \varrho^m$ (i.e. λ belongs to the straight line part of Γ) or $|\lambda| = \varrho^m$ and

$$L^m |\xi|^m \geq L^m \left(\frac{1}{4} - |\lambda|^{2/m}\right)^{m/2} = L^m \left(\frac{1}{4} - \varrho^2\right)^{m/2} > \varrho^m = |\lambda|$$

(the last inequality is equivalent to (12.22)). In this way, and in view of the fact that $\theta(\xi, \lambda) = 0$ for $|\xi|^2 + |\lambda|^{2/m} \leq 1/2$, we may always assume that $b^0_{-m-j}(x, \xi, \lambda)$ is holomorphic inside the curved parts of Γ.

The kernels $'B^{(z)}_{mz-j}(x, y)$ are defined and holomorphic for $\operatorname{Re} z < j/m$ (for these z-values, the integral in (12.21) converges absolutely). Our present aim is to analytically continue these kernels to the entire complex z-plane.

Let us show first of all, that for $x \neq y$, the kernel $'B^{(z)}_{mz-j}(x, y)$ may be continued to an entire function of z (and furthermore, if K_1 and K_2 are disjoint compact subsets of X, then the kernel $'B^{(z)}_{mz-j}(x, y)$ may be continued to an entire function of z with values in $C(K_1 \times K_2)$). Indeed by the standard integrating by parts

$$'B_{mz-j}^{(z)}(x,y)$$

$$= \frac{i}{2\pi} \int d\xi \int_{\Gamma} \lambda^z e^{i(x-y)\cdot\xi} |x-y|^{-2M} \Delta_\xi^M(b_{-m-j}(x,\xi,\lambda)) \, d\lambda, \qquad (12.23)$$

where M is a positive integer, and from this expression the holomorphy of $'B_{mz-j}^{(z)}(x,y)$ for z such that $\operatorname{Re} z < (j+2M)/m$ is obvious.

6. We now demonstrate that $'B_{mz-j}^{(z)}(x,y) = 0$ for $x \neq y$ and $z \in \mathbb{Z}_+$. If $z \in \mathbb{Z}_+$, then λ^z is a single-valued function and the integrals along the straight line parts of Γ in (12.23) cancel. But also the integral along the curved part of Γ equals 0 by the Cauchy theorem, since

$$\Delta_\xi^M(b_{-m-j}(x,\xi,\lambda)) = \sum_{|\alpha|+|\beta|=2M} C_{\alpha\beta} [\partial_\xi^\alpha b_{-m-j}^0(x,\xi,\lambda)] [\partial_\xi^\beta \theta(\xi,\lambda)],$$

the function $\partial_\xi^\alpha b_{-m-j}^0(x,\xi,\lambda)$ is holomorphic in λ inside the curved part of Γ and any derivative $\partial_\xi^\beta \theta(\xi,\lambda)$ is constant in λ for $|\lambda|=\mathrm{const}$.

7. Let us now study the analytic continuation of the integral

$$'B_{mz-j}^{(z)}(x,x) = \frac{i}{2\pi} \int d\xi \int_{\Gamma} \lambda^z \theta(\xi,\lambda) \, b_{-m-j}^0(x,\xi,\lambda) \, d\lambda. \qquad (12.24)$$

Let Γ^1 be the part of Γ, in $|\xi|^2 + |\lambda|^{2/m} > 1$ and Γ^2 the part where $|\xi|^2 + |\lambda|^{2/m} \leqq 1$ (for $|\xi| > 1$ this set is empty). We then clearly have

$$\int_{\Gamma} \lambda^z \theta b_{-m-j}^0 \, d\lambda = \int_{\Gamma^1} \lambda^z b_{-m-j}^0 \, d\lambda + \int_{\Gamma^2} \lambda^z \theta b_{-m-j}^0 \, d\lambda.$$

It is obvious that the integral

$$\frac{i}{2\pi} \int d\xi \int_{\Gamma^2} \lambda^z \theta(\xi,\lambda) \, b_{-m-j}^0(x,\xi,\lambda) \, d\lambda$$

is an entire function of z. This entire function equals 0 for $z = 0, 1, 2, \ldots$, since then the integrals along the straight line parts of Γ^2 cancel due to the single-valuedness of λ^z, and the integral along the curved part is 0 by Cauchy's theorem ($\theta(\xi,\lambda)$ is constant for $|\xi|=\mathrm{const}$ and $|\lambda|=\mathrm{const}$ and $b_{-m-j}^0(x,\xi,\lambda)$ is holomorphic in λ inside the curved part of Γ^2). We may therefore prove all the statements on continuation for the integral

$$I(z) = \frac{i}{2\pi} \int d\xi \int_{\Gamma^1} \lambda^z b_{-m-j}^0(x,\xi,\lambda) \, d\lambda. \qquad (12.25)$$

It is obvious that $\Gamma^1 = \Gamma$ for $|\xi| \geqq \sqrt{1-\varrho^2}$ and Γ^1 consists of two rays for $|\xi| < \sqrt{1-\varrho^2}$. Let us put $\Gamma_\xi^1 = \Gamma^1$ for $|\xi| < \sqrt{1-\varrho^2}$ and

$\Gamma_\xi^1 = |\xi|^m (1 - \varrho^2)^{-m/2} \Gamma$ for $|\xi| \geqq \sqrt{1 - \varrho^2}$. Since it follows from (12.22) that $\varrho / \sqrt{1 - \varrho^2} < L$ then $b^0_{-m-j}(x, \xi, \lambda)$ has no singularities for $|\lambda| < (\varrho |\xi| / \sqrt{1 - \varrho^2})^m$ and by the Cauchy theorem

$$\int_{\Gamma^1} \lambda^z b^0_{-m-j}(x, \xi, \lambda)\, d\lambda = \int_{\Gamma_\xi^1} \lambda^z b^0_{-m-j}(x, \xi, \lambda)\, d\lambda. \tag{12.26}$$

Let us now make a change of coordinates, putting

$\lambda = e^{\pm i\pi} t^m$ on the straight line parts of Γ_ξ;

$\lambda = \alpha^m |\xi|^m e^{i\varphi}$, where $\alpha = \dfrac{\varrho}{\sqrt{1 - \varrho^2}}$, $-\pi \leqq \varphi \leqq \pi$ on the curved part of Γ_ξ^1.

The purpose of this change is to derive from (12.25) an integral of a homogeneous function in (ξ, t) and to proceed as in part 3. of this proof. After the change, we obtain

$$\frac{i}{2\pi} \int_{\Gamma_\xi^1} \lambda^z b^0_{-m-j}\, d\lambda = I_1 \quad \text{if} \quad |\xi| < \sqrt{1 - \varrho^2},$$

$$\frac{i}{2\pi} \int_{\Gamma_\xi^1} \lambda^z b^0_{-m-j}\, d\lambda = I_2 + I_3 \quad \text{if} \quad |\xi| \geqq \sqrt{1 - \varrho^2},$$

where

$$I_1 = m\, \frac{\sin \pi z}{\pi} \int_{\sqrt{1 - |\xi|^2}}^{+\infty} t^{mz + m - 1}\, b^0_{-m-j}(x, \xi, -t^m)\, dt,$$

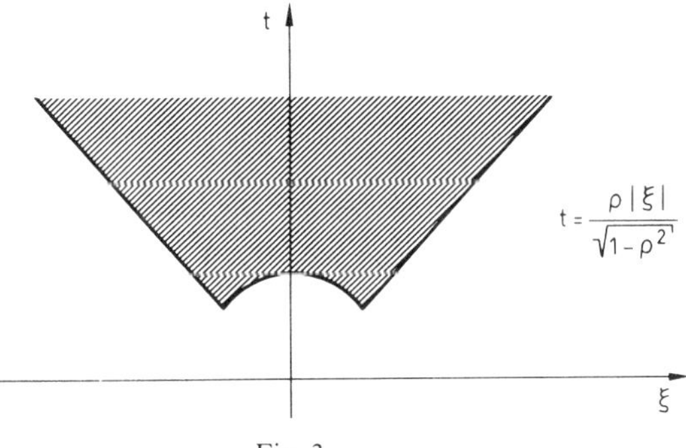

Fig. 3

$$I_2 = m\, \frac{\sin \pi z}{\pi} \int_{\alpha |\xi|}^{+\infty} t^{mz + m - 1}\, b^0_{-m-j}(x, \xi, -t^m)\, dt,$$

$$I_3 = \frac{1}{2\pi} \int_{-\pi}^{\pi} (\alpha |\xi|)^{mz + m}\, e^{i\varphi(z+1)}\, b^0_{-m-j}(x, \xi, \alpha^m |\xi|^m e^{i\varphi})\, d\varphi.$$

Let us remark that the sum

$$\int\limits_{|\xi| < \sqrt{1-\varrho^2}} I_1 \, d\xi + \int\limits_{|\xi| \geq \sqrt{1-\varrho^2}} I_2 \, d\xi \tag{12.27}$$

is an integral of a function in (ξ, t), homogenous of order $mz + m - 1 - m - j$ in the domain consisting of the intersection of a conic domain and the complement of a ball (Fig. 3).

Introducing spherical coordinates in the space (ξ, t), we see that the integral (12.27) converges absolutely for $\mathrm{Re}\,(mz - j - 1) < -n - 1$ and that, in view of (12.17), it can be written in the form

$$-\frac{m \sin \pi z}{\pi (mz - j + n)} \int\limits_{\substack{|\xi|^2 + t^2 = 1 \\ t > \varrho}} t^{mz + m - 1} \, b^0_{-m-j}(x, \xi, -t^m) \, d(\xi, t)', \tag{12.28}$$

where $d(\xi, t)'$ is the surface element on the unit sphere S^n in the (ξ, t)-space. Since $\varrho > 0$ and $t > \varrho$, the integral (12.28) is well-defined as an entire function of z. Further this whole expression can have only one pole for $z = z_j = \dfrac{j - n}{m}$; if $z_j \neq 0$, $1, 2, \ldots$, it vanishes for all $z \in \mathbb{Z}_+$; if $z_j = l \in \mathbb{Z}_+$ it vanishes for all integer $z \neq l$ and there is no longer a pole at $z = l$. Here one can, of course, write down the value of (12.28) for $z = z_j = l \in \mathbb{Z}_+$, but it will be more convenient to do this later for the whole integral (12.25).

Let us consider the remaining term

$$\int\limits_{|\xi| > \sqrt{1-\varrho^2}} I_3 \, d\xi.$$

Here it is convenient to go over to spherical coordinates in the ξ-space. We then obtain

$$\int\limits_{|\xi| > \sqrt{1-\varrho^2}} I_3 \, d\xi = \frac{C}{mz - j + n} \int\limits_{-\pi}^{\pi} e^{i\varphi(z+1)} \int\limits_{|\xi| = 1} b^0_{-m-j}(x, \xi, \alpha^m e^{i\varphi}) \, d\xi' \, d\varphi, \tag{12.29}$$

where $C = \mathrm{const}$. From this the fact that the integral is meromorphic is obvious and the only possible pole is at $z = z_j$. If $z = l \in \mathbb{Z}_+$ we have

$$\int\limits_{-\pi}^{\pi} e^{i\varphi(z+1)} b^0_{-m-j}(x, \xi, \alpha^m e^{i\varphi}) \, d\varphi = i^{-1} \int\limits_{|w| = 1} w^l b^0_{-m-j}(x, \xi, \alpha^m w) \, dw = 0,$$

since the function $b^0_{-m-j}(x, \xi, \lambda)$ is holomorphic in λ for $|\lambda| < \alpha^m$. From this it is also clear that for $z = l \in \mathbb{Z}_+$ the integral (12.29) vanishes, except maybe at $z = z_j$ (if $z_j \in \mathbb{Z}$), where in this case there is no pole.

8. In this way we have demonstrated that the integral (12.24) is meromorphically extendable to the whole complex plane, having no more than

one simple pole, which is only possible for $z = z_j = \dfrac{j-n}{m}$. Further, the value of this integral for $z = l \in \mathbb{Z}_+$ is zero, with the possible exception of the point $z = l = z_j$, but then we have no pole at z_j. Let us consider now just this case, $z = l = z_j \in \mathbb{Z}_+$, and compute the value of the analytic continuation (12.24) at $z = l = z_j$.

Note that the integral with respect to λ in (12.24) is convergent for $z = z_j$, since $z_j = \dfrac{j-n}{m} < j/m$. Decompose the ξ-integral into a sum of the integrals over the ball $|\xi| \leq 1$ and over its complement $|\xi| > 1$. Standard arguments, already used before, show that the integral over the ball $|\xi| \leq 1$ is 0 for $z = z_j$. For $|\xi| \geq 1$ we have $\theta(\xi, \lambda) = 1$ and instead of (12.24) it is enough to consider the integral

$$I = \frac{i}{2\pi} \int\limits_{|\xi| \geq 1} d\xi \int\limits_{\Gamma} \lambda^z\, b^0_{-m-j}(x, \xi, \lambda)\, d\lambda . \tag{12.30}$$

Changing to spherical coordinates in the ξ-space, we obtain

$$I = -\frac{1}{mz - j + n} \int\limits_{|\xi|=1} \frac{i}{2\pi} \int\limits_{\Gamma} \lambda^z\, b^0_{-m-j}(x, \xi, \lambda)\, d\lambda\, d\xi' . \tag{12.31}$$

Using the Cauchy theorem for $\operatorname{Re} z > -1$ we may contract the curved part of Γ to 0. Then we obtain for (12.31)

$$I = \frac{\sin \pi z}{\pi (mz - j + n)} \int\limits_{|\xi|=1} \int\limits_0^\infty r^z\, b^0_{-m-j}(x, \xi, -r)\, dr\, d\xi' .$$

The value of this expression for $z = z_j = l \in \mathbb{Z}_+$ equals exactly $\varkappa_l(x)$, where $\varkappa_l(x)$ is given by the formula (12.6) or (12.9). Therefore $'B^{(l)}_{-n}(x, x) = \varkappa_l(x)$.

9. Let us now note that the difference

$$'R^{(z)}_{(N)}(x, y) = A_z(x, y) - {}'B^{(z)}_{(N)}(x, y)$$

$$= \frac{i}{2\pi} \int d\xi \int\limits_{\Gamma} \lambda^z\, e^{i(x-y)\cdot\xi}\, r_{(N)}(x, \xi, \lambda)\, d\lambda \tag{12.32}$$

can be extended to a holomorphic function of z for $\operatorname{Re} z < \dfrac{N-n}{m}$ with values in $C(K \times K)$, where K is any compact set in X. From this one can see that in the half-plane $\operatorname{Re} z < \dfrac{N-n}{m}$ the functions $A_z(x, y)$ and $'B^{(z)}_{(N)}(x, y)$ have the same poles with identical residues. Further, if $z = l \in \mathbb{Z}_+$ and $x \neq y$, then $A_l(x, y) = {}'B^{(l)}_{(N)}(x, y) = 0$ (cf. parts 4 and 6 of this proof). It is therefore clear, that by continuity $A_l(x, y) = {}'B^{(l)}_{(N)}(x, y)$ for all x, y $\left(\text{if } l < \dfrac{N-n}{m}\right)$. In particular, $A_l(x, x) = {}'B^{(l)}_{(N)}(x, x)$, which together with the result of part 8. completes the proof of Theorem 12.1. $\square$

§13. The ζ-function of an Elliptic Operator and Formal Asymptotic Behaviour of the Spectrum

13.1 Definition and the continuation theorem. Let A be an elliptic operator on a closed manifold M, satisfying the same conditions as in the foregoing section. Let $A_z(x, y)\, dy$ be the kernel of A^z. For $x = y$ we obtain from this kernel the density $A_z(x, x)\, dx$ which is well-defined on the whole manifold M and which can be integrated over M.

Definition 13.1. The function

$$\zeta_A(z) = \int_M A_z(x, x)\, dx \tag{13.1}$$

is called the ζ-*function* of the elliptic operator A.

In the next section we show that $\zeta_A(z)$ can be expressed via the eigenvalues of A allowing us in the self-adjoint case to obtain the simplest theorem on the asymptotic behaviour of the eigenvalues. For now, we shall be content with the formal Definition 13.1 and will formulate a theorem on the analytic continuation of the ζ-function.

Theorem 13.1. *The function $\zeta_A(z)$ defined by the formula* (13.1) *for* $\operatorname{Re} z < -n/m$ *can be continued to a meromorphic function in the entire complex z-plane with at most simple poles, which can be situated only in the points of the arithmetic progression* $z_j = \dfrac{j-n}{m}$, $j = 0,\ 1,\ 2,\ \ldots$, *except for the points* $z_j = l = 0,\ 1,\ 2,\ \ldots$, *and where the residue γ_j at z_j and the value $\varkappa_l = \zeta(l)$, in the notations of Theorem 12.1, are given by the formulae*

$$\gamma_j = \int_M \gamma_j(x)\, dx = -\frac{1}{m} \int_M dx \int_{|\xi|=1} b^{(z_j),\, 0}_{-n}(x, \xi)\, d\xi'$$

$$= -\frac{i}{2\pi m} \int_M dx \int_{|\xi|=1} d\xi' \int_\Gamma \lambda^{\frac{j-n}{m}}\, b^0_{-m-j}(x, \xi, \lambda)\, d\lambda, \tag{13.2}$$

$$\varkappa_l = \int_M \varkappa_l(x)\, dx = (-1)^l \frac{1}{m} \int_M dx \int_{|\xi|=1} d^{(l),\, 0}_{-n}(x, \xi)\, d\xi'$$

$$= (-1)^l \frac{1}{m} \int_M dx \int_{|\xi|=1} d\xi' \int_0^\infty r^l\, b^0_{-m(l+1)-n}(x, \xi, -r)\, dr. \tag{13.3}$$

Proof. Follows from Theorem 12.1 by integrating (12.5) and (12.6) with respect to x, taking into account the remarks following Theorem 12.1. $\square$

13.2 The spectral meaning of the ζ-function. In this section we shall assume that on M there is given a smooth positive density, which by abuse of notation is denoted dx. Then the kernel of an operator may be identified with an ordinary function on $M \times M$. In addition, the self-adjointness of an elliptic operator on M is a meaningful concept.

Theorem 13.2. *Let A be a self-adjoint positive elliptic differential operator on M and let $\lambda_j (j=1, 2, \ldots)$ be its eigenvalues. Then*

$$\zeta_A(z) = \sum_{j=1}^{\infty} \lambda_j^z, \qquad \mathrm{Re}\, z < -n/m, \tag{13.4}$$

where the right-hand side converges absolutely for the indicated z-values. This convergence is uniform in z in the half-plane $\mathrm{Re}\, z < -n/m - \varepsilon$ *for arbitrary* $\varepsilon > 0$.

Proof. Let $\mathrm{Re}\, z < -n/m$ and let $A_z(x, y)$ be the kernel of A^z, which is a continuous function on $M \times M$. Let $\{\varphi_j(x)\}_{j=1}^{\infty}$ be complete orthonormal system of eigenfunctions for A. Decomposing $A_z(x, y)$ into a Fourier series in the complete orthonormal system of functions $\{\varphi_j(x)\overline{\varphi_k(y)}\}_{j,k=1}^{\infty}$ we obtain

$$A_z(x, y) = \sum_{j=1}^{\infty} \lambda_j^z \varphi_j(x)\,\overline{\varphi_j(y)}, \tag{13.5}$$

where the series converges in $L^2(M \times M)$. If z is real, then by the Mercer theorem (cf. Riesz and Sz.-Nagy [1], §98) the series (13.5) converges absolutely and uniformly. Putting $x = y$ in (13.5) and integrating over x, we obtain the identity (13.4). In the case of a non-real z it is only necessary to note that $|\lambda_j^z| = \lambda_j^{\mathrm{Re}\, z}$, from which it follows that the series in (13.4) and (13.5) converge absolutely and uniformly. The last statement of the theorem follows from the fact that if $s_0 \in \mathrm{IR}$, $s_0 < -n/m$, then the series (13.5) for $\mathrm{Re}\, z < s_0$ is majorized in absolute value by the sums of the series

$$\frac{1}{2} \sum_{j=1}^{\infty} \lambda_j^{s_0} |\varphi_j(x)|^2 + \frac{1}{2} \sum_{j=1}^{\infty} \lambda_j^{s_0} |\varphi_j(y)|^2,$$

which are themselves absolutely and uniformly converging series with positive terms. $\square$

Remark 1. One may also prove Theorem 13.2 without using the Mercer theorem, noting that for $\mathrm{Re}\, z < -n/(2m)$ the operator A^z has a kernel $A_z(x, y) \in L^2(M \times M)$ (i.e. A^z is a Hilbert-Schmidt operator) and in view of the Parceval identity we have

$$\sum_{j=1}^{\infty} \lambda_j^{2s} = \int_{M \times M} |A_s(x, y)|^2 \, dx\, dy, \qquad s < -n/(2m). \tag{13.6}$$

From the group property it follows that for $s < -n$

$$A_s(x, y) = \int A_{s/2}(x, z)\, A_{s/2}(z, y)\, dz\,,$$

or

$$A_s(x, y) = \int A_{s/2}(x, z)\, \overline{A_{s/2}(y, z)}\, dz$$

using the fact that the kernel $A_{s/2}(x, y)$ is hermitean. Putting now $x = y$ and integrating in x we obtain from (13.6) that (13.4) holds for real $z < -n/m$. The transition to complex z is accomplished in the same way as above or by using analytic continuation.

Let us remark however that from this proof it is hard to get exact information on the decomposition (13.5) (in particular about the uniform convergence of the series there).

Remark 2. The equality (13.4) is valid also without the assumption on self-adjointness of A. The proof is easily obtained from the theorem of V. B. Lidskii (cf. Gohberg I.C. and Krein M.G. [1], Theorem 8.4). However, in the non-selfadjoint case it is not possible to extract any kind of interesting information about the eigenvalues from (13.4). The only exception is the case of a normal operator, where in fact the results may be deduced from the corresponding results in the self-adjoint case.

13.3 Formal asymptotic behaviour of the function $N(t)$ in the self-adjoint case. The function $V(t)$. Let A be as in Theorem 13.2. Set

$$N(t) = \sum_{\lambda_j \leq t} 1 \tag{13.7}$$

for arbitrary $t \in \mathbb{R}$, i.e. $N(t)$ is the number of eigenvalues of A not exceeding t (counting multiplicity). It is clear that $N(t)$ is a non-decreasing function of t which equals 0 for $t < \lambda_1$. We assume here for convenience that the eigenvalues have been arranged in increasing order:

$$0 < \lambda_1 \leq \lambda_2 \leq \lambda_3 \leq \ldots \tag{13.8}$$

We then have an obvious formula, expressing $\zeta_A(z)$ in terms of $N(t)$ in the form of a Stieltjes integral:

$$\zeta_A(z) = \int_0^\infty t^z\, dN(t). \tag{13.9}$$

Assume now that $N(t)$ admits the following asymptotic expansion as $t \to +\infty$:

$$N(t) = c_1 t^{\alpha_1} + c_2 t^{\alpha_2} + \ldots + c_k t^{\alpha_k} + O(t^{\alpha_{k+1}}), \tag{13.10}$$

where $\operatorname{Re}\alpha_1 > \operatorname{Re}\alpha_2 > \ldots > \operatorname{Re}\alpha_k > \operatorname{Re}\alpha_{k+1}$, then

$$\zeta_A(z) = \sum_{l=1}^{k} c_l \int_1^\infty t^z \, d(t^{\alpha_l}) + f_k(z), \tag{13.11}$$

where $f_k(z)$ is holomorphic for $\operatorname{Re} z < -\operatorname{Re}\alpha_{k+1}$. Since

$$\int_1^\infty t^z \, d(t^{\alpha_l}) = \int_1^\infty \alpha_l t^{z+\alpha_l-1} \, dt = -\frac{\alpha_l}{z+\alpha_l},$$

it follows that (13.11) may be rewritten as

$$\zeta_A(z) = -\sum_{l=1}^{k} \frac{c_l\alpha_l}{z+\alpha_l} + f_k(z). \tag{13.12}$$

Hence for $\operatorname{Re} z < -\operatorname{Re}\alpha_{k+1}$ the function $\zeta_A(z)$ has simple poles at $-\alpha_l$ with residues $-c_l\alpha_l$, $l=1,\ldots,k$. Therefore knowing the poles of $\zeta_A(s)$ and the residues at these poles allows us to write down a formal asymptotic expansion for $N(t)$. In reality however, only the computation of the first term of the asymptotic expansion works out well. This term, dictated by comparison of (13.10) and (13.12), must have the form $-\dfrac{r_0}{s_0} t^{s_0}$ if $\zeta_A(s)$ has its left-most pole at $-s_0 < 0$ with residue r_0.

The formula

$$N(t) \sim -\frac{r_0}{s_0} t^{s_0} \tag{13.13}$$

will be rigorously proved in the following section using the Tauberian theorem of Ikehara, while for the present we shall occupy ourselves with the specification of its coefficients.

By Theorem 13.1 we have $s_0 = n/m$ and

$$r_0 = -\frac{1}{m} \int_M dx \int_{|\xi|=1} a_m^{-n/m}(x,\xi) \, d\xi'. \tag{13.14}$$

Since in the situation under consideration $a_m(x,\xi) > 0$ for $\xi \neq 0$, then $r_0 \neq 0$, so that the pole at $-n/m$ really exists. Formula (13.13) may now be written

$$N(t) \sim \frac{1}{n} \int_M dx \int_{|\xi|=1} a_m^{-n/m}(x,\xi) \, d\xi' \cdot t^{n/m}. \tag{13.15}$$

Now rewrite the right-hand side of (13.15) in a more natural form. For this, introduce the function

$$V(t) = (2\pi)^{-n} \int\limits_{a_m(x,\,\xi)<t} dx\,d\xi. \tag{13.16}$$

Note that this function has an invariant meaning: it is the volume in T^*M of all points (x, ξ) such that $a_m(x, \xi) < t$, multiplied by $(2\pi)^{-n}$. Here the volume is given in T^*M by the measure, induced by the canonical symplectic structure (cf. Arnol'd [1]).

Lemma 13.1. *We have the following formula*

$$V(t) = \frac{1}{n} \int\limits_M dx \int\limits_{|\xi|=1} a_m^{-n/m}(x, \xi)\,d\xi' \cdot t^{n/m}, \tag{13.17}$$

and may, therefore, instead of (13.15) *write*

$$N(t) \sim V(t). \tag{13.18}$$

Proof. Let us remark to begin with, that the condition $a_m(x, \xi) < t$, in view of the homogeneity of a_m, is equivalent to $a_m(x, t^{-1/m}\xi) < 1$. Thus, changing variables in (13.16), $\eta = t^{-1/m}\xi$, we obtain

$$V(t) = \int\limits_{a_m(x,\,\eta)<1} dx\,d\eta \cdot t^{n/m}.$$

The later arguments will take place for a fixed value of x, and we shall write $a_m(\xi)$ instead of $a_m(x, \xi)$.

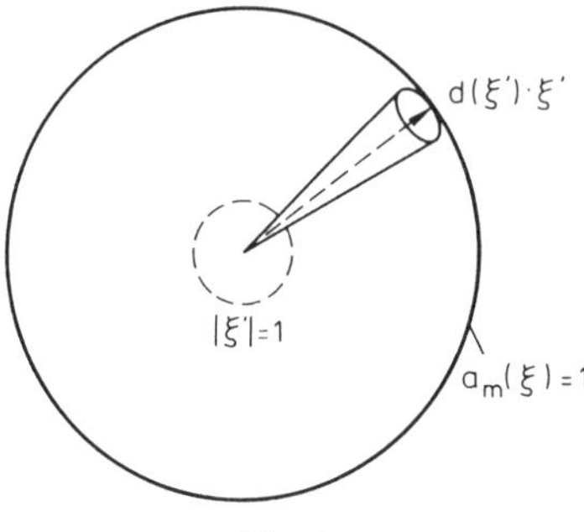

Fig. 4

We have to show that

$$\int\limits_{a_m(\xi)<1} d\xi = \frac{1}{n} \int\limits_{|\xi|=1} a_m^{-n/m}(\xi')\,d\xi'. \tag{13.19}$$

Let $|\xi'| = 1$ and $d(\xi') = \sup\limits_{t>0} \{t: a_m(t\xi') \leq 1\}$ (Fig. 4), so that $d(\xi')$ is the distance of 0 from the surface $a_m(\xi) = 1$ in the direction ξ'. Let us consider the infinitesimal cone with vertex in 0 and height $d(\xi') \cdot \xi'$, cutting out the area $d\xi'$ on the surface $|\xi'| = 1$. The volume of this cone equals $\frac{1}{n}(d(\xi'))^n \cdot d\xi'$.

Therefore

$$\int\limits_{a_m(\xi) < 1} d\xi = \frac{1}{n} \int\limits_{|\xi'| = 1} (d(\xi'))^n \, d\xi' . \tag{13.20}$$

But $a_m(\xi) = a_m\left(|\xi| \cdot \dfrac{\xi}{|\xi|}\right) = |\xi|^m a_m\left(\dfrac{\xi}{|\xi|}\right)$, in view of the homogeneity of $a_m(\xi)$; hence if $a_m(\xi) = 1$ then $a_m\left(\dfrac{\xi}{|\xi|}\right) = |\xi|^{-m}$, from which $d(\xi') = |\xi|$ $= a_m(\xi')^{-1/m}$. Substituting this expression for $d(\xi')$ in (13.20) we obtain (13.19). $\square$

13.4 Asymptotic behaviour of the eigenvalues. We shall now state an asymptotic formula for λ_k as $k \to +\infty$, equivalent to the asymptotic formula for $N(t)$ ((13.13), (13.15) or (13.18)), and infer from this that, essentially, $N(t)$ as a function of t, and λ_k as a function of k, are mutually inverse functions.

Denote by V_1 the coefficient of the term $t^{n/m}$ in (13.17), so that $V_1 = V(1)$. Then the desired formula is of the form

$$\lambda_k \sim V_1^{-m/n} \cdot k^{m/n} \quad as \quad k \to +\infty . \tag{13.21}$$

Proposition 13.1. *The asymptotic formulae* (13.21) *and* (13.18) *are equivalent (i.e. each implies the other).*

Proof. 1. Suppose (13.18) holds, i.e. for any $\varepsilon > 0$ there exists $t_0 > 0$ such that

$$1 - \varepsilon \leq N(t)\, V_1^{-1}\, t^{-n/m} \leq 1 + \varepsilon \tag{13.22}$$

for $t > t_0$. Choose an integer $k_0 > 0$, such that $\lambda_{k_0} > t_0$ and $\lambda_{k_0+1} > \lambda_{k_0}$. Then show that

$$1 - \varepsilon \leq k V_1^{-1} \lambda_k^{-n/m} \leq 1 + \varepsilon \tag{13.23}$$

for $k \geq k_0$. Indeed, for any $k \geq k_0$ there exist integers k_1 and k_2 such that $k_0 \leq k_1 < k \leq k_2$ and $\lambda_{k_1} < \lambda_{k_1+1} = \lambda_{k_2} < \lambda_{k_2+1}$. In particular, we have $N(\lambda_{k_1}) = k_1$ and $N(\lambda_{k_2}) = k_2$, so that from (13.22) it follows that

$$1 - \varepsilon \leq k_1 V_1^{-1} \lambda_{k_1}^{-n/m} \leq 1 + \varepsilon , \tag{13.24}$$

$$1 - \varepsilon \leq k_2 V_1^{-1} \lambda_{k_2}^{-n/m} \leq 1 + \varepsilon . \tag{13.25}$$

Further, $N(t) = k_1$ for $\lambda_{k_1} \leq t < \lambda_{k_2}$ so that for these t

$$1 - \varepsilon \leq k_1 V_1^{-1} t^{-n/m} \leq 1 + \varepsilon$$

and by continuity

$$1 - \varepsilon \leq k_1 V_1^{-1} \lambda_{k_2}^{-n/m} \leq 1 + \varepsilon. \tag{13.26}$$

It follows from (13.25) and (13.26) that

$$1 - \varepsilon \leq kV_1^{-1} \lambda_k^{-n/m} \leq 1 + \varepsilon, \tag{13.27}$$

and it remains to note that $\lambda_k = \lambda_{k_2}$. Hence (13.23) is proven. But from this it follows that

$$(1 + \varepsilon)^{-m/n} V_1^{-m/n} k^{m/n} \leq \lambda_k \leq (1 - \varepsilon)^{-m/n} V_1^{-m/n} k^{m/n}, \tag{13.28}$$

and this implies (13.21).

2. Now let (13.21) hold. Choose $\varepsilon > 0$ and let k_0 be an integer such that for any integer $k > k_0$, the inequality (13.28) holds. Let $\lambda_{k_1} \leq t < \lambda_{k_2}$, where k_1 and k_2 are the same as in part 1. of this proof. It follows from (13.28) that, in particular,

$$1 - \varepsilon \leq k_1 V_1^{-1} \lambda_{k_1}^{-n/m} \leq 1 + \varepsilon, \tag{13.29}$$

$$1 - \varepsilon \leq (k_1 + 1) V_1^{-1} \lambda_{k_2}^{-n/m} \leq 1 + \varepsilon, \tag{13.30}$$

since $\lambda_{k_1+1} = \lambda_{k_2}$. Choose now a number k_0 so large that $V_1^{-1} \lambda_k^{-n/m} \leq \varepsilon$ for $k \geq k_0$. We then obtain from (13.30) that

$$1 - 2\varepsilon \leq k_1 V_1^{-1} \lambda_{k_2}^{-n/m} \leq 1 + 2\varepsilon. \tag{13.31}$$

But $N(t) = k_1$ and therefore from (13.29) and (13.31) we get

$$1 - 2\varepsilon \leq N(t) V_1^{-1} \lambda_{k_1}^{-n/m} \leq 1 + 2\varepsilon,$$

$$1 - 2\varepsilon \leq N(t) V_1^{-1} \lambda_{k_2}^{-n/m} \leq 1 + 2\varepsilon,$$

from which it follows that

$$1 - 2\varepsilon \leq N(t) V_1^{-1} t^{-n/m} \leq 1 + 2\varepsilon,$$

in view of the fact that $\lambda_{k_1} \leq t < \lambda_{k_2}$, as required. $\quad\square$

13.5 Problems

Problem 13.1. Find the poles, residues and values at the non-negative integer points of the ζ-function of the operator $A = -\dfrac{d^2}{dx^2}$ on the circle $\mathbb{R}/2\pi\,\mathbb{Z}$. Express the classical Riemann ζ-function

$$\zeta(z) = \sum_{n=1}^{\infty} \frac{1}{n^z} \tag{13.32}$$

in terms of $\zeta_A(z)$ and find all the poles, residues and values of $\zeta(z)$ at $z = 0,\ -2,\ -4,\ \dots$.

Problem 13.2. Consider the Schrödinger operator $A = -\Delta + q(x)$ on the torus $\mathbb{T}^2 = \mathbb{R}^2/2\pi\,\mathbb{Z}^2$. Here $q(x) \in C^{\infty}(\mathbb{T}^2)$. Express $\zeta_A(0)$ in terms of $q(x)$.

Problem 13.3. Let A be an elliptic differential operator, mapping $C^{\infty}(M, E)$ into $C^{\infty}(M, F)$, where E and F are smooth vector bundles on M. Let there be given a smooth density on M and hermitean structures on E and F (a hermitean metric on each fiber). Show that

$$\text{index } A = \zeta_{I + A^*A}(z) - \zeta_{I + AA^*}(z), \tag{13.33}$$

where the right-hand side does not depend on z.

(This formula allows us, in principle to write down the index A in terms of the symbol of A, using Theorem 13.1 which gives the possibility of computing $\zeta_B(0)$ for $B = I + A^*A$ or $B = I + AA^*$.)

Hint: All non-zero eigenvalues of AA^* and A^*A have the same multiplicities, since A maps an eigensubspace of A^*A into an eigensubspace of AA^* and A^* acts in the opposite direction.

Problem 13.4. Show that the kernel $K(t, x, y)$ of the operator e^{At} from Problem 10.1 is infinitely differentiable in t, x and y for $t > 0$ and for all $x \in M$, $y \in M$. As $t \to +0$, we have the following asymptotic properties of $K(t, x, y)$:

a) If $x \neq y$, then $K(t, x, y) = 0\,(t^N)$ for any $N > 0$.

b) $K(t, x, x)$ has the following asymptotic expansion as $t \to +0$:

$$K(t, x, x) \sim \sum_{j=0}^{\infty} \alpha_j(x)\, t^{\frac{j-n}{m}}, \tag{13.34}$$

where $\alpha_j(x) \in C^{\infty}(M)$. Express $\alpha_j(x)$ in terms of $\gamma_j(x)$ and $\varkappa_l(x)$ (cf. Theorem 12.1) and write down an expression for $\alpha_j(x)$ in terms of the symbol of A.

Verify that if $A^* = A$, then

$$K(t, x, y) = \sum_{j=0}^{\infty} e^{-\lambda_j t} \varphi_j(x) \overline{\varphi_j(y)} \tag{13.35}$$

and the θ-function

$$\theta(t) = \sum_{j=0}^{\infty} e^{-\lambda_j t} = \int_M K(t, x, x)\, dx \tag{13.36}$$

has the asymptotic expansion

$$\theta(t) \sim \sum_{j=0}^{\infty} \alpha_j\, t^{\frac{j-n}{m}}. \tag{13.37}$$

Express index A in terms of the θ-functions of the operators A^*A and AA^*.

Problem 13.5. Let E be a hermitean vector bundle on a closed manifold M with smooth positive density and let A be an elliptic self-adjoint differential operator mapping $C^\infty(M, E)$ into $C^\infty(M, E)$ (not necessarily semibounded). Consider the function

$$\eta_A(z) = \sum_{\lambda} (\operatorname{sign} \lambda) \cdot |\lambda|^z, \tag{13.38}$$

where the sum runs over all the eigenvalues of A. Show that the series (13.38) converges absolutely for $\operatorname{Re} z < -n/m$ and the function defined by it, $\eta_A(z)$, may be continued to the whole complex z-plane as a meromorphic function with simple poles at $z_j = \dfrac{j-n}{m}, j = 0, 1, 2, \ldots$. Express the residues at these poles via the symbol of A.

Hint. Express $\eta_A(z)$ in terms of $\zeta_A'(z)$ and $\zeta_A''(z)$ where $\zeta_A'(z)$ and $\zeta_A''(z)$ are two ζ-functions of A, obtained by different choices of the branch for λ^z with cuts along the upper and lower semi axes of the imaginary axis.

§14. The Tauberian Theorem of Ikehara

14.1 Formulation. The Tauberian theorem of Ikehara allows us to deduce from the fact that the ζ-function is meromorphic asymptotic formulae for $N(t)$ as $t \to +\infty$ or for λ_k as $k \to +\infty$ (cf. §13). Let us give its exact formulation.

Theorem 14.1. *Let $N(t)$ be a non-decreasing function equal to 0 for $t \leq 1$ and such that the integral*

$$\zeta(z) = \int_1^{\infty} t^z\, dN(t) \tag{14.1}$$

converges for $\mathrm{Re}\, z < -k_0$, *where* $k_0 > 0$ *and the function*

$$\zeta(z) + \frac{A}{z + k_0}$$

can be extended by continuity to the closed half-plane $\mathrm{Re}\, z \leqq -k_0$. *We will assume that* $A \neq 0$. *Then, as* $t \to +\infty$ *we have*

$$N(t) \sim \frac{A}{k_0}\, t^{k_0} \tag{14.2}$$

(recall that $f_1(t) \sim f_2(t)$ *as* $t \to +\infty$ *means that* $\lim\limits_{t \to +\infty} f_1(t)/f_2(t) = 1$*).*

(The convergence of the integral in (14.1) for $\mathrm{Re}\, z < -k_0$ easily follows from a weaker condition. Namely, it suffices to suppose that the integral converges for $\mathrm{Re}\, z < -k_1$ for some k_1 and the function $\zeta(z)$ thus expressed can be holomorphically continued to the half-plane $\mathrm{Re}\, z < -k_0$).

Corollary 14.1. *Suppose that the function* $\zeta(z)$, *defined for* $\mathrm{Re}\, z < -k_0$ *by* (14.1), *can be meromorphically continued into the larger half-plane* $\mathrm{Re}\, z < -k_0 + \varepsilon$, *where* $\varepsilon > 0$, *so that on the line* $\mathrm{Re}\, z = -k_0$ *there is a single and moreover simple pole at* $-k_0$ *with residue* $-A$. *Then the asymptotic formula* (14.2) *holds.*

14.2 Beginning of the proof of Theorem 14.1: The reductions.

1st reduction. It is convenient to consider instead of $\zeta(z)$ the function $f(z) = \zeta(-z)$. We then obtain

$$f(z) = \int\limits_1^\infty t^{-z}\, dN(t), \tag{14.3}$$

where the integral converges for $\mathrm{Re}\, z > k_0$ and the function $f(z) - \dfrac{A}{z - k_0}$ is continuous for $\mathrm{Re}\, z \geqq k_0$.

2nd reduction: Reduction to the case $k_0 = 1$. By introducing the function $f_1(z) = f(k_0 z)$, we obtain

$$f_1(z) = \int\limits_1^\infty t^{-k_0 z}\, dN(t) = \int\limits_1^\infty \tau^{-z}\, dN_1(\tau),$$

where $N_1(\tau) = N(\tau^{1/k_0})$. Since

$$f(k_0 z) - \frac{A}{k_0 z - k_0} = f_1(z) - \frac{A}{k_0} \cdot \frac{1}{z - 1}$$

and since $N(t) \sim \dfrac{A}{k_0}\, t^{k_0}$ is equivalent to $N_1(\tau) \sim \dfrac{A}{k_0}\, \tau$, then Theorem 14.1 reduces to the following statement:

Let $N(t)$ be a non-decreasing function and let the integral

$$f(z) = \int\limits_1^\infty t^{-z}\, dN(t) \tag{14.4}$$

be convergent for $\operatorname{Re} z > 1$, where $f(z) - \dfrac{A}{z-1}$ is continuous for $\operatorname{Re} z \geq 1$. Then

$$N(t) \sim At \quad \text{as} \quad t \to +\infty . \tag{14.5}$$

Note that from the continuity of $f(z) - \dfrac{A}{z-1}$ for $\operatorname{Re} z \geq 1$ and the fact that $f(z) \geq 0$ for real $z \geq 1$, it follows that $A > 0$. Changing $N(t)$ for $A^{-1} N(t)$, which results in changing $f(z)$ for $A^{-1} f(z)$, we see that it suffices to show the statement for $A = 1$.

3rd reduction. Let us pass from the Melin transformation to the Laplace transformation, i.e. make a change of variables $t = e^x$. Put $N(e^x) = \varphi(x)$. We then see that $\varphi(x)$ is a non-decreasing function, equal to zero for $x < 0$ and that the integral

$$f(z) = \int\limits_0^\infty e^{-zx}\, d\varphi(x) \tag{14.6}$$

converges for $\operatorname{Re} z > 1$ and $f(z) - \dfrac{1}{z-1}$ is continuous for $\operatorname{Re} z \geq 1$. We must show that

$$\lim_{x \to +\infty} e^{-x} \varphi(x) = 1 . \tag{14.7}$$

4th reduction. Denote $H(x) = e^{-x}\varphi(x)$. The $\varphi(x)$ is non-decreasing if and only if

$$H(y) \geq H(x)\, e^{x-y} \quad \text{for} \quad y \geq x . \tag{14.8}$$

Integrating by parts in (14.6) gives, for $\operatorname{Re} z > 1$

$$f(z) = z \int\limits_0^\infty e^{-zx}\, \varphi(x)\, dx = z \int\limits_0^\infty e^{-(z-1)x}\, H(x)\, dx \tag{14.9}$$

Now put $z = 1 + \varepsilon + it$, where $\varepsilon > 0$ and t is real. Note that

$$\int\limits_0^\infty e^{-(z-1)x}\, dx = \frac{1}{z-1},$$

therefore, (14.9) implies

$$\frac{f(z)}{z} - \frac{1}{z-1} = \int\limits_{0}^{\infty} e^{-(z-1)x}\,(H(x)-1)\,dx\,.$$

Since

$$\frac{f(z)}{z} - \frac{1}{z-1} = \frac{1}{z}\left(f(z) - \frac{1}{z-1} - 1\right),$$

then, putting

$$h_{\varepsilon}(t) = \frac{1}{z}\left(f(z) - \frac{1}{z-1} - 1\right)\Bigg|_{z=1+\varepsilon+it}\,, \tag{14.10}$$

we obtain

$$h_{\varepsilon}(t) = \int\limits_{0}^{\infty} e^{-\varepsilon x - itx}\,(H(x)-1)\,dx\,. \tag{14.11}$$

We may now give the following reformulation of Theorem 14.1.

Theorem 14.1′. *Let $H(x)$ be a function, equal to 0 for $x < 0$ and satisfying (14.8) for all real x and y. Assume that the integral (14.11) converges absolutely for any $\varepsilon > 0$ and the function $h_{\varepsilon}(t)$ defined by it, is such that the limit*

$$\lim_{\varepsilon \to 0} h_{\varepsilon}(t) = h(t) \tag{14.12}$$

exists and is uniform on any finite segment $|t| \leqq 2\lambda$. Then

$$\lim_{x \to +\infty} H(x) = 1\,. \tag{14.13}$$

Remark. If $H(x)$ tends to 1 sufficiently quickly (if e.g. $H(x) - 1 \in L^{1}([0, +\infty)))$, then we obtain (14.12) from (14.13) by passing to the limit under the integral sign, which one may do in view of the dominated convergence theorem (the function $h(t)$ then equals the Fourier transform of $\theta(x)(H(x)-1)$, $\theta(x)$ the Heaviside function). In some sense, the Tauberian condition (14.8) allows one to invert this statement.

14.3 The basic lemma. It is clear that in order to prove Theorem 14.1′ we have to somehow express $H(x) - 1$ in terms of $h(t)$ which, formally, is possible by the inverse Fourier transformation. However, we know nothing about the behaviour of $h(t)$ as $t \to +\infty$ or about the nature of the convergence of $h_{\varepsilon}(t)$ to $h(t)$ on the whole line and it is therefore necessary, to begin with, to multiply the limit equality (14.12) with a finite cut-off function $\tilde{\varrho}(t)$. These considerations, linked to the convenience of having transformations with positive kernels (of the Fejér type), demonstrate that it is convenient to consider $\tilde{\varrho}(t)$ to be the Fourier transform of a non-negative function $\varrho(v) \in L^{1}(\mathbb{R})$:

$$\tilde{\varrho}(t) = \int e^{-itv}\,\varrho(v)\,dv\,. \tag{14.14}$$

We shall assume that $\tilde{\varrho}(t)$ is a continuous function with compact support such that $\tilde{\varrho}(0) = 1$, $\varrho(v) \geq 0$ and $\varrho(v) \in L^1(\mathbb{R})$. From this it follows that

$$\int_{-\infty}^{+\infty} \varrho(v)\, dv = 1 . \tag{14.15}$$

The existence of a function $\tilde{\varrho}(t)$ of the type described may be shown in the same way as in 6.3 (at the beginning of the proof of Theorem 6.3 a function $\tilde{\varrho}(t) \in C_0^\infty(\mathbb{R}^1)$ is constructed which satisfies all these requirements). We can also explicitly define $\tilde{\varrho}(t)$, putting

$$\tilde{\varrho}(t) = \begin{cases} 1 - \dfrac{|t|}{2}; & |t| \leq 2, \\[2mm] 0 & ; \quad |t| > 2 . \end{cases}$$

Then indeed, for a fixed $v \neq 0$ we have

$$\varrho(v) = \int_{-2}^{2} e^{itv}\left(1 - \frac{|t|}{2}\right) dt = -\int_{-2}^{2} \frac{e^{itv}}{iv}\, d\left(1 - \frac{|t|}{2}\right) + \frac{e^{itv}}{2\pi i v}\left(1 - \frac{|t|}{2}\right)\Bigg|_{-2}^{2}$$

$$= \int_{-2}^{2} \frac{e^{itv}}{2iv}\operatorname{sign} t\, dt = -\frac{e^{itv}}{4\pi v^2}\Bigg|_{0}^{2} - \frac{e^{itv}}{4\pi v^2}\Bigg|_{0}^{-2} = \frac{1 - \cos 2v}{2\pi v^2} = \frac{1}{\pi}\frac{\sin^2 v}{v^2} ,$$

from which all the necessary properties of $\varrho(v)$ are obvious.

Lemma 14.1. *For any fixed $\lambda > 0$*

$$\lim_{y \to +\infty} \int_{-\infty}^{+\infty} H\left(y - \frac{v}{\lambda}\right) \varrho(v)\, dv = 1 . \tag{14.16}$$

Proof. 1. Put $\tilde{\varrho}_\lambda(t) = \tilde{\varrho}(t/\lambda)$ and $\varrho_\lambda(v) = \lambda \varrho(\lambda v)$ so that $\tilde{\varrho}_\lambda(t)$ is the Fourier transform of $\varrho_\lambda(v)$. It is clear that

$$\int_{-\infty}^{+\infty} H\left(y - \frac{v}{\lambda}\right) \varrho(v)\, dv = \int_{-\infty}^{+\infty} H(y - v)\, \varrho_\lambda(v)\, dv , \tag{14.17}$$

and since $\varrho_\lambda(v)$ possesses the same properties as $\varrho(v)$, it suffices to prove (14.16) for $\lambda = 1$.

2. Putting $F_\varepsilon(t) = \tilde{\varrho}(t) h_\varepsilon(t)$, we compute the inverse Fourier transform of the function $F_\varepsilon(t)$ with compact support, taking into account that $\tilde{\varrho}(t)$ and $h_\varepsilon(t)$ are the Fourier transforms of the absolutely integrable functions $\varrho(v)$ and $\theta(v)(H(v) - 1)e^{-\varepsilon v}$:

$$\int_{-\infty}^{+\infty} e^{ity} F_\varepsilon(t)\, dt = \int_{-\infty}^{+\infty} e^{ity} \tilde{\varrho}(t)\left[\int_{0}^{\infty} (H(x) - 1)e^{-\varepsilon x - itx}\, dx\right] dt$$

$$= \int_0^{+\infty} (H(x) - 1)\, e^{-\varepsilon x} \left[\int_{-\infty}^{+\infty} \tilde{\varrho}(t)\, e^{it(y-x)}\, dt \right] dx \qquad (14.18)$$

$$= \int_0^{+\infty} (H(x) - 1)\, e^{-\varepsilon x}\, \varrho(y-x)\, dx.$$

As a result, as one might have anticipated, we obtain a convolution and we have made sure that (14.18) holds everywhere and in the usual sense (the change in the order of integration is permitted by the Fubini's theorem).

Let us now rewrite (14.18) in the form

$$\int_{-\infty}^{+\infty} e^{ity} F_\varepsilon(t)\, dt + \int_0^{\infty} e^{-\varepsilon x}\, \varrho(y-x)\, dx = \int_0^{\infty} H(x)\, e^{-\varepsilon x}\, \varrho(y-x)\, dx \qquad (14.19)$$

and take the limit as $\varepsilon \to +0$. Since $\operatorname{supp} F_\varepsilon \subset \operatorname{supp} \tilde{\varrho}$ and $F_\varepsilon(t) \to F(t)$ uniformly in $t \in \operatorname{supp} \tilde{\varrho}$ (here $F(t) = \tilde{\varrho}(t)\, h(t)$), then the first integral on the left-hand side has a limit as $\varepsilon \to +0$ for any y. The same also holds for the second integral (e.g. by the dominated convergence theorem). Therefore, the integral on the right-hand side of (14.19) has for any y a limit as $\varepsilon \to +0$. Since $H(x)\, e^{-\varepsilon x}\, \varrho(y-x)$ converges monotonely as $\varepsilon \to +0$ to $H(x)\, \varrho(y-x)$, we get

$$\int_{-\infty}^{+\infty} e^{ity} F(t)\, dt + \int_0^{\infty} \varrho(y-x)\, dx = \int_0^{\infty} H(x)\, \varrho(y-x)\, dx. \qquad (14.20)$$

Now let y tend to $+\infty$. By the Riemann lemma

$$\lim_{y \to +\infty} \int_{-\infty}^{+\infty} e^{ity} F(t)\, dt = 0.$$

In addition, it is clear that $\displaystyle\lim_{y \to +\infty} \int_0^{\infty} \varrho(y-x)\, dx = 1$. Therefore, it follows from (14.20) that

$$\lim_{y \to +\infty} \int_0^{\infty} H(x)\, \varrho(y-x)\, dx = 1. \qquad (14.21)$$

But

$$\int_0^{\infty} H(x)\, \varrho(y-x)\, dx = \int_{-\infty}^{+\infty} H(x)\, \varrho(y-x)\, dx = \int_{-\infty}^{+\infty} H(y-v)\, \varrho(v)\, dv,$$

so that (14.21) implies the statement of the lemma. $\square$

14.4 Proof of Theorem 14.1′. 1. First, we show that

$$\varlimsup_{y \to +\infty} H(y) \le 1 . \tag{14.22}$$

Since

$$\int_{-a}^{a} H\left(y - \frac{v}{\lambda}\right) \varrho(v)\, dv \le \int_{-\infty}^{+\infty} H\left(y - \frac{v}{\lambda}\right) \varrho(v)\, dv$$

for any $a > 0$, it follows from Lemma 14.1 that

$$\varlimsup_{y \to +\infty} \int_{-a}^{a} H\left(y - \frac{v}{\lambda}\right) \varrho(v)\, dv \le 1 . \tag{14.23}$$

Now, in view of the Tauberian condition (14.8) we have

$$H\left(y - \frac{v}{\lambda}\right) \ge H\left(y - \frac{a}{\lambda}\right) e^{-\frac{2a}{\lambda}} \quad \text{for} \quad v \in [-a, a] .$$

Now, it follows from (14.23) that

$$\varlimsup_{y \to +\infty} H\left(y - \frac{a}{\lambda}\right) e^{-\frac{2a}{\lambda}} \int_{-a}^{a} \varrho(v)\, dv \le 1 ,$$

or

$$\varlimsup_{y \to +\infty} H(y) \le e^{\frac{2a}{\lambda}} \left(\int_{-a}^{a} \varrho(v)\, dv \right)^{-1} . \tag{14.24}$$

Inequality (14.24) holds for any $a > 0$ and $\lambda > 0$. Let $a \to +\infty$ and $\lambda \to +\infty$ in this inequality in such a way that $a/\lambda \to 0$. Then we obtain the required estimate (14.22) from (14.24).

2. We will now verify that

$$\varliminf_{y \to +\infty} H(y) \ge 1 . \tag{14.25}$$

To begin with, note that (14.22) implies the boundedness of $H(y)$:

$$|H(y)| \le M , \quad y \in \mathbb{R}^1 , \tag{14.26}$$

in view of which

$$\int_{|v| \ge b} H\left(y - \frac{v}{\lambda}\right) \varrho(v)\, dv \le \alpha(b) , \tag{14.27}$$

where $\alpha(b) \to 0$ as $b \to +\infty$ (this means, in particular that the integral on the left-hand side of (14.27) approaches 0 as $b \to +\infty$, uniformly in y and λ).

Since

$$\int_{-\infty}^{+\infty} H\left(y - \frac{v}{\lambda}\right) \varrho(v)\, dv = \int_{-b}^{b} \ldots + \int_{|v| \geq b} \ldots\,,$$

then, by (14.27) and Lemma 14.1 we obtain for arbitrary $b > 0$

$$\lim_{y \to +\infty} \int_{|v| \leq b} H\left(y - \frac{v}{\lambda}\right) \varrho(v)\, dv \geq 1 - \alpha(b). \tag{14.28}$$

Let us again use condition (14.8). We have

$$H\left(y + \frac{b}{\lambda}\right) \geq H\left(y - \frac{v}{\lambda}\right) e^{-\frac{2b}{\lambda}}; \quad v \in [-b, b],$$

from which, in view of (14.28), it follows that

$$\lim_{y \to +\infty} H\left(y + \frac{b}{\lambda}\right) e^{\frac{2b}{\lambda}} \int_{-b}^{b} \varrho(v)\, dv \geq 1 - \alpha(b),$$

or

$$\lim_{y \to +\infty} H(y) \geq (1 - \alpha(b))\, e^{-\frac{2b}{\lambda}} \left(\int_{-b}^{b} \varrho(v)\, dv\right)^{-1}. \tag{14.29}$$

Now let $b \to +\infty$ and $\lambda \to +\infty$, so that $b/\lambda \to 0$. Then from (14.29) we obtain the desired inequality (14.25). $\square$

Problem 14.1. Let $N(t)$ be a non-decreasing function, equal to 0 for $t \leq 1$ and let the integral (14.1) be convergent for $\mathrm{Re}\, z < -k_0$, some $k_0 > 0$. Assume furthermore, that the function $\zeta(z)$, defined by (14.1), can be meromorphically continued to larger half-space $\mathrm{Re}\, z < -k_0 + \varepsilon$, where $\varepsilon > 0$ so that on the line $\mathrm{Re}\, z = -k_0$ there is a single pole at $-k_0$ with principal part $A(z + k_0)^{-l}$ in the Laurent expansion (here l is a positive integer, equal to the order of the pole at $-k_0$). Show that

$$N(t) \sim \frac{(-1)^{l-1} A}{(l-1)!} \cdot t^{k_0} (\ln t)^{l-1}$$

as $t \to +\infty$.

Problem 14.2. Prove the Karamata Tauberian theorem:

Let $N(t)$ be a non-decreasing function of $t \in \mathbb{R}^1$, equal to 0 for $t < 1$ and such that the integral

$$\theta(z) = \int_0^\infty e^{-zt}\, dN(t) \tag{14.30}$$

converges for all $z > 0$ and

$$\theta(z) \sim Az^{-\alpha} \quad \text{as} \quad z \to +0 \tag{14.31}$$

(here $A > 0$ and $\alpha > 0$ are constants). Then

$$N(t) \sim \frac{A}{\Gamma(\alpha+1)}\, t^\alpha \quad \text{as} \quad t \to +\infty. \tag{14.32}$$

§15. Asymptotic Behaviour of the Spectral Function and the Eigenvalues (Rough Theorem)

15.1 The spectral function and its asymptotic behaviour on the diagonal. Let M be a closed n-dimensional manifold on which there is given a smooth positive density and let A be a self-adjoint, elliptic operator on M such that

$$a_m(x, \xi) > 0; \quad \xi \neq 0. \tag{15.1}$$

Then A is semibounded. Denote by λ_j its eigenvalues, enumerated in increasing order (counting multiplicities):

$$\lambda_1 \leqq \lambda_2 \leqq \lambda_3 \leqq \ldots,$$

By $\varphi_j(x)$ we denote the corresponding eigenfunctions, which constitute an orthonormal system.

Let E_t be the spectral projection of A (the orthogonal projection onto the linear hull of all eigenvectors with eigenvalues not exceeding t). It is clear that

$$E_t u = \sum_{\lambda_j \leqq t} (u, \varphi_j)\, \varphi_j. \tag{15.2}$$

Definition 15.1. The *spectral function* of A is the kernel (in the sense of L. Schwartz) of the operator E_t.

Taking into account that on M there is a correspondence between functions and densities, we may assume that the spectral function is a function, not a density. From (15.2) it is obvious that this function, $e(x, y, t)$, is given by the formula

$$e(x, y, t) = \sum_{\lambda_j \leqq t} \varphi_j(x)\, \overline{\varphi_j(y)} \tag{15.3}$$

and, in particular, belongs to $C^\infty(M \times M)$ for every fixed t. Let us note immediately the following properties of $e(x, y, t)$:

1) $e(x, x, t)$ is a non-decreasing function of t for any fixed $x \in M$;

2) the function $N(t)$ introduced in section 13.3, can be expressed in terms of $e(x, x, t)$ by the formula

$$N(t) = \int_M e(x, x, t)\, dx, \qquad (15.4)$$

where dx is a fixed density on M.

Now assume that local coordinates in a neighbourhood of x are so chosen, that the density coincides with the Lebesgue measure in these coordinates and put

$$V_x(t) = \int_{a_m(x,\xi)<t} d\xi. \qquad (15.5)$$

Theorem 15.1. *For any $x \in M$ the following holds:*

$$e(x, x, t) \sim V_x(t) \qquad \text{as} \qquad t \to +\infty \qquad (15.6)$$

Proof. 1. To begin with, note that without loss of generality we may assume $\lambda_1 \geq 1$. Indeed this is satisfied by the operator $A_1 = A + MI$ for sufficiently large M. Now, if $e_1(x, y, t)$ is the spectral function of A_1, we have $e(x, y, t) = e_1(x, y, t+M)$. Therefore, the asymptotic formula $e_1(x, x, t) \sim V_x(t)$ implies $e(x, x, t) \sim V_x(t+M)$. But

$$V_x(t+M) = V_x(1)\,(t+M)^{n/m} = V_x(1)\,t^{n/m}(1+O(t^{-1})) \sim V_x(1)\,t^{n/m} = V_x(t),$$

which implies (15.6).

2. Thus let $\lambda_1 \geq 1$. We may then define complex powers A^z of A in accordance with the scheme of §10. Using (13.5), we may for $x = y$ express the kernel $A_z(x, y)$ of A^z in terms of the spectral function as follows

$$A_z(x, x) = \int_0^\infty t^z\, de(x, x, t), \qquad (15.7)$$

where d signifies the differentiation with respect to t (for a fixed x this is simply a Stieltjes integral). In view of Theorems 12.1 and 14.1 we obtain now for $e(x, x, t)$ the asymptotic formula

$$e(x, x, t) \sim \left[\frac{1}{n} \int_{|\xi|=1} a_m^{-n/m}(x, \xi')\, d\xi'\right] \cdot t^{n/m}. \qquad (15.8)$$

An elementary transformation of the right-hand side of this formula, carried out in the proof of Lemma 13.1, shows that it equals $V_x(t)$, implying (15.6). □

15.2 Asymptotic behaviour of the Eigenvalues

Theorem 15.2. *Let A satisfy the conditions described at the beginning of this section. Then one has the following asymptotic relations*

$$N(t) \sim V(t), \qquad\qquad t \to +\infty, \tag{15.9}$$

$$\lambda_k \sim V(1)^{-m/n} k^{m/n}, \quad k \to +\infty, \tag{15.10}$$

where $V(t)$ is defined by the formula (13.16).

Proof. In §13 we showed the equivalence of (15.9) and (15.10). (Proposition 13.1). Let us prove (15.9). This is done on the basis of the Tauberian theorem of Ikehara, by analogy with the proof of Theorem 15.1. Indeed, again we may assume that $\lambda_1 \geq 1$. Then for $\operatorname{Re} z < -n/m$, we clearly have the formula

$$\zeta_A(z) = \int_1^\infty t^z \, dN(t). \tag{15.11}$$

It remains to use Theorems 13.1 and 14.1 and Lemma 13.1.   □

Remark. One can derive (15.9) from (15.6) by integration over x. To justify this integration, it is necessary, however, to prove the uniformity in x of (15.6), which requires in several places (in particular, in the proof of the Ikehara theorem) the verification of uniformity in the parameter. To avoid this cumbersome verification, we have preferred to give an independent proof.

15.3 Problems

Problem 15.1. In the situation of this section prove the estimate

$$|e(x, y, t)| \leq Ct^{n/m},$$

where $x, y \in M$ and the constant $C > 0$ does not depend on x, y and $t \, (t \geq 1)$.

Problem 15.2. Let A be an elliptic differential operator, on closed manifold M with smooth positive density, which is normal, i.e.

$$A^*A = AA^*. \tag{15.12}$$

a) Show that A has an orthonormal basis of smooth eigenfunctions $\varphi_j(x)$, $j = 1, 2, \ldots$, with eigenvalues $\lambda_j \in \mathbb{C}$, such that

$$|\lambda_j| \to +\infty \quad \text{as } j \to +\infty. \tag{15.13}$$

b) Show that if $N(t)$ denotes the number of λ_j, such that $|\lambda_j| \leq t$, and if $V(t)$ is defined by the formula

$$V(t) = (2\pi)^{-n} \int\limits_{|a_m(x,\xi)| < t} dx\, d\xi, \qquad (15.14)$$

then

$$N(t) \sim V(t) \qquad \text{as} \quad t \to +\infty. \qquad (15.15)$$

Problem 15.3. Deduce Theorems 15.1 and 15.2 from the results of Problem 13.4 and the Tauberian theorem of Karamata (Problem 14.2).

Problem 15.4. Let M be a closed manifold, A an elliptic differential operator on M, such that $a_m(x, \xi) > 0$ for $\xi \neq 0$. Let λ_j be its eigenvalues and $N_1(t)$ the number of eigenvalues with $\operatorname{Re} \lambda_j \leq t$ (here we take for the multiplicity of an eigenvalue λ_0 the dimension of the root subspace E_{λ_0}, cf. Theorem 8.4), $N_2(t)$ the number of eigenvalues with $|\lambda_j| \leq t$. Show that

$$N_1(t) \sim N_2(t) \sim V(t) \qquad \text{as } t \to +\infty, \qquad (15.16)$$

where $V(t)$ is defined as before. Show that

$$\lambda_k \sim V(1)^{-m/n} k^{m/n} \qquad k \to +\infty \qquad (15.17)$$

(this means, in particular, that $\operatorname{Im} \lambda_k$ has a lower degree of growth than $\operatorname{Re} \lambda_k$).

Chapter III
Asymptotic Behaviour of the Spectral Function

§16. Formulation of the Hörmander Theorem and Comments

16.1 Formulation and an example. Let M be a closed n-dimensional manifold on which there is given a smooth positive density dx and let A be an elliptic, self-adjoint operator of degree m on M such that $a_m(x, \xi) > 0$ for $\xi \neq 0$. We will use the notations $e(x, y, \lambda)$, $N(\lambda)$, $V_x(\lambda)$ and $V(\lambda)$ introduced in §15. The following theorem refines Theorems 15.1 and 15.2.

Theorem 16.1 (L. Hörmander). *The following estimate holds*

$$|e(x, x, \lambda) - V_x(\lambda)| \leq C\lambda^{(n-1)/m}, \qquad \lambda \geq 1, \qquad x \in M, \tag{16.1}$$

where the constant $C > 0$ is independent of x and λ.

Corollary 16.1. *The following asymptotic formula holds*

$$N(\lambda) = V(\lambda)\,(1 + O(\lambda^{-1/m})) \qquad \text{as} \quad \lambda \to +\infty \tag{16.2}$$

Remark 16.1. In general the estimate of the remainder in (16.1), (16.2) cannot be improved. This can be seen by looking, for instance, at the operator $A = -\dfrac{d^2}{dx^2}$ on the circle $S^1 - \mathbb{R}/2\pi\mathbb{Z}$. The corresponding eigenfunctions are of the form

$$\psi_k(x) = \frac{1}{\sqrt{2\pi}}\, e^{ikx}, \qquad k = 0, \quad \pm 1, \quad \pm 2, \ldots,$$

and the eigenvalues are $\lambda_k = k^2$, $k = 0, \pm 1, \pm 2, \ldots$.

Further, since $|\psi_k(x)|^2 = (2\pi)^{-1}$, then clearly $e(x, x, \lambda) = (2\pi)^{-1} N(\lambda)$. Since $V_x(\lambda) = (2\pi)^{-1} V(\lambda)$ then (16.1) and (16.2) are equivalent. So it suffices to show that the estimate of remainder in (16.2) can not be improved. But in this example (16.2) has the form $N(\lambda) = V(\lambda)\,(1 + O(\lambda^{-1/2}))$ or $N(\lambda) = 2\sqrt{\lambda} + O(1)$. The estimate $O(1)$ can not be improved because $N(\lambda)$ has only integer values.

Later, in §22, we will study a more interesting example, which is a generalization of the present one (the Laplace operator on the sphere) and shows that (16.1) cannot be improved in the case of arbitrary n and m.

16.2 Sketch of the proof. First of all, the theory of complex powers of operators, allows a reduction to the case when A is a ΨDO of order 1. In this situation we will show, that for small t, e^{itA} is itself an FIO, with a phase function which is a solution of a certain first order non-linear equation. Let us now remark, that the kernel of e^{itA} is the Fourier transform (in λ) of the spectral function of A. From this the asymptotic (16.1) is obtained, by invoking Tauberian type arguments for the Fourier transformation.

The remainder of this chapter is as follows: §17 contains some indispensible information on first order non-linear equations; in §18 an important theorem on the action of ΨDO on exponents is proved, from which, in particular, the composition formula for a ΨDO with an FIO follows; in §19 the class of phase functions corresponding to ΨDO is studied; in §20 we construct the operator e^{itA} in the form of an FIO for a first order operator A; in §21 Theorem 16.1 is proved in the general case (there is also information about $e(x, y, \lambda)$ for $x \neq y$); finally, §22 contains the definition of the Laplace operator on a Riemannian manifold and the computation of its spectral function in the case of a sphere.

Problem 16.1. Compute $N(\lambda)$ and $e(x, x, \lambda)$ for the operator

$$A = -\Delta = -\left(\frac{\partial^2}{\partial x_1^2} + \frac{\partial^2}{\partial x_2^2} + \ldots + \frac{\partial^2}{\partial x_n^2} \right)$$

on the torus $\mathbb{T}^n = \mathbb{R}^n/2\pi\mathbb{Z}^n$ and verify that the asymptotic formulae (16.1) and (16.2) hold.

§17. Non-linear First Order Equations

17.1 Bicharacteristics. Let M be an n-dimensional manifold and $a(x, \xi)$ a smooth real-valued function, defined on an open subset of T^*M. Consider the Hamiltonian system on T^*M, generated by $a(x, \xi)$ as Hamiltonian:

$$\begin{cases} \dot{x} = a_\xi \\ \dot{\xi} = -a_x, \end{cases} \tag{17.1}$$

where $a_\xi = \left(\dfrac{\partial a}{\partial \xi_1}, \ldots, \dfrac{\partial a}{\partial \xi_n} \right)$, $a_x = \left(\dfrac{\partial a}{\partial x_1}, \ldots, \dfrac{\partial a}{\partial x_n} \right)$ and (x, ξ) are the coordinates on T^*M, induced by a local coordinate system on M. It is well-kown, that the vector field on T^*M, defined by the right-hand side of (17.1), is independent of the choice of local coordinates on M (cf. e.g. V.I. Arnol'd [1]).

Definition 17.1. A solution curve $(x(t), \xi(t))$ of (17.1) is called a *bicharacteristic* of the function $a(x, \xi)$.

A bicharacteristic is not necessarily defined for all $t \in \mathbb{R}$. In this case we assume that it is defined on the maximal possible interval (concerning this consult also Problems 17.1 and 17.2).

Proposition 17.1. *The function $a(x, \xi)$ is a first integral of the system* (17.1), *i.e. if $(x(t), \xi(t))$ is a bicharacteristic of the function $a(x, \xi)$, then* $a(x(t), \xi(t)) = \text{const}$.

Proof. We have

$$\frac{d}{dt} a(x(t), \xi(t)) = a_x \dot{x} + a_\xi \dot{\xi} = a_x a_\xi - a_x a_\xi = 0 . \qquad \square$$

Proposition 17.1 makes sense of the following definition:

Definition 17.2. A bicharacteristic $(x(t), \xi(t))$ of the function $a(x, \xi)$ is called a *null-bicharacteristic* if $a(x(t), \xi(t)) = 0$.

17.2 The Hamilton-Jacobi equation. Consider the first order partial differential equation

$$a(x, \varphi_x(x)) = 0 , \tag{17.2}$$

where φ is a smooth, real-valued function, defined on an open subset of M and φ_x its gradient. Such an equation is called a Hamilton-Jacobi equation. For its treatment, it is convenient to introduce the graph of φ_x, i.e. the set

$$\Gamma_\varphi = \{(x, \varphi_x(x)), \ x \in M\} \subset T^*M . \tag{17.3}$$

Proposition 17.2. *If φ is a solution of (17.2), then the manifold Γ_φ is invariant under the phase flow of the system (17.1), i.e. if $(x(t), \xi(t))$ is a bicharacteristic of a, $x(t)$ for $t \in [0, b]$ belongs to the domain of φ and $(x(0), \xi(0)) \in \Gamma_\varphi$ then $(x(t), \xi(t)) \in \Gamma_\varphi$ for all $t \in [0, b]$.*

Proof. In view of the uniqueness theorem, it sufficies to verify that the Hamiltonian vector field $(a_\xi, -a_x)$ is tangent to Γ_φ at all its points. This is equivalent to the following: if $(x(t), \xi(t))$ is a bicharacteristic and $(x(0), \xi(0)) \in \Gamma_\varphi$ (i.e. $\xi(0) = \varphi_x(x(0))$), then $\left\{ \dfrac{d}{dt} [\xi(t) - \varphi_x(x(t))] \right\} \Big|_{t=0} = 0$. But this follows from the computation:

$$\left\{ \frac{d}{dt} [\xi(t) - \varphi_x(x(t))] \right\} \Big|_{t=0} = (\dot{\xi} - \varphi_{xx} \cdot \dot{x}) \Big|_{t=0}$$

$$= -a_x(x(0), \xi(0)) - \varphi_{xx}(x(0)) \, a_\xi(x(0), \xi(0))$$

$$= -\frac{\partial}{\partial x} [a(x, \varphi_x(x))]_{x=x(0)} = 0 . \qquad \square$$

In what follows the only important case for us is when $a(x, \xi)$ is positively homogeneous with respect to ξ of degree m, i.e.

$$a(x, t\xi) = t^m a(x, \xi), \qquad t > 0, \ \xi \neq 0, \tag{17.4}$$

where m is any real number. Such functions are characterized by the Euler theorem:

$$\xi \cdot a_\xi = ma. \tag{17.5}$$

Proposition 17.3. *Let $a(x, \xi)$ be homogeneous of degree m and $\varphi(x)$ a solution of (17.2). Then $\varphi(x)$ is constant along the projections of the null-bicharacteristics of the function $a(x, \xi)$ belonging to Γ_φ, i.e. if $(x(t), \xi(t))$ is a null-bicharacteristic and $\xi(0) = \varphi_x(x(0))$, then $\varphi(x(t)) = \mathrm{const}$.*

Proof. We have

$$\frac{d}{dt} \varphi(x(t)) = \varphi_x \dot{x} = \varphi_x a_\xi = \varphi_x(x(t))\, a_\xi(x(t), \xi(t))$$

$$= \varphi_x(x(t))\, a_\xi(x(t), \varphi_x(x(t))) = ma(x(t), \varphi_x(x(t))) = 0. \quad \square$$

17.3 The Cauchy problem. The Cauchy problem for the Hamilton-Jacobi equation (17.2) consists in finding a solution $\varphi(x)$ of this equation, subject to the condition

$$\varphi|_S = \psi, \tag{17.6}$$

where S is a hypersurface (submanifold of codimension 1) in M and $\psi \in C^\infty(S)$. Locally, we may consider the hypersurface as a hyperplane, i.e. by choosing the local coordinate system in a neighbourhood of a point $x_0 \in S$, we may achieve that

$$S = \{x: \ x_n = 0\} \tag{17.7}$$

so that $\psi = \psi(x')$, where $x' = (x_1, \ldots, x_{n-1})$. In this coordinate system, it is convenient to formulate the condition of being *non-characteristic*, guaranteeing local solvability of the Cauchy problem in a neighbourhood of the point $x' \in S$: the equation

$$a(x', 0, \psi_{x'}(x'), \lambda) = 0 \tag{17.8}$$

has a simple root λ, i.e. a root $\lambda \in \mathbb{R}$, which in addition to (17.8) satisfies

$$\frac{\partial a}{\partial \xi_n}(x', 0, \psi_{x'}(x'), \lambda) \neq 0. \tag{17.9}$$

Let a point $0 \in S$ be fixed. Then by the implicit function theorem, the equation

$$a(x, \xi', \lambda) = 0 \tag{17.10}$$

for $|x| < \varepsilon$ and $|\xi' - \psi_{x'}(0)| < \varepsilon$ has a solution $\lambda = a'(x, \xi')$, which is a smooth function of x and ξ'. It is easy to verify that $a'(x, \xi')$ is homogeneous of the first order in ξ', so we may assume that it is defined for $|x| < \varepsilon$ and for all $\xi' \neq 0$ in a conical neighbourhood of $\psi_{x'}(0)$. Equation (17.10) for $|x| < \varepsilon$ and for a vector (ξ', λ) close to the direction of $(\psi'_{x'}(0), a'(0, \psi'_{x'}(0)))$, may be represented in the form

$$\lambda - a'(x, \xi') = 0. \tag{17.11}$$

Therefore the local Cauchy problem takes the following form: find a solution $\varphi = \varphi(x)$ of (17.2), which satisfies (17.6) and, additionally, satisfies

$$\frac{\partial \varphi}{\partial x_n}(0, 0) = a'(0, \psi'_{x'}(0)). \tag{17.12}$$

Since in this situation it is possible to pass from (17.10) to (17.11), our problem may be written in the following form

$$\frac{\partial \varphi}{\partial x_n} - a'\left(x, \frac{\partial \varphi}{\partial x'}\right) = 0, \tag{17.13}$$

$$\varphi\,|_{x_n = 0} = \psi(x'), \tag{17.14}$$

i.e. the matter reduces to the case

$$a(x, \xi) = \xi_n - a'(x, \xi'). \tag{17.15}$$

Let us consider the bicharacteristics of $a(x, \xi)$ of the form (17.15). Their equations are

$$\begin{cases} \dot{x}' = -a'_{\xi'}(x, \zeta'), \\ \dot{x}_n = 1, \\ \dot{\xi} = a'_x(x, \xi'). \end{cases} \tag{17.16}$$

Consider a null-bicharacteristics $(x(t), \xi(t))$ belonging to Γ_φ and starting in S, i.e. such that $x_n(0) = 0$. Then it is obvious from (17.16) that $x_n(t) = t$. Fix another point $x' = x'(0) \in S$. It is clear that the condition $(x(0), \xi(0)) \in \Gamma_\varphi$ means the following

$$\xi'(0) = \psi'_{x'}(x'), \quad \xi_n(0) = \frac{\partial \varphi}{\partial x_n}(x', 0), \tag{17.17}$$

and the condition $a(x(0), \xi(0)) = 0$ gives

$$\xi_n(0) = a'(x', 0, \psi'_{x'}(x')). \tag{17.18}$$

Therefore, the null-bicharacteristic belonging to Γ_φ and such that $x_n(0) = 0$ and $x'(0) = x'$, is uniquely defined. From (17.17) and (17.18) the smooth dependence on x' is clear. In addition, if we consider the transformation

$$g: (x', x_n) \to (x'(x_n), x_n), \tag{17.19}$$

defined for $|x| < \varepsilon$, then from the initial condition $x'(0) = x'$, it follows that its Jacobian is 1 for $x_n = 0$, so that g is a local diffeomorphism. Now, from Proposition 17.3, it necessarily follows that

$$\varphi(x) = \psi([g^{-1}(x)]'), \tag{17.20}$$

where $[g^{-1}(x)]'$ is the vector, obtained from $g^{-1}(x)$ by neglecting the last component (corresponding to the notation x' for $x = (x', x_n)$).

Therefore, we have shown the uniqueness of the solution of the local Cauchy problem and obtained a formula, (17.20), for this solution. The existence of this solution is a simple verification. We recommend the reader to do the following exercise.

Exercise 17.1. Show that formula (17.20) actually gives a solution of the local Cauchy problem as described above.

17.4 Global formulation. We would like to formulate sufficient conditions for the existence of a solution of the Cauchy problem in a neighbourhood of S without restricting to a small neighbourhood of a point on S (although the neighbourhood of the hypersurface S may be very small, in the sense of, for example, some distance from S). First, these conditions must of course, guarantee the existence of solutions of the local problem at any point $x \in S$ and secondly, roughly speaking, provide continuous dependence of the root λ of equation (17.8) on x. This means, that on S we may define a covector field $\xi = \xi(x') \in T^*_x M$, continuously depending on $x' \in S$ and such that

1) $i^* \xi(x') = \psi_{x'}(x')$, where $i: S \to M$ is the natural inclusion map and $\psi_{x'}(x')$ is the gradient of $\psi(x')$ at $x' \in S$, viewed as a covector on S (an element of $T^*_{x'} S$);

2) Introduce local coordinates as described in 17.3 in a neighbourhood of any point $x' \in S$. Then

$$\xi(x') = (\psi_{x'}(x'), \lambda(x')),$$

where $\lambda(x')$ is a root of (17.8), satisfying (17.9), i.e. satisfying all the conditions for the local solvability of the Cauchy problem. Let us note that (17.12) may be written, here without local coordinates as

$$\varphi_x(x') = \xi(x'), \quad x' \in S \tag{17.21}$$

Therefore, the final statement of the Cauchy problem goes as follows: find a solution of (17.2), defined on a connected neighbourhood of the hypersurface S, satisfying the initial condition (17.6) and the additional condition (17.21). In this form, the problem has a unique solution, depending smoothly on the parameters (if any), provided that the given quantities a, S, ψ and ξ also depend smoothly on these parameters.

Remark 17.1. Condition 1) is obviously necessary (assuming the rest is also fulfilled) for the solvability of the Cauchy problem and signifies simply the absence of topological obstructions to the global existence of a field $\xi(x')$, the local existence and smoothness of which is ensured by solvability conditions of the local problems at the points $x' \in S$.

17.5 Linear homogeneous equations. Equation (17.2) is called *linear homogenous* if $a(x, \xi)$ is linear in ξ, i.e.

$$a(x, \xi) = \mathbb{V}(x) \cdot \xi, \tag{17.22}$$

where $\mathbb{V}(x)$ is a vector field on M. The projections on M of the bicharacteristics, are in this case the solutions of the system

$$\dot{x} = \mathbb{V}(x), \tag{17.23}$$

and the solutions of (17.2) are simply the first integrals of the system (17.23). The same system (17.1) contains also, along with (17.23), the equations

$$\dot{\xi} = -\mathbb{V}_x(x) \cdot \xi \tag{17.24}$$

which are linear in ξ. A standard growth estimate for $|\xi(t)|$ shows that if $x(t) \in K$, where K is a compact set in M, then $|\xi(t)|$ is bounded on any finite interval on the t-axis. Therefore a bicharacteristic is either defined for all t or its projection $x(t)$ will leave any compact set $K \subset M$. The condition that S is noncharacteristic means, that $\mathbb{V}(x)$ is everywhere transversal to S.

Let us consider the map g mapping (x', t) into $x(t)$ with $x(t)$ a solution of the system (17.23) with the initial value $x(0) = x'$. If there exists $\varepsilon > 0$ such that $x(t)$ is defined for any x' for all $|t| < \varepsilon$, then g determines a map

$$g \colon S \times (-\varepsilon, \varepsilon) \to M. \tag{17.25}$$

If g is a diffeomorphism, then the solution of the Cauchy problem with initial data on S is defined on the image of g. It is therefore important to be able to estimate from below the number $\varepsilon > 0$, for which the map (17.25) is a diffeomorphism. One important case, where such an estimate is possible will be shown below.

17.6 Non-homogeneous linear equations. These are equations of the form

$$\mathbb{V}(x) \cdot \varphi_x(x) + b(x)\,\varphi(x) = f(x), \tag{17.26}$$

where $b(x), f(x) \in C^\infty(M)$, $\mathbb{V}(x)$ is a vector field on M, $\varphi(x)$ is an unknown function and φ_x its gradient. If $x(t)$ is a solution of the system (17.23) then obviously

$$\frac{d}{dt}\,\varphi(x(t)) + b(x(t))\,\varphi(x(t)) = f(x(t)),$$

from which $\varphi(x(t))$ can be found as a solution of an ordinary first order linear differential equation, provided that $\varphi(x(0))$ is known. The basic feature following from this is that the domain, on which a solution of the Cauchy problem exists, depends only on $\mathbb{V}(x)$ and S and is independent of the right-hand side $f(x)$ and the initial value $\psi \in C^\infty(S)$.

In particular, in what follows, we will need an equation of the special form

$$\frac{\partial \varphi}{\partial x_n} - \sum_{j=1}^{n-1} a_j(x)\,\frac{\partial \varphi}{\partial x_j} + b(x)\,\varphi = f(x), \tag{17.27}$$

where $x = (x', x_n)$, $x' \in M'$ for some $(n-1)$-dimensional closed manifold M' and $x_n \in (-a, a)$ with $a > 0$. The system (17.23) (for the corresponding homogeneous equation) is of the form

$$\begin{cases} \dot{x}' = \mathbb{V}'(x), \\ \dot{x}_n = 1. \end{cases} \tag{17.28}$$

The solutions $x(t)$ of this system which start at $x_n = 0$, are defined for $t \in (-a, a)$ and if we put $S = M' = \{x : x_n = 0\}$, then the map g of the preceding section becomes a diffeomorphism $g \colon M \to M$, where $M = M' \times (-a, a)$ and where it is clear from (17.28) that the "fiber" $M' \times x_0$ is mapped onto itself diffeomorphically. Because of this, the Cauchy problem for (17.27) with initial condition

$$\varphi\big|_{x_n = 0} = \psi(x'), \qquad x' \in M', \tag{17.29}$$

has a solution $\varphi \in C^\infty(M)$.

In a number of cases one can carry out similar arguments also for non-compact M'.

Problem 17.1. Let $a(x, \xi)$ be defined for $x \in M$, $\xi \neq 0$ with degree of homogeneity 1 in ξ. Show that if $(x(t), \xi(t))$ is a bicharacteristic, then it is either defined for all t or $x(t)$ will leave any compact set $K \subset M$. In particular, if M is compact, then all bicharacteristics are defined for all $t \in \mathbb{R}$.

Problem 17.2. Show that the same holds for an arbitrary degree of homogeneity of $a(x, \xi)$, if condition of ellipticity holds:

$$a(x, \xi) \neq 0 \qquad \text{for} \quad \xi \neq 0, \; x \in M.$$

§18. The Action of a Pseudodifferential Operator on an Exponent

18.1 Formulation of the result. Here we describe the asymptotic behaviour as $\lambda \to +\infty$ of the expression $A(e^{i\lambda\psi(x)})$, with A a ΨDO and ψ a smooth function without critical points.

Theorem 18.1. *Let X be an open set in $\mathbb{R}^n$, $A \in L^m_{\varrho,\delta}(X)$, $1 - \varrho \leq \delta < \varrho$, A properly supported and with symbol $a(x, \xi)$. Let $\psi(x) \in C^\infty(X)$ and $\psi'_x(x) \neq 0$ for $x \in X$ (here ψ'_x denotes the gradient of ψ). Then for any function $f \in C^\infty(X)$ and arbitrary integer $N \geq 0$, for $\lambda \geq 1$ we have*

$$A(f e^{i\lambda\psi}) = e^{i\lambda\psi} \left[\sum_{|\alpha| < N} a^{(\alpha)}(x, \lambda\psi'_x(x)) \, \frac{D_z^\alpha(f(z) e^{i\lambda\varrho_x(z)})}{\alpha!} \Bigg|_{z=x} + \right.$$

$$\left. + \lambda^{m-(\varrho-1/2)N} R_N(x, \lambda) \right], \tag{18.1}$$

where $\varrho_x(y) = \psi(y) - \psi(x) - (y-x) \cdot \psi'_x(x)$, $a^{(\alpha)}(x, \xi) = \partial_\xi^\alpha a(x, \xi)$, *and for* $R_N(x, \lambda)$ *the following estimate holds*

$$|\partial_x^\gamma R_N(x, \lambda)| \leq C_{\gamma, N, K} \lambda^{\delta|\gamma|}, \qquad x \in K, \tag{18.2}$$

where K is compact in X and the constants $C_{\gamma, N, K}$ do not depend on λ. If there are families of functions $f(x)$, $\psi(x)$, bounded in $C^\infty(X)$ (i.e. the derivatives $\partial^\gamma f(x)$, $\partial^\gamma \psi(x)$ are uniformly bounded on any compact set for arbitrary fixed γ) and if the gradients ψ'_x are uniformly separated (in absolute value) from 0 on any compact set, then the constants $C_{\gamma, N, K}$ in the estimates (18.2) are independent of the choice of functions f, ψ of these families.

Remark 18.1. The statement of this theorem is similar to that of Theorem 4.2 on the transformation of the symbol under diffeomorphisms and, as will become clear in what follows, the theorems are actually equivalent. However, in view of the importance of Theorem 18.1, we shall give two proofs for it: one derived from Theorem 4.2 and an independent one, essentially based on the stationary phase method. The latter allows us to deduce from Theorem 18.1 the invariance of the class of ΨDO under diffeomorphisms and the formulae for change of coordinates. The reader is recommended to pursue this in the form of an exercise (historically this was the first method by which the invariance of the class of ΨDO under diffeomorphisms was demonstrated).

Remark 18.2. It is useful to note right from the beginning, in what sense (18.1) is asymptotic as $\lambda \to +\infty$.

It is clear that the condition $\psi'_x(x) \neq 0$ will ensure the estimate

$$|a^{(\alpha)}(x, \lambda \psi'_x(x))| \leq C_{\alpha, K} \lambda^{m - \varrho |\alpha|}, \qquad x \in K. \tag{18.3}$$

Let us now verify that $D^\alpha_z(f(z) e^{i\lambda \varrho_x(z)})|_{z=x}$ is a polynomial of degree not higher than $|\alpha|/2$. Indeed we have

$$D^\alpha_z(f(z) e^{i\lambda \varrho_x(z)}) \tag{18.4}$$

$$= \sum_{\gamma_0 + \gamma_1 + \ldots + \gamma_k = \alpha} c_{\gamma_0 \ldots \gamma_k} (D^{\gamma_0}_z f(z)) \cdot \lambda^k (D^{\gamma_1}_z \varrho_x(z)) \ldots (D^{\gamma_k}_z \varrho_x(z)) e^{i\lambda \varrho_x(z)},$$

where in this sum $|\gamma_j| \geq 1$ for $j = 1, 2, \ldots, k$. Since $\varrho_x(z)$ has a zero of second order for $x = z$, then in (18.4) for $x = z$ only terms in which $|\gamma_j| \geq 2, j = 1, \ldots, k$, remain. But $\sum_{j=1}^{k} |\gamma_j| \leq |\alpha|$, so we obtain $2k \leq |\alpha|$ as required.

Now, taking (18.3) into account, we see that

$$\left| a^{(\alpha)}(x, \lambda \psi'_x(x)) \left. \frac{D^\alpha_z(f(z) e^{i\lambda \varrho_x(z)})}{\alpha!} \right|_{z=x} \right| \leq C_{\alpha, K} \lambda^{m - (\varrho - 1/2)|\alpha|}, \qquad x \in K,$$

giving the required decrease in the degrees of growth of the finite terms in formula (18.4), since $\varrho > 1/2$.

18.2 First proof of Theorem 18.1. Let us make a change of coordinates $y = \varkappa(x) = (x_1, \ldots, x_{n-1}, \psi(x))$. This change has Jacobian ψ'_{x_n} and may be made, therefore, only where $\psi'_{x_n} \neq 0$. The general case however can be easily reduced to this case using a partition of unity and a rearrangement of the coordinate axes. Indeed A is properly supported and consequently for any compact set K_1 there exists a compact set K_2 such that $Au|_{K_1}$ only depends on $u|_{K_2}$. Therefore $A(fe^{i\lambda\psi})|_{K_1}$ may be written as a finite sum of terms of the form $A(f_j e^{i\lambda\psi})$, with $f_j \in C^\infty_0(X)$ and where to any j there exists an integer k, $1 \leq k \leq n$, such that $\psi'_{x_k}(x) \neq 0$ for $x \in \mathrm{supp} f_j$.

Thus, let $\psi'_{x_n} \neq 0$, $x \in X$ and let A_1 be the operator A written in the coordinates $y = \varkappa(x)$ (cf. §4). Let $a_1(y, \eta)$ be the symbol of A_1. By Theorem 4.2 we have

$$a_1(y, \eta)|_{y=\varkappa(x)} = e^{-i\varkappa(x) \cdot \eta} A(e^{i\varkappa(x) \cdot \eta})$$

$$= \sum_{|\alpha| < N} \frac{1}{\alpha!} a^{(\alpha)}(x, {}^t\varkappa'(x)\eta)(D^\alpha_z e^{i\varkappa''_x(z) \cdot \eta})|_{z=x} + r_N(x, \eta),$$

where

$$\varkappa''_x(z) = \varkappa(z) - \varkappa(x) - \varkappa'(x)(z-x), \qquad r_N \in S^{m - (\varrho - 1/2)N}_{\varrho, \delta}(X \times \mathbb{R}^n).$$

In particular, putting $\eta = (0, \ldots, 0, \lambda)$, we obtain precisely formula (18.1) for $f \equiv 1$ and where R_N can be estimated as in (18.2). The statements about uniformity and the case of arbitrary f are obtained by repeating the proof of Theorem 4.2; we leave this to the reader as an exercise. $\quad\square$

18.3 Second proof of Theorem 18.1. For simplicity, we shall assume in this proof that $\delta = 0$ and $\varrho = 1$.

Note, once again, that since A is properly supported, we may assume that $f \in C_0^\infty(X)$. Put, for brevity, $I(\lambda) = e^{-i\lambda\psi(x)} A\left(fe^{i\lambda\psi(x)}\right)(x)$ (for a fixed x). We have

$$I(\lambda) = \int a(x, \xi)\, f(y)\, e^{i\lambda(\psi(y) - \psi(x)) + i(x - y)\cdot\xi}\, dy\, d\xi. \tag{18.5}$$

Let us now make the change of coordinates $\xi = \lambda\zeta$:

$$I(\lambda) = \lambda^n \int a(x, \lambda\zeta)\, f(y)\, e^{i\lambda[\psi(y) - \psi(x) - (y-x)\cdot\zeta]}\, dy\, d\zeta. \tag{18.6}$$

We want to find the asymptotic behaviour of this integral as $\lambda \to +\infty$. We shall see that a major role (and this is the point of the stationary phase method) is played by neighbourhoods of the critical points of the function

$$g(y, \zeta) = \psi(y) - \psi(x) - (y - x)\cdot\zeta. \tag{18.7}$$

Now, since $g'_\zeta = x - y$, $g'_y = \psi'_y(y) - \zeta$, there exists exactly one critical point of this function, the point $y = x$, $\zeta = \psi'_x(x)$. For short we put $\xi_x = \psi'_x(x)$. Introduce the cut-off function $\chi \in C_0^\infty(\mathbb{R}^n)$, $\chi(z) = 1$ for $|z| < \varepsilon/2$, $\chi(z) = 0$ for $|z| > \varepsilon$, where $\varepsilon > 0$ and consider the integral

$$\tilde{I}(\lambda) = \lambda^n \int a(x, \lambda\zeta)\, \chi(y - x)\, \chi(\zeta - \xi_x)\, f(y)\, e^{i\lambda g(y,\zeta)}\, dy\, d\zeta. \tag{18.8}$$

Then for any $N > 0$

$$|I(\lambda) - \tilde{I}(\lambda)| \leqq C_N \lambda^{-N}. \tag{18.9}$$

Indeed, putting

$${}^tL = \lambda^{-1}[(\psi'(y) - \zeta)^2 + (x - y)^2]^{-1}[(\psi'(y) - \zeta)\cdot D_y + (x - y)\cdot D_\zeta],$$

we see that ${}^tLe^{i\lambda g(y,\zeta)} = e^{i\lambda g(y,\zeta)}$. Integrating by parts in the oscillatory integral $I(\lambda) - \tilde{I}(\lambda)$, and considering that

$$|(\psi'(y) - \zeta)^2 + (x - y)^2| \geqq \varepsilon_1 > 0 \quad \text{on} \quad \text{supp}\,[1 - \chi(y - x)\,\chi(\zeta - \xi_x)],$$

we see that

$$I(\lambda) - \tilde{I}(\lambda) = \lambda^n \int e^{i\lambda g(y,\zeta)}\, L^N[a(x, \lambda\zeta)(1 - \chi(y - x)\,\chi(\zeta - \xi_x))f(y)]\, dy\, d\zeta,$$

and transforming this oscillatory integral into an absolutely convergent one (cf. §1) we easily obtain the estimate (18.9) due to the factor λ^{-1}, in the expression for L. Analogously, one also obtains estimates for the x-derivatives from the difference $I(\lambda) - \tilde{I}(\lambda)$. However, note that they follow from the estimates of $I(\lambda) - \tilde{I}(\lambda)$ with arguments similar to the proof of Proposition 3.6. In the sequel we shall omit estimates of the derivatives, leaving them to the reader.

Thus, instead of $I(\lambda)$, we may consider $\tilde{I}(\lambda)$. Making yet another change of coordinates $\zeta = \xi_x + \lambda^{-1}\eta$, we obtain

$$\tilde{I}(\lambda) = \int e^{i(x-y)\cdot\eta}\, a(x, \lambda\xi_x + \eta)\, \chi\!\left(\frac{\eta}{\lambda}\right) \chi(y-x)\, f(y)\, e^{i\lambda\varrho_x(y)}\, dy\, d\eta\,. \qquad (18.10)$$

Expand $a(x, \lambda\xi_x + \eta)$ in a Taylor series at $\eta = 0$:

$$a(x, \lambda\xi_x + \eta) = \sum_{|\alpha| < N} a^{(\alpha)}(x, \lambda\xi_x)\, \frac{\eta^\alpha}{\alpha!} + r_N(x, \eta, \lambda)\,,$$

where

$$r_N(x, \eta, \lambda) = \sum_{|\alpha| = N} c_\alpha \int_0^1 (1-t)^{N-1} \eta^\alpha a^{(\alpha)}(x, \lambda\xi_x + t\eta)\, dt\,.$$

Multiplying this expansion with the cut-off function $\chi(\eta/\lambda)\, \chi(y-x)$ and substituting the result into (18.10) we obtain

$$\tilde{I}(\lambda) = \sum_{|\alpha| < N} \int e^{i(x-y)\cdot\eta}\, a^{(\alpha)}(x, \lambda\xi_x)$$

$$\times \frac{\eta^\alpha}{\alpha!}\, \chi\!\left(\frac{\eta}{\lambda}\right) \chi(y-x)\, f(y)\, e^{i\lambda\varrho_x(y)}\, dy\, d\eta + R_N'(x, \lambda)\,, \qquad (18.11)$$

where

$$R_N'(x, \lambda) = \sum_{|\alpha| = N} c_\alpha \int dy\, d\eta \int_0^1 (1-t)^{N-1} \eta^\alpha a^{(\alpha)}(x, \lambda\xi_x + t\eta)$$

$$\times \chi(y-x)\, \chi\!\left(\frac{\eta}{\lambda}\right) e^{i(x-y)\cdot\eta}\, f(y)\, e^{i\lambda\varrho_x(y)}\, dt\,. \qquad (18.12)$$

As follows from the arguments above, the asymptotic behaviour of the finite parts in formula (18.11) does not change if we remove the cut-off function $\chi(\eta/\lambda)$. But then these terms can be easily transformed, by the Fourier inversion formula:

$$\int e^{i(x-y)\cdot\eta}\, a^{(\alpha)}(x, \lambda\xi_x)\, \frac{\eta^\alpha}{\lambda!}\, \chi(y-x)\, f(y)\, e^{i\lambda\varrho_x(y)}\, dy\, d\eta$$

$$= a^{(\alpha)}(x, \lambda\xi_x)\, D_y^\alpha \big(f(y)\, e^{i\lambda\varrho_x(y)}\big)\big|_{y=x}\,,$$

so that it only remains to estimate the remainder R'_N or to estimate, uniformly in t $(0 < t \leqq 1)$, the integral

$$r_\alpha(x, \lambda, t) = \int \eta^\alpha \, a^{(\alpha)}(x, \lambda\xi_x + t\eta) \, \chi(y - x) \, \chi\left(\frac{\eta}{\lambda}\right) e^{i(x-y)\cdot\eta} \left(f(y) \, e^{i\lambda\varrho_x(y)}\right) dy \, d\eta$$

$$= \sum_{\alpha' + \alpha'' = \alpha} c_{\alpha'\alpha''} \int e^{i(x-y)\cdot\eta} \, a^{(\alpha)}(x, \lambda\xi_x + t\eta) \, \chi^{(\alpha')}(y - x) \, \chi\left(\frac{\eta}{\lambda}\right) D^{\alpha''}\left(f(y) \, e^{i\lambda\varrho_x(y)}\right) dy \, d\eta$$

$$\tag{18.13}$$

where $|\alpha| = N$.

Let us introduce the notation

$$\tilde{a}_\alpha(x, \eta, \lambda, t) = \chi\left(\frac{\eta}{\lambda}\right) a^{(\alpha)}(x, \lambda\xi_x + t\eta).$$

If the number ε in the definition of $\chi(z)$ is chosen so that $\varepsilon < |\xi_x|/2$, then for $\tilde{a}_\alpha$ one has the estimates

$$|\partial_\eta^\gamma \partial_x^\beta \tilde{a}_\alpha(x, \eta, \lambda, t)| \leqq C_{\alpha\beta\gamma} \lambda^{m - N - |\gamma|}. \tag{18.14}$$

Now let us use (18.4) and substitute into (18.13) the expression obtained from this for $D_y^{\alpha''}(f(y)e^{i\lambda\varrho_x(y)})$. In this expression all the terms contain products

$$\lambda^k (D_y^{\gamma_1} \varrho_x(y)) \, \ldots \, (D_y^{\gamma_k} \varrho_x(y)), \tag{18.15}$$

in which $|\gamma_1| + \ldots + |\gamma_k| \leqq N$. If $k \leqq N/2$, we do not transform this product. If $k > N/2$, then by the Dirichlet principle, in (18.15) there are no less than $k - N/2$ indices γ_j such that $|\gamma_j| = 1$. But then, by the Hadamard lemma

$$(D_y^{\gamma_1} \varrho_x(y)) \, \ldots \, (D_y^{\gamma_k} \varrho_x(y)) = \sum_{|\gamma| \geqq k - N/2} g_\gamma(y, x) \, (x - y)^\gamma,$$

where $g_\gamma(y, x)$ is a smooth function (in x and y), defined for y sufficiently close to x. Inserting this expression into (18.13) and integrating by parts (utilizing the exponent $e^{i(x-y)\cdot\eta}$, allowing us to change $(x - y)^\gamma$ into $(\ \ D_\eta)^\gamma$), we see that $r_\alpha(x, \lambda, t)$ is a linear combination of terms of the form

$$I_1(\lambda) = \lambda^k \int \tilde{a}_\alpha^{(\gamma)}(x, \eta, \lambda, t) \, e^{i(x-y)\cdot\eta} \, e^{i\lambda\varrho_x(y)} \tilde{f}(y, x) \, dy \, d\eta, \tag{18.16}$$

where $\tilde{a}_\alpha^{(\gamma)} = \partial_\eta^\gamma \tilde{a}_\alpha$ and $\tilde{f}(y, x)$ is smooth (in x and y) and supported in $|y - x| \leqq \varepsilon$. The indices k and γ are related by $|\gamma| \geqq k - N/2$. Taking into account that the volume of the domain of integration in η in (18.16) does not exceed $C\lambda^n$, and using (18.14) we obtain for $I_1(\lambda)$ the estimate

$$|I_1(\lambda)| \leqq C \lambda^{k + m - (N + |\gamma|) + n} \leqq C \lambda^{m + n - N/2},$$

which allows us to conclude the proof by applying the type of arguments used in the proof of Proposition 3.6. $\Box$

18.4 The product of a pseudodifferential operator and a Fourier integral operator. Let X, Y be open sets in $\mathbb{R}^{n_x}$ and $\mathbb{R}^{n_y}$ and let P be an FIO of the form

$$Pu(x) = \int p(x, y, \theta) \, e^{i\varphi(x,y,\theta)} \, u(y) \, dy \, d\theta \,, \qquad (18.17)$$

where $p(x, y, \theta) \in S^{m'}(X \times Y \times \mathbb{R}^N)$ and $\varphi(x, y, \theta)$ is an operator phase function (cf. §2, Definition 2.3). Let there also be on X a properly supported PDO $A \in L^m_{\varrho,\delta}(X)$ with symbol $a(x, \xi)$. Since P maps $C_0^\infty(Y)$ into $C^\infty(X)$ and $\mathscr{E}'(Y)$ into $\mathscr{D}'(X)$ and A maps the spaces $C^\infty(X)$ and $\mathscr{D}'(X)$ into themselves, then the composition $A \cdot P$ is defined as an operator, mapping $C_0^\infty(Y)$ into $C^\infty(X)$ and $\mathscr{E}'(Y)$ into $\mathscr{D}'(X)$.

Theorem 18.2. *Let $1 - \varrho \leqq \delta < \varrho \leqq 1$. Then the composition $Q = A \cdot P$ is also of the form (18.17) with the same phase function $\varphi(x, y, \theta)$ as P and with an amplitude of the form*

$$q(x, y, \theta) = e^{-i\varphi(x,y,\theta)} \, a(x, D_x) \, [p(x, y, \theta) \, e^{i\varphi(x,y,\theta)}], \qquad (18.18)$$

with the asymptotic formula

$$q(x, y, \theta) \sim \sum_\alpha a^{(\alpha)}(x, \varphi_x(x, y, \theta)) \, \frac{D_z^\alpha(p(z, y, \theta) \, e^{i\varrho(z,x,y,\theta)})}{\alpha!} \Bigg|_{z=x}$$

$$\text{as} \quad |\theta| \to +\infty \qquad (18.19)$$

where $\varrho(z, x, y, \theta) = \varphi(z, y, \theta) - \varphi(x, y, \theta) - (z - x) \cdot \varphi_x(x, y, \theta)$.

Remark 18.3. Since $\varphi(x, y, \theta)$ is not smooth for $\theta = 0$, it is not immediately clear from (18.18) that Q is an FIO. This is the case however, since adding an operator with smooth kernel to P we may assume that $p(x, y, \theta) = 0$ for $|\theta| < 1$. Then (18.18) defines a smooth function in all the variables and the same holds for all terms in the expansion (18.19), which has the usual meaning (cf. Definition 3.4). However, an operator with smooth kernel may always be written in the form (18.17) with an amplitude $p(x, y, \theta)$ which has compact support in θ and equal 0 for $|\theta| < 1$ (cf. the hint to Exercise 2.4). Therefore Q is an FIO with phase function φ.

Proof of Theorem 18.2. Let us introduce the set

$$C_\varphi = \{(x, y, \theta): \varphi'_\theta(x, y, \theta) = 0\} \,.$$

used in §1 and §2. Note that $\varphi'_x(x, y, \theta) \neq 0$ for $(x, y, \theta) \in C_\varphi$ by the definition of an operator phase function. Changing P by adding an operator with a smooth kernel, we may assume that $\operatorname{supp} p(x, y, \theta)$ lies in an arbitrarily small conical neighbourhood of the set C_φ (cf. Proposition 2.1) and, in particular that $\varphi'_x \neq 0$ on $\operatorname{supp} p$. In addition and in accordance with Remark 18.3, assume that $p(x, y, \theta) = 0$ for $|\theta| < 1$.

Now note, that since A continuously maps $C^\infty(X)$ into $C^\infty(X)$, we may apply it under the integral sign in (18.17) (for this it is necessary to begin by transforming the integral into an absolutely convergent one, as in §1; note that the variable x is not involved in this change, being just a parameter). Now it is only necessary to verify (18.19), understood in the sense of Definition 3.4 (cf. also Remark 18.3). But this is a trivial consequence of Theorem 18.1, putting $\lambda = |\theta|$ and viewing y and θ as parameters. Indeed, we have

$$q(x, y, \theta) = \lambda^{m'} \, e^{-i\lambda\varphi(x, y, \theta')} \, a(x, D_x) \, [\lambda^{-m'} \, p(x, y, \theta) \, e^{i\lambda\varphi(x, y, \theta')}],$$

where $\theta' = \theta/|\theta|$. Noting now that by varying the parameters y, θ the functions $\varphi(x, y, \theta')$, $\lambda^{-m'} p(x, y, \theta)$ belong to a bounded subset of $C^\infty(X)$, we see that Theorem 18.1 applies. $\square$

Exercise 18.1. Obtain from Theorem 18.2 the composition formula for two properly supported ΨDO of the type $L^m(X)$ (cf. Theorem 3.4).

Exercise 18.2. Let A and P be as in Theorem 18.2. Prove the result, similar to Theorem 18.2, for the operator $Q_1 = P \cdot A$.

Hint. Use transposition.

Problem 18.1. Obtain from Theorem 18.1 the change of variable formula for PDO (cf. Theorem 4.2).

§19. Phase Functions Defining the Class of Pseudodifferential Operators

19.1 In the formulation of Theorem 4.1 there is an example of a class of phase functions, for which the corresponding class of FIO coincides with the class of ΨDO. In the sequel, we shall need the following variant of this theorem for non-linear phase functions.

Theorem 19.1. *Let X be an open set in $\mathbb{R}^n$ and $\varphi(x, y, \xi)$ a phase function on $X \times X \times \mathbb{R}^n$ such that*

1) $\varphi'_\xi(x, y, \xi) = 0 \Leftrightarrow x = y$;

2) $\varphi'_x(x, x, \xi) = \xi$.

Then, if $1 - \varrho \leq \delta < \varrho$, the class of FIO with phase function φ and amplitude $p(x, y, \xi) \in S^m_{\varrho, \delta}(X \times X \times \mathbb{R}^n)$ coincides with the class $L^m_{\varrho, \delta}(X)$. The class of FIO with phase function φ and with an amplitude $a(x, y, \xi)$ which is classical, coincides with the class of classical ΨDO.

Remark 19.1. Conditions 1) and 2) for nearby x and y, may be expressed in one single condition

$$\varphi(x, y, \xi) = (x - y) \cdot \xi + O(|x - y|^2 |\xi|). \tag{19.1}$$

19.2 Proof of Theorem 19.1. 1. Since the kernels of FIO with phase functions satisfying condition 1) of Theorem 19.1 are smooth for $x \neq y$ by Proposition 2.1, then we may assume that all amplitudes are supported on an arbitrarily small neighbourhood of the diagonal $x = y$.

2. Let φ, φ_1 be two phase functions, satisfying the conditions of Theorem 19.1. Denoting by $L^m_{\varrho, \delta}(X, \varphi)$ the class of FIO with phase function φ and with amplitudes in the class $S^m_{\varrho, \delta}(X \times X \times \mathbb{R}^n)$, we see that it suffices to verify the inclusion

$$L^m_{\varrho, \delta}(X, \varphi) \subset L^m_{\varrho, \delta}(X, \varphi_1), \tag{19.2}$$

because it clearly implies that all the classes $L^m_{\varrho, \delta}(X, \varphi)$ coincide and, in particular, that they coincide with $L^m_{\varrho, \delta}(X)$.

3. Denote by $\mathscr{H}_k$ the class of all functions $\psi(x, y, \xi)$, which are positively homogenous of degree k in ξ, smooth for $\xi \neq 0$ and defined for nearby x and y. Note that $\mathscr{H}_0$ is an algebra, containing the smooth functions of x and y as a subalgebra, and that $\mathscr{H}_1$ is a $\mathscr{H}_0$-module.

For us it is essential that for nearby x and y, the difference $\varphi_1 - \varphi$ can be written in the form

$$\varphi_1 - \varphi = \sum_{j, k = 1}^n b_{jk} \frac{\partial \varphi}{\partial \xi_j} \frac{\partial \varphi}{\partial \xi_k}, \qquad b_{jk} \in \mathscr{H}_1. \tag{19.3}$$

Let us verify this. From the Taylor formula it follows that

$$\frac{\partial \varphi}{\partial \xi_j} = (x_j - y_j) + \sum_{k = 1}^n a_{jk}(x_k - y_k),$$

where $a_{jk} \in \mathscr{H}_0$, $a_{jk}(x, x, \xi) = 0$. This can also be written in the form

$$\varphi'_\xi = (I + A)(x - y),$$

where I is the unit matrix and A is a matrix with elements in $\mathscr{H}_0$, equal to 0 for $x = y$. But then, for nearby x and y the matrix $(I + A)^{-1}$ exists and has elements from $\mathscr{H}_0$. This means that we may write

$$x_j - y_j = \sum_{k = 1}^n \tilde{a}_{jk} \frac{\partial \varphi}{\partial \xi_k}, \qquad \tilde{a}_{jk} \in \mathscr{H}_0. \tag{19.4}$$

Now, using (19.1) for φ and φ_1, we see that on the diagonal (as $x = y$) $\varphi_1 - \varphi$ has a zero of order two and by the Taylor formula

$$\varphi_1 - \varphi = \sum_{j,\,k=1}^{n} \tilde{b}_{jk}(x_j - y_j)(x_k - y_k), \qquad \tilde{b}_{jk} \in \mathscr{H}_1.$$

Inserting here the expression for $x_j - y_j$ from (19.4), we obtain (19.3).

4. Consider now the homotopy

$$\varphi_t = \varphi + t(\varphi_1 - \varphi), \qquad 0 \le t \le 1. \tag{19.5}$$

Each of the functions φ_t satisfies (19.1). A trivial repetition of the above argument shows that instead of (19.3) we may write

$$\varphi_1 - \varphi = \sum_{j,\,k=1}^{n} b_{jk}^{t}\, \frac{\partial \varphi_t}{\partial \xi_j}\, \frac{\partial \varphi_t}{\partial \xi_k}, \tag{19.6}$$

where $b_{jk}^{t} \in \mathscr{H}_1$ and depends smoothly on t.

Now let P_t be the FIO given by the formula

$$P_t u(x) = \int e^{i\varphi_t(x,y,\xi)}\, p(x,y,\xi)\, u(y)\, dy\, d\xi.$$

Then

$$\frac{d^r}{dt^r}(P_t u) = \iint p(x,y,\xi)\, i^r (\varphi_1 - \varphi)^r\, e^{i\varphi_t}\, u(y)\, dy\, d\xi$$

$$= \sum_{j_1,\,\ldots,\,j_{2r}} \iint d_{j_1,\,\ldots,\,j_{2r}}\, \frac{\partial \varphi_t}{\partial \xi_{j_1}} \cdots \frac{\partial \varphi_t}{\partial \xi_{j_{2r}}}\, e^{i\varphi_t}\, pu\, dy\, d\xi, \tag{19.7}$$

where $d_{j_1,\,\ldots,\,j_{2r}} \in \mathscr{H}_r$. Without loss of generality, we may assume that $p(x,y,\xi) = 0$ for $|\xi| < 1$, so that $d_{j_1,\,\ldots,\,j_{2r}} p \in S^{m+r}(X \times X \times \mathbb{R}^n)$. We now integrate by parts in (19.7), using the formula

$$\frac{\partial \varphi_t}{\partial \xi_j}\, e^{i\varphi_t} = i^{-1}\, \frac{\partial}{\partial \xi_j}\, e^{i\varphi_t}.$$

This integration demonstrates that

$$\frac{\overset{\bullet}{d^r}}{dt^r}\, P_t \in L_{\varrho,\,\delta}^{m+r(1-2\varrho)}(X, \varphi_t), \tag{19.8}$$

where all estimates are uniform in t. But if we now put

$$Q_j = \frac{(-1)^j}{j!}\, \frac{d^j P_t}{dt^j}\bigg|_{t=1} \in L_{\varrho,\,\delta}^{m+j(1-2\varrho)}(X, \varphi_1),$$

then, by the Taylor formula,

$$P_0 = \sum_{j=0}^{k-1} Q_j + (-1)^k \int_0^1 \frac{t^{k-1}}{(k-1)!} \frac{d^k P_t}{dt^k} \, dt. \tag{19.9}$$

The remainder in (19.9) has a kernel with increasing smoothness as $k \to +\infty$. It is therefore clear that if $Q \sim \sum_{j=0}^{\infty} Q_j$ (adding the amplitudes asymptotically), then

$$Q \in L_{\varrho,\delta}^m(X, \varphi_1) \quad \text{and} \quad P_0 - Q \in L^{-\infty}(X),$$

which shows (19.2).

The fact that classical amplitudes remain classical under this procedure, is clear from the construction. $\square$

Problem 19.1. Let φ_1, φ_2 be two phase functions such that

$$\varphi_1 - \varphi_2 = \sum_{j,k=1}^{n} b_{jk} \frac{\partial \varphi_2}{\partial \xi_j} \frac{\partial \varphi_2}{\partial \xi_k},$$

where $b_{jk} \in C^\infty(X \times X \times (\mathbb{R}^n \setminus 0))$ and b_{jk} homogenous in ξ of degree 1. Show that $L_{\varrho,\delta}^m(X, \varphi_2) \subset L_{\varrho,\delta}^m(X, \varphi_1)$.

§20. The Operator $\exp(-itA)$

20.1 Definition and formulation of results. Let M be a closed n-dimensional manifold with a smooth density and let A be a self-adjoint, classical ΨDO of degree 1 on M with principal symbol $a_1(x, \xi)$ satisfying the condition

$$a_1(x, \xi) > 0 \quad \text{for} \quad \xi \neq 0 \tag{20.1}$$

(in particular, A is elliptic). Let $\{\varphi_k\}_{k=1,2,\ldots}$ be a complete orthonormal system of eigenvectors for A and λ_k the corresponding eigenvalues. If $u(x) \in C^\infty(M)$ we denote by u_k the Fourier coefficients of $u(x)$ with respect to the system $\{\varphi_k(x)\}$:

$$u_k = (u, \varphi_k). \tag{20.2}$$

Proposition 20.1. *If $u(x) \in C^\infty(M)$, then*

$$u(x) = \sum_{k=1}^{\infty} u_k \varphi_k(x), \tag{20.3}$$

where the series converges in the topology of $C^\infty(M)$.

Proof. We clearly have

$$u_k \lambda_k^N = (u, A^N \varphi_k) = (A^N u, \varphi_k),$$

and since $A^N u \in C^\infty(M)$, we may apply A^N termwise to the series (20.3) for any $N \in \mathbb{Z}_+$, obtaining each time a series converging in $L^2(M)$. But from this it immediately follows that for any $s \in \mathbb{R}$, the series (20.3) converges in norm in $H^s(M)$, since for $s = N \in \mathbb{Z}_+$ this norm is equivalent to $\|u\| + \|A^N u\|$, with $\| \cdot \|$ the $L^2(M)$-norm. From this and the embedding theorem 7.6, the required statement follows. $\square$

This proposition allows us to make the following

Definition 20.1. The operator $\exp(-itA)$ for $t \in \mathbb{R}$ is defined by the formula

$$\exp(-itA) u(x) = \sum_{k=1}^\infty \exp(-it\lambda_k) u_k \varphi_k(x). \qquad (20.4)$$

Clearly the series (20.4) for $u \in C^\infty(M)$ converges in the topology of $C^\infty(M)$. Further, if $u \in H^s(M)$, s an integer, then this series converges in norm in $H^s(M)$ (cf. the proof of Proposition 10.2). The operator $\exp(-itA)$ is for integer s a bounded operator on $H^s(M)$. Note, that it is also a unitary operator on $L^2(M)$.

Another definition consists in considering the Cauchy problem

$$\begin{cases} \dfrac{\partial u}{\partial t} + iAu = 0, & (20.5) \\[2mm] u|_{t=0} = u_0, & (20.6) \end{cases}$$

where $u = u(t, x) \in C^\infty(\mathbb{R} \times M)$, $u_0 \in C^\infty(M)$ and the operator A in (20.5) acts on x for any fixed t. Solving this problem by the "Fourier method", we see that the solution is given by (20.4), with u replaced by u_0, i.e.

$$u(t, x) = \exp(-itA) u_0(x). \qquad (20.7)$$

The solution of (20.5)–(20.6) is also unique. This can be seen, for instance by writing for a solution $u(t, x)$ the expansion

$$u(t, x) = \sum_{k=1}^\infty c_k(t) \varphi_k(x),$$

which, being inserted into (20.5)–(20.6) gives the equations $c_k'(t) + i\lambda_k c_k(t) = 0$, with the initial values $c_k(0) = (u_0, \varphi_k)$, from which one can uniquely recover $c_k(t)$.

Theorem 20.1. *If $\varepsilon > 0$ is sufficiently small, then for $|t| < \varepsilon$ one can represent $U(t) = \exp(-itA)$ in the form of a sum of an operator with a smooth kernel in t, x and y and an FIO, given by the phase function*

$$\varphi(t, x, y, \xi) = \psi(x, y, \xi) - ta_1(y, \xi) \qquad (20.8)$$

linear in t, and by an amplitude $p(t, x, y, \xi)$ which is a classical symbol of order 0, smooth in t and such that the following estimate is fulfilled

$$|\partial_\xi^\alpha \partial_{x,y,t}^\beta \, p(t, x, y, \xi)| \leqq C_{\alpha\beta} \langle \xi \rangle^{-|\alpha|}. \tag{20.9}$$

Example 20.1. Consider in $\mathbb{R}^n$ the operator $A = \sqrt{-\Delta}$ with symbol $|\xi|$. The Cauchy problem (20.5)–(20.6) for functions decreasing as $|x| \to +\infty$ can be solved using the Fourier transformation in x. Indeed, since $Au(x) = F_{\xi \to x}^{-1}(|\xi| \tilde{u}(\xi))$, then

$$u(t, x) = \exp(-itA)\, u_0(x) = F_{\xi \to x}^{-1}(e^{-it|\xi|} \tilde{u}_0(\xi))$$

$$= \int\int e^{i(x-y)\cdot\xi - it|\xi|}\, u(y)\, dy\, d\xi.$$

We see that in this case (formally this does not follow from the theorem however) the operator $\exp(-itA)$ is an FIO with phase function $(x-y)\cdot\xi - t|\xi|$.

20.2 Proof of Theorem 20.1. 1. Let us construct the operator $Q(t)$, which approximates $U(t)$ and is an FIO of the form

$$Q(t)f(x) = \int\int q(t, x, y, \xi)\, e^{i\varphi(t, x, y, \xi)} f(y)\, dy\, d\xi. \tag{20.10}$$

The operator $U(t)$ satisfies the conditions

$$\begin{cases} (D_t + A)\, U(t) = 0, & \tag{20.11} \\ U(0) = I. & \tag{20.12} \end{cases}$$

We will try to choose $Q(t)$ satisfying the conditions

$$\begin{cases} (D_t + A)\, Q(t) \in L^{-\infty}(M), & \tag{20.13} \\ Q(0) - I \in L^{-\infty}(M). & \tag{20.14} \end{cases}$$

More precisely, the left side of (20.13) will also be a smooth function of t with values in $L^{-\infty}(M)$.

The linearity of the problem, allows a reduction, using a partition of unity, to the case of constructing q and φ in local coordinates.

In view of Theorem 18.2, (20.13) will be satisfied if

$$e^{-i\varphi}[D_t + A]\,(qe^{i\varphi}) \in S^{-\infty}, \tag{20.15}$$

where A operates on x.

Writing down an asymptotic expansion for the symbol $a(x, \xi)$ and for $q(t, x, y, \xi)$ in terms of homogeneous functions, we have

$$a \sim a_1 + a_0 + a_{-1} + \ldots,$$

$$q \sim q_0 + q_{-1} + q_{-2} + \ldots.$$

Then by Theorem 18.2

$$e^{-i\varphi}[D_t + A](qe^{i\varphi}) \sim (\varphi_t + a_1(x, \varphi_x)) q_0 + r_0 + r_{-1} + \cdots ,$$

where r_j is the sum of terms of degree of homogeneity j.

We require

$$\varphi_t + a_1(x, \varphi_x) = 0 . \qquad (20.16)$$

The initial values

$$\varphi|_{t=0} = \psi(x, y, \xi) \qquad (20.17)$$

must be chosen, so that for $t = 0$ we may guarantee (20.14). But for this we have to require that ψ be a phase function, corresponding to the class of ΨDO (cf. §19), i.e. for nearby x and y

$$\psi(x, y, \xi) = (x - y) \cdot \xi + O(|x - y|^2 |\xi|) . \qquad (20.18)$$

We will look for this function in a neighbourhood of the diagonal $x = y$. The term $O(|x - y|^2 |\xi|)$ in (20.18) is necessary in order to achieve linearity in t for the function $\varphi(t, x, y, \xi)$, which – in its turn – is useful since later on we will take the Fourier transformation in t. Thus, we look for φ in the form

$$\varphi(t, x, y, \xi) = \psi(x, y, \xi) - ta'(y, \xi) .$$

This is yet another requirement on φ. We will see that it can be satisfied. Putting this expression for φ into (20.16) yields:

$$-a'(y, \xi) + a_1(x, \psi_x(x, y, \xi)) = 0 .$$

Setting $x = y$, we obtain using (20.18) that $a'(y, \xi) = a_1(y, \xi)$. Hence

$$\varphi(t, x, y, \xi) = \psi(x, y, \xi) - ta_1(y, \xi), \qquad (20.19)$$

where $\psi(x, y, \xi)$ satisfies the equation

$$a_1(x, \psi_x(x, y, \xi)) = a_1(y, \xi). \qquad (20.20)$$

Instead of (20.18) we require

$$\psi(x, y, \xi)|_{(x-y)\cdot\xi=0} = 0, \qquad (20.21)$$

$$\psi_x(x, y, \xi)|_{x=y} = \xi, \qquad (20.22)$$

from which (20.18) immediately follows. The relations (20.20)–(20.22) define a Cauchy problem for ψ, in which y and ξ are parameters. In this, we may assume that $|\xi| = 1$, since solving this problem for $|\xi| = 1$ we can afterwards extend ψ by homogeneity (of degree 1) to all values of ξ.

Let us note, that the plane $(x - y) \cdot \xi = 0$ is non-characteristic, because $\partial_\xi a_1(x, \xi) \neq 0$ in view of the Euler formula $\xi \cdot \partial_\xi a_1(x, \xi) = a_1(x, \xi)$. From the results in §17 it is clear that the solution exists for x which are close to y and, furthermore, that the corresponding neighbourhood of y in which the solution exists, may be chosen uniformly in y, ξ so that the function $\psi(x, y, \xi)$ is defined in some neighbourhood of the diagonal $x = y$.

2. It follows from Theorem 19.1 that there exists a classical symbol $I(x, y, \xi) \in CS^0$, which is supported on an arbitrarily small neighbourhood of the diagonal $x = y$ and

$$\int\int I(x, y, \xi) e^{i\psi(x,y,\xi)} f(y) dy \, d\xi - f(x) = kf(x), \qquad (20.23)$$

where k is an operator with smooth kernel.

From (20.23) and (20.14) it now follows that we must have the following initial condition for q:

$$q(0, x, y, \xi) = I(x, y, \xi) \qquad (\mathrm{mod} \, S^{-\infty}). \qquad (20.24)$$

If we introduce the decomposition of $I(x, y, \xi)$ into homogeneous components

$$I \sim I_0 + I_{-1} + \dots,$$

then (20.24) may be rewritten in the form

$$q_{-j}(0, x, y, \xi) = I_{-j}(x, y, \xi), \qquad j = 0, 1, 2, \dots \qquad (20.25)$$

We now write out the equations for the functions q_{-j}, given by (20.13) and (20.25). The first order terms equal 0 in view of (20.16). For the 0-th order terms we obtain

$$\begin{cases} \partial_t q_0 + \sum_{|\alpha|=1} a_1^{(\alpha)}(x, \varphi_x) \partial_x^\alpha q_0 + \sum_{|\alpha|=2} a_1^{(\alpha)}(x, \varphi_x) \dfrac{\partial_x^\alpha \varphi}{\alpha!} q_0 + i a_0(x, \varphi_x) q_0 = 0, \\[2mm] q_0|_{t=0} = I_0, \end{cases} \qquad (20.26)$$

where $a_1^{(\alpha)} = \partial_\xi^\alpha a_1$. This Cauchy problem allows us to define $q_0(t, x, y, \xi)$ for small t. Furthermore, for any integer $j \geq 0$ we obtain the following equations, called transport equations:

$$\begin{cases} \partial_t q_{-j} + \sum_{|\alpha|=1} a_1^{(\alpha)}(x, \varphi_x)\, \partial_x^\alpha q_{-j} \\ \quad + \sum_{|\alpha|=2} a_1^{(\alpha)}(x, \varphi_x)\, \dfrac{\partial_x^\alpha \varphi}{\alpha!}\, q_{-j} + i a_0(x, \varphi_x)\, q_{-j} + R_j = 0\,, \\ q_{-j}\big|_{t=0} = I_{-j}\,, \end{cases} \qquad (20.27)$$

where R_j only depends on $q_0, q_{-1}, \ldots, q_{-(j-1)}$.

In view of the reasoning in section 17.6, the solution of (20.27) may be defined in the same t-interval as the solution of (20.26).

Hence, we finally obtain an operator $Q(t)$, defined for $|t| < \varepsilon$, which is an FIO and for which

$$(D_t + A)\, Q(t) = K(t)\,, \qquad (20.28)$$

$$Q(0) = I + k\,, \qquad (20.29)$$

where $K(t)$ has a smooth kernel with a smooth dependence on t and where k is an operator with a smooth kernel.

3. Let us now demonstrate that $[U(t) - Q(t)]$ is an operator with infinitely differentiable kernel in t, x, y (for $|t| < \varepsilon$). For this consider the operator

$$R(t) = U(-t)\, Q(t) - I \qquad (20.30)$$

and differentiate it with respect to t:

$$\begin{aligned} D_t R(t) &= -(D_t U)(-t)\, Q(t) + U(-t)\,(D_t Q)(t) \\ &= A U(-t)\, Q(t) - U(-t)\, A Q(t) + U(-t)\, K(t)\,. \end{aligned} \qquad (20.31)$$

The validity of this computation follows from the described structure of $Q(t)$ and from the remarks on $U(t)$, made at the beginning of this section (the derivative is of course taken, after having applied $R(t)$ to some function $u(x) \in C^\infty(M)$). Since $U(t)A = AU(t)$, it follows from (20.31) that

$$D_t R(t) = U(-t)\, K(t)\,. \qquad (20.32)$$

On the other hand, it is clear from (20.29) that

$$R(0) = k\,. \qquad (20.33)$$

Since $U(-t)\, K(t)$ is an operator with smooth kernel in t, x, y, integrating (20.32) and taking (20.33) into account, we see that $R(t)$ has the same property and then from (20.30) it is clear that the operator $U(t) - Q(t) = -U(t)\, R(t)$ also has a kernel which is smooth in t, x, y for $|t| < \varepsilon$. $\square$

Problem 20.1. Let A be a classical first order ΨDO on M, satisfying (20.1) but not necessarily self-adjoint. Carry out the construction of the parametrix

$Q(t)$ for this operator and show that the Cauchy problem (20.5)–(20.6) has a unique solution and that for the operator $U(t) = \exp(-itA)$, defined by this problem, the statement of Theorem 20.1 is also true.

Hint: Obtain an integral equation of the Volterra type for $U(t)$, using the operator $U_0(t) = \exp(-itA_0)$, where $A_0 = \frac{1}{2}(A + A^*)$.

§21. Precise Formulation and Proof of the Hörmander Theorem

21.1 The singularities of the Fourier transformed spectral function near zero and estimates of the averaged spectral function. Let $e(x, y, \lambda)$ be the spectral function of the same (first order) operator A as in §20, $U(t, x, y)$ the kernel of the operator $\exp(-itA)$. If $\varphi_j(x)$ are the eigenfunctions of A with eigenvalues λ_j, we have

$$e(x, y, \lambda) = \sum_{\lambda_j \leq \lambda} \varphi_j(x)\, \overline{\varphi_j(y)}, \tag{21.1}$$

$$U(t, x, y) = \sum_{j} e^{-i\lambda_j t}\, \varphi_j(x)\, \overline{\varphi_j(y)}, \tag{21.2}$$

where the latter series is summed in the sense of e.g. distributions on $M \times M$, depending smoothly on t (this is easy to prove by arguments similar to those used after Definition 20.1). It follows from (21.1) and (21.2) that

$$U(t, x, y) = \int e^{-i\lambda t} d_\lambda e(x, y, \lambda), \tag{21.3}$$

where the integral is understood as a Fourier transformation (from λ to t) in the distribution sense.

Let $\varrho(\lambda) \in S(\mathbb{R}^1)$ and let $\hat{\varrho}(t) = F_{\lambda \to t}\, \varrho(\lambda)$ be the Fourier transform. Then from the known properties of the Fourier transformation it follows that

$$\hat{\varrho}(t)\, U(t, x, y) = F_{\lambda \to t} \int \varrho(\lambda - \mu)\, d_\mu e(x, y, \mu). \tag{21.4}$$

Now choose $\varrho(\lambda)$ such that

   1) $\varrho(\lambda) > 0$ for all $\lambda \in \mathbb{R}^1$;

   2) $\hat{\varrho}(0) = 1$;

   3) $\operatorname{supp} \hat{\varrho}(t) \subset (-\varepsilon, \varepsilon)$, where $\varepsilon > 0$ is sufficiently small.

Exercise 21.1. Prove the existence of such a function $\varrho(\lambda)$.

Hint: This is done by analogy with the construction of $\chi(x)$ at the beginning of the proof of Theorem 6.3.

Let now $Q(t, x, y)$ be the distribution kernel of $Q(t)$, constructed (for small $|t|$) in §20. In view of Theorem 20.1, we have

$$U(t, x, y) - Q(t, x, y) \in C^\infty((-\varepsilon, \varepsilon) \times M \times M). \tag{21.5}$$

Thus the functions $U(t, x, y)$ and $Q(t, x, y)$ have the same singularities in a neighbourhood of the point $t = 0$.

Now (21.4) implies

Lemma 21.1. *The function*

$$\int \varrho(\lambda - \mu)\, d_\mu e(x, y, \mu) - F^{-1}_{t \to \lambda}(\hat{\varrho}(t)\, Q(t, x, y)) \tag{21.6}$$

is a smooth function in all variables, tending to 0 faster than any power of λ as $\lambda \to +\infty$, uniformly in $x, y \in M$.

Let us now compute the second term in (21.6). This can be easily done, thanks to the linearity in t of the phase function $\varphi(t, x, y, \xi)$, in the definition of the operator $Q(t)$.

First we compute formally, not worrying about convergence of the integrals. In the notation of §20 we have

$$Q(t, x, y) = \int q(t, x, y, \xi)\, e^{i(\psi(x, y, \xi) - a_1(y, \xi)t)}\, d\xi, \tag{21.7}$$

$$F^{-1}_{t \to \lambda}(\hat{\varrho}(t)\, Q(t, x, y))\,(\lambda)$$
$$= (2\pi)^{-1} \int \hat{\varrho}(t)\, q(t, x, y, \xi)\, e^{it\lambda - ita_1(y, \xi) + i\psi(x, y, \xi)}\, dt\, d\xi. \tag{21.8}$$

Set

$$R(\lambda, x, y, \xi) = (2\pi)^{-1} \int \hat{\varrho}(t)\, q(t, x, y, \xi)\, e^{it\lambda}\, dt. \tag{21.9}$$

Then

$$F^{-1}_{t \to \lambda}(\hat{\varrho}(t)\, Q(t, x, y))\,(\lambda) = \int R(\lambda - a_1(y, \xi), x, y, \xi)\, e^{i\psi(x, y, \xi)}\, d\xi. \tag{21.10}$$

Let us now note that $R(\lambda, x, y, \xi)$ is a smooth function of all variables and, furthermore since $q(t, x, y, \xi)$ is a smooth function of all variables (including t), then R rapidly tends to 0 as $|\lambda| \to +\infty$ and we have for any $N > 0$ the estimates

$$|\partial_x^\alpha \partial_y^\beta \partial_\xi^\gamma \partial_\lambda^\delta R(\lambda, x, y, \xi)| \leqq C_{\alpha\beta\gamma\delta N} \langle \xi \rangle^{-|\gamma|} \langle \lambda \rangle^{-N}. \tag{21.11}$$

Here R admits an asymptotic expansion into homogeneous functions in ξ. From (21.11) it is clear in view of the ellipticity of $a_1(y, \xi)$, that $R(\lambda - a_1(y, \xi), x, y, \xi)$ rapidly tends to 0 as $|\xi| \to +\infty$, hence the integral in (21.10) is absolutely convergent.

As for the justification of the transition from (21.8) to (21.10), which was done formally, it remains to note that it is valid for symbols $q(t, x, y, \xi)$ which have compact support in ξ, and perform the standard passage to the limit, as in the definition of oscillatory integrals (cf. §1).

Taking Lemma 21.1 into account, we see that we have proven

Lemma 21.2. *If R is defined via the formula (21.9), then for any $N > 0$*

$$|\int \varrho(\lambda - \mu)\, d_\mu e(x, y, \mu) - \int R(\lambda - a_1(y, \xi), x, y, \xi)\, e^{i\psi(x, y, \xi)}\, d\xi|$$
$$\leqq C_N (1 + |\lambda|)^{-N}, \tag{21.12}$$

where C_N is independent of x and y.

Let us now estimate the second term in (21.12).

Lemma 21.3. *We have*

$$\left| \int R\left(\lambda - a_1\left(y, \xi\right), x, y, \xi\right) e^{i\psi(x,y,\xi)} d\xi \right| \leq C\left(1 + |\lambda|\right)^{n-1}, \tag{21.13}$$

where C does not depend on x and y.

Proof. Let us denote, as in §15:

$$V_y(\lambda) = \int\limits_{a_1(y,\xi)<\lambda} d\xi. \tag{21.14}$$

Since the function $a_1(x, \xi)$ is homogenous of degree 1 in ξ, then

$$V_y(\lambda) = V_y(1)\,\lambda^n. \tag{21.15}$$

Let us now utilize the obvious identity

$$\int R(\lambda - a_1(y,\xi), x, y, \xi)\, e^{i\psi(x,y,\xi)} d\xi = \int R(\lambda - \sigma, x, y, \xi)\, e^{i\psi(x,y,\xi)} dV_y(\sigma),$$

both sides of which are defined due to the estimate (21.11). Taking (21.15) into account, we see now that the left-hand side of (21.13) can be estimated via

$$C_N \int\limits_0^\infty \left(1 + |\lambda - \sigma|\right)^{-N} dV_y(\sigma) = C_N' \int\limits_0^\infty \left(1 + |\lambda - \sigma|\right)^{-N} \sigma^{n-1}\, d\sigma.$$

But in view of the obvious inequality $1 + |\sigma| \leq (1 + |\lambda - \sigma|)(1 + |\lambda|)$ we have

$$\left(1 + |\lambda - \sigma|\right)^{-N} \sigma^{n-1} \leq \left(1 + |\lambda - \sigma|\right)^{-N+n-1} \left(1 + |\lambda|\right)^{n-1}.$$

Therefore

$$\int\limits_0^\infty \left(1 + |\lambda - \sigma|\right)^{-N} \sigma^{n-1}\, d\sigma$$

$$\leq \left(1 + |\lambda|\right)^{n-1} \int\limits_{-\infty}^\infty \left(1 + |\lambda - \sigma|\right)^{-N+n-1} d\sigma = C\left(1 + |\lambda|\right)^{n-1},$$

from which (21.13) follows. $\square$

Corollary 21.1. *The following estimate holds*

$$\left| \int \varrho\left(\lambda - \mu\right) d_\mu e\left(x, y, \mu\right) \right| \leq C\left(1 + |\lambda|\right)^{n-1}, \tag{21.16}$$

where C is independent of x and y.

21.2 Passage to estimates of the spectral function

Lemma 21.4. *The following estimate holds*

$$|e(x, x, \lambda+1) - e(x, x, \lambda)| \leqq C(1+|\lambda|)^{n-1}, \tag{21.17}$$

where C does not depend on x and λ.

Proof. Since $\varrho(\lambda)$ is positive and non-zero on $[-2, 2]$, we have

$$\int \varrho(\lambda-\mu)\, d_\mu e(x, x, \mu) \geq C \int_{\lambda-1}^{\lambda+2} d_\mu e(x, x, \mu) \geq C\,[e(x, x, \lambda+1) - e(x, x, \lambda)],$$

where $C > 0$; the statement of the lemma follows now from Corollary 21.1. $\square$

Lemma 21.5. *The following estimate holds*

$$|e(x, y, \lambda+1) - e(x, y, \lambda)| \leqq C(1+|\lambda|)^{n-1}, \tag{21.18}$$

where C does not depend on x, y and λ.

Proof. It follows from (21.1) that

$$e(x, y, \lambda+1) - e(x, y, \lambda) = \sum_{\lambda < \lambda_j \leqq \lambda+1} \varphi_j(x)\, \overline{\varphi_j(y)},$$

and from the Cauchy-Bunyakovskij-Schwarz inequality we get

$$|e(x, y, \lambda+1) - e(x, y, \lambda)|$$

$$\leqq \left[\sum_{\lambda < \lambda_j \leqq \lambda+1} |\varphi_j(x)|^2\right]^{1/2} \left[\sum_{\lambda < \lambda_j \leqq \lambda+1} |\varphi_j(y)|^2\right]^{1/2}$$

$$= (e(x, x, \lambda+1) - e(x, x, \lambda))^{1/2}\, (e(y, y, \lambda+1) - e(y, y, \lambda))^{1/2}$$

$$\leqq C(1+|\lambda|)^{n-1}$$

by Lemma 21.4 (with the same constant C). $\square$

Lemma 21.6. *The following estimate holds*

$$|e(x, y, \lambda+\mu) - e(x, y, \lambda)| \leqq C(1+|\lambda|+|\mu|)^{n-1}(1+|\mu|), \tag{21.19}$$

where C is independent of x, y, λ, μ.

Proof. We can prove (21.19) by partitioning $e(x, y, \lambda+\mu) - e(x, y, \lambda)$ into a sum of at most $1 + |\mu|$ terms, estimated as in Lemma 21.5. $\square$

Lemma 21.7. *The following estimate holds*

$$\left| \int \varrho(\lambda-\mu)\, e(x,y,\mu)\, d\mu - e(x,y,\lambda) \right| \leq C(1+|\lambda|)^{n-1}, \tag{21.20}$$

where C does not depend on $x,\ y,\ \lambda.$

Proof. We use the fact that $\varrho \in S(\mathbb{R}^1)$ and $\int \varrho(\lambda)\, d\lambda = 1$. With the help of the obvious inequality

$$1 + |\lambda| + |\mu| \leq (1+|\lambda|)\,(1+|\mu|)$$

we obtain

$$\left| \int \varrho(\lambda-\mu)\, e(x,y,\mu)\, d\mu - e(x,y,\lambda) \right|$$
$$= \left| \int \varrho(\lambda-\mu)\, [e(x,y,\mu) - e(x,y,\lambda)]\, d\mu \right|$$
$$= \left| \int \varrho(\mu)[e(x,y,\lambda+\mu) - e(x,y,\lambda)]\, d\mu \right|$$
$$\leq \int (1+|\lambda|+|\mu|)^{n-1}\,(1+|\mu|)^{-N}\, d\mu \leq C(1+|\lambda|)^{n-1},$$

as required. $\square$

Lemma 21.8. *We have*

$$\left| \int_{-\infty}^{\lambda} d\lambda \int \varrho(\lambda-\mu)\, d_\mu e(x,y,\mu) - e(x,y,\lambda) \right| \leq C(1+|\lambda|)^{n-1}. \tag{21.21}$$

Proof. Integrating by parts, we have

$$\int \varrho(\lambda-\mu)\, d_\mu e(x,y,\mu) = \int \varrho'(\lambda-\mu)\, e(x,y,\mu)\, d\mu,$$

from which

$$\int_{-\infty}^{\lambda} \left(\int \varrho(\lambda-\mu)\, d_\mu e(x,y,\mu) \right) d\lambda = \int_{-\infty}^{\lambda} \left(\int \varrho'(\lambda-\mu)\, e(x,y,\mu)\, d\mu \right) d\lambda$$

$$= \int \left(\int_{-\infty}^{\lambda} \varrho'(\lambda-\mu)\, d\lambda \right) e(x,y,\mu)\, d\mu = \int \varrho(\lambda-\mu)\, e(x,y,\mu)\, d\mu,$$

so that (21.21) now follows from (21.20). $\square$

Taking Lemma 21.2 into account, we derive the following

Proposition 21.1. *We have*

$$\left| e(x,y,\lambda) - \iint_{\sigma<\lambda} R(\sigma - a_1(y,\xi), x, y, \xi)\, e^{i\psi(x,y,\xi)}\, d\sigma\, d\xi \right|$$
$$\leq C(1+|\lambda|)^{n-1}, \tag{21.22}$$

where C is independent of $x,\ y,\ \lambda.$

21.3 The Hörmander theorem for first order operators. We would like now to rewrite the estimate (21.22) in a simpler form. Note, that in view of (21.9)

$$\int_{-\infty}^{+\infty} R(\lambda, x, y, \xi)\, d\lambda = q(0, x, y, \xi) = I(x, y, \xi). \tag{21.23}$$

Therefore, putting

$$R_1(\tau, x, y, \xi) = \begin{cases} \displaystyle\int_{-\infty}^{\tau} R(\sigma, x, y, \xi)\, d\sigma, & \tau < 0, \\[2ex] \displaystyle -\int_{\tau}^{\infty} R(\sigma, x, y, \xi)\, d\sigma \\[2ex] \displaystyle = \int_{-\infty}^{\tau} R(\sigma, x, y, \xi)\, d\sigma - I(x, y, \xi), & \tau > 0, \end{cases}$$

the following estimates for R_1 follow from (21.11):

$$|\partial_x^\alpha \partial_y^\beta \partial_\xi^\gamma R_1(\lambda, x, y, \xi)| \leq C_{\alpha\beta\gamma N} \langle \xi \rangle^{-|\gamma|} \langle \lambda \rangle^{-N}. \tag{21.24}$$

We clearly have,

$$\iint_{\sigma < \lambda} R(\sigma - a_1(y, \xi), x, y, \xi)\, e^{i\psi(x, y, \xi)}\, d\sigma\, d\xi$$
$$= \int_{a_1(y, \xi) < \lambda} I(x, y, \xi)\, e^{i\psi(x, y, \xi)}\, d\xi + \int R_1(\lambda - a_1(y, \xi), x, y, \xi)\, e^{i\psi(x, y, \xi)}\, d\xi.$$

One can show by analogy with Lemma 21.3 that

$$\left| \int R_1(\lambda - a_1(y, \xi), x, y, \xi)\, e^{i\psi(x, y, \xi)}\, d\xi \right| \leq C(1 + |\lambda|)^{n-1}.$$

Therefore we see that the following holds

Lemma 21.9. *We have*

$$\left| e(x, y, \lambda) - \int_{a_1(y, \xi) < \lambda} I(x, y, \xi)\, e^{i\psi(x, y, \xi)}\, d\xi \right| \leq C(1 + |\lambda|)^{n-1}, \tag{21.25}$$

where C is independent of x, y, λ.

For further simplification we need

Lemma 21.10. *For nearby x and y we have*

$$\left| \int_{a_1(y, \xi) < \lambda} I(x, y, \xi)\, e^{i\psi(x, y, \xi)}\, d\xi - \int_{a_1(y, \xi) < \lambda} e^{i\psi(x, y, \xi)}\, d\xi \right| \leq C(1 + |\lambda|)^{n-1},$$

where C does not depend on x, y, λ.

Proof. From the proof of Theorem 19.1, it is easily deduced that for nearby x and y

$$|I(x, y, \xi) - 1| \leqq C(1 + |\xi|)^{-1}. \tag{21.26}$$

Therefore

$$\left| \int\limits_{a_1(y,\xi)<\lambda} (I(x, y, \xi) - 1)\, e^{i\psi(x,y,\xi)}\, d\xi \right| \leqq C \int\limits_{a_1(y,\xi)<\lambda} (1 + |\xi|)^{-1}\, d\xi$$

$$\leqq C \int\limits_{a_1<\lambda} (1 + |a_1|)^{-1}\, d\xi = C \int\limits_{\mu<\lambda} (1 + |\mu|)^{-1}\, dV_y(\mu)$$

$$\leqq C \int\limits_0^\lambda (1 + |\mu|)^{-1}\, \mu^{n-1}\, d\mu \leqq C \int\limits_0^\lambda (1 + |\mu|)^{n-2}\, d\mu \leqq C(1 + |\lambda|)^{n-1},$$

as required. $\square$

Note, that if the pair x, y belongs to some compact set in $M \times M$, disjoint from the diagonal, then we may assume that $I(x, y, \xi) = 0$ and, in this case, Lemma 21.9 implies that $|e(x, y, \lambda)| \leqq C(1 + |\lambda|)^{n-1}$.

We summarize these results in the form of a theorem.

Theorem 21.1. 1) *Let* $\psi(x, y, \xi)$ *be defined for nearby* x *and* y *and let*

$$a_1(x, \psi_x(x, y, \xi)) = a_1(y, \xi), \quad \psi|_{(x-y)\cdot\xi=0} = 0, \quad \psi_x|_{x=y} = \xi. \tag{21.27}$$

Then for nearby x *and* y *we have*

$$\left| e(x, y, \lambda) - \int\limits_{a_1(y,\xi)<\lambda} e^{i\psi(x,y,\xi)}\, d\xi \right| \leqq C(1 + |\lambda|)^{n-1}, \tag{21.28}$$

where C *does not depend on* x, y *and* λ. *In particular, we have*

$$|e(x, x, \lambda) - V_x(\lambda)| \leqq C(1 + |\lambda|)^{n-1}. \tag{21.29}$$

2) *If the pair* x, y *belongs to a compact set in* $M \times M$ *disjoint from the diagonal, then*

$$|e(x, y, \lambda)| \leqq C(1 + |\lambda|)^{n-1}. \tag{21.30}$$

Corollary 21.2. *The following asymptotic formula for the number* $N(\lambda)$ *of eigenvalues of the operator* A *smaller than* λ, *holds*

$$\left| N(\lambda) - \int\limits_{a_1(x,\xi)<\lambda} d\xi\, dx \right| \leqq C(1 + |\lambda|)^{n-1}.$$

21.4 The case of higher order operators. Let A be a self-adjoint elliptic differential operator of order m on a closed manifold M with principal symbol $a_m(x, \xi) \geq 0$. Let us construct for nearby x and y a function $\psi(x, y, \xi)$ such that

$$a_m(x, \psi_x(x, y, \xi)) = a_m(y, \xi), \tag{21.31}$$

$$\psi|_{(x-y)\cdot\xi=0} = 0, \quad \psi_x|_{x=y} = \xi, \tag{21.32}$$

from which it follows that ψ is homogeneous of degree one in ξ and such that

$$\psi(x, y, \xi) = (x-y)\cdot\xi + O(|x-y|^2 |\xi|) \quad \text{as} \quad x \to y. \tag{21.33}$$

Theorem 21.2. *For the operator A of order m described above we have:*
1) For nearby x and y

$$\left| e(x, y, \lambda) - \int\limits_{a_m(y,\xi)<\lambda} e^{i\psi(x,y,\xi)}\, d\xi \right| \leq C(1+|\lambda|)^{\frac{n-1}{m}}, \tag{21.34}$$

where C does not depend on x and y.
In particular,

$$\left| e(x, x, \lambda) - \int\limits_{a_m(x,\xi)<\lambda} d\xi \right| \leq C(1+|\lambda|)^{\frac{n-1}{m}} \tag{21.35}$$

and consequently,

$$\left| N(\lambda) - \int\limits_{a_m(x,\xi)<\lambda} dx\, d\xi \right| \leq C(1+|\lambda|)^{\frac{n-1}{m}}. \tag{21.36}$$

2) If the pair x, y belongs to some compact set in $M \times M$ disjoint from the diagonal then

$$|e(x, y, \lambda)| \leq C(1+|\lambda|)^{\frac{n-1}{m}}. \tag{21.37}$$

Proof. Let us first note that, without loss of generality, we may assume $A > 0$ (if this is not true for A, then, in view of Corollary 9.3, it is true for $A + cI$, for sufficiently large $c > 0$). Introduce the operator $A_1 = A^{1/m}$. This is an elliptic pseudodifferential operator of order 1 with principal symbol $a_1(x, \xi) = a_m(x, \xi)^{1/m}$. It is clear that (21.31), (21.32) for $\psi(x, y, \xi)$ simply coincide with the equations (21.27). In addition it is clear that

$$e(x, y, \lambda) = \sum_{\lambda_j \leq \lambda} \varphi_j(x)\, \overline{\varphi_j(y)} = \sum_{\lambda_j^{1/m} \leq \lambda^{1/m}} \varphi_j(x)\, \overline{\varphi_j(y)} = e_1(x, y, \lambda^{1/m}),$$

where $e_1(x, y, \lambda)$ is the spectral function of A_1. All statements in Theorem 21.2 now follow from the corresponding statements in Theorem 21.1. $\square$

Problem 21.1. Consider the case of a homogeneous operator $a(D)$ with constant coefficients in $\mathbb{R}^n$. Write down $e(x, y, \lambda)$ as an integral and verify the estimates (21.34), (21.35) and (21.37) directly.

§22. The Laplace Operator on the Sphere

22.1 The Laplace operator on a Riemannian manifold. 1) Let M be a manifold with a Riemannian metric, i.e. in any tangent space $T_x M$ there is given a bilinear form $\langle \cdot, \cdot \rangle$ with a positive definite quadratic form. If $x^1, \ldots, x^n$ are the local coordinates in an open set $U \subset M$, then $\left(\dfrac{\partial}{\partial x^1}, \ldots, \dfrac{\partial}{\partial x^n} \right)$ is a basis for the tangent space at all points in U. Denoting

$$g_{ij}(x) = \left\langle \left(\frac{\partial}{\partial x^i} \right)_x, \left(\frac{\partial}{\partial x^j} \right)_x \right\rangle, \tag{22.1}$$

we obtain a positive definite matrix $g_{ij}(x)$. If now $v = \sum\limits_{j=1}^{n} v^j \dfrac{\partial}{\partial x^j} \in T_x M$, then

$$\langle v, v \rangle = \sum\limits_{i,j=1}^{n} g_{ij}(x) v^i v^j. \tag{22.2}$$

The cotangent space $T_x^* M$ is, by definition, the dual of $T_x M$. A basis of it (for $x \in U$) consists of the 1-forms dx^i, defined by the relations

$$\left\langle dx^i, \frac{\partial}{\partial x^j} \right\rangle = \delta^i_j.$$

A metric on a vector space E induces an isomorphism of E with E^*. With the help of this isomorphism we may transfer the metric from E to E^*. Fixing $x \in U$, let us compute $\langle dx^i, dx^j \rangle$ at x. Which tangent vector corresponds to dx^i? Denote its coordinates by a^{ik}, $k = 1, \ldots, n$; then we have to satisfy the condition

$$\delta^i_j = \left\langle dx^i, \frac{\partial}{\partial x^j} \right\rangle = \left\langle \{a^{ik}\}, \frac{\partial}{\partial x^j} \right\rangle = \sum\limits_{k=1}^{n} g_{kj} a^{ik},$$

from which $a^{ik} = g^{ik}$, the elements of the matrix inverse to $\| g_{ik} \|$. We now have

$$\langle dx^i, dx^j \rangle = \left\langle \sum\limits_{k=1}^{n} g^{ik} \frac{\partial}{\partial x^k}, dx^j \right\rangle = g^{ij}, \tag{22.3}$$

i.e. for any cotangent vector $a = \sum\limits_{i=1}^{n} a_i \, dx^i \in T_x^* M$

$$\langle a, a \rangle = \sum_{i,j=1}^{n} g^{ij} a_i a_j. \tag{22.4}$$

2) Now introduce on the Riemannian manifold a smooth density (volume) such that in the tangent space the volume of a parallelepiped in an orthonormal system equals 1. For an arbitrary parallelepiped, defined by the vectors $e_1, \ldots, e_n$, the volume equals

$$\mathrm{vol}\,\{e_1, \ldots, e_n\} = |\det\,\{e_1^{\mathrm{col}}, \ldots, e_n^{\mathrm{col}}\}|, \tag{22.5}$$

where e_i^{col} is the column of coordinates of e_i in the orthonormal basis. Formula (22.5) follows from the fact that the volume has to be an additive, non-negative invariant of parallelepipeds.

What about the case where the coordinate basis is not orthonormal? Let us then consider the operator A, mapping an orthonormal basis $e_1', \ldots, e_n'$ into the basis $e_1, \ldots, e_n$. The columns of its matrix (in the basis $e_1', \ldots, e_n'$), will be the coordinates of the vectors e_i in the orthonormal basis e_i', so that $\mathrm{vol}\,\{e_1, \ldots, e_n\} = |\det A| = \sqrt{\det (A^*A)}$.

Now, the matrix elements of A^*A in the basis $\{e_i'\}$ are of the form

$$\langle A^*Ae_i', e_j' \rangle = \langle Ae_i', Ae_j' \rangle = \langle e_i, e_j \rangle. \tag{22.6}$$

Therefore, the volume of the parallelepiped $\left(\dfrac{\partial}{\partial x^1}, \ldots, \dfrac{\partial}{\partial x^n} \right)$ equals $\sqrt{g}$, where $g = \det \| g_{ij}(x) \|$ and the volume of an arbitrary set F in the local coordinates of U is defined as

$$\mathrm{vol}\,(F) - \int_{F} \sqrt{g(x)} \, dx^1 \ldots dx^n. \tag{22.7}$$

It follows directly from the change of variable formula for an integral that the integral (22.7) is independent of the choice of local coordinates. We may therefore take it immediately as a definition of $\mathrm{vol}\,(F)$, defining a smooth positive density dv on M by the formula

$$dv = \sqrt{g(x)} \, dx^1 \ldots dx^n. \tag{22.8}$$

3) For any differential operator

$$A\colon C^\infty(M) \to C^\infty(M)$$

there exists a unique operator A^*, such that

$$(Af, g) = (f, A^*g), \quad f, \ g \in C_0^\infty(M), \tag{22.9}$$

where

$$(f, g) = \int f(x)\, \overline{g(x)}\, dv. \tag{22.10}$$

Analogous arguments hold also for operators on sections of vector bundles over M, if in each fiber of the bundle we have a hermitean metric (positive definite hermitean form), which replaces $f(x)\,\overline{g(x)}$ in (22.10).

Let us define a scalar product on the space of 1-forms $\Lambda^1(M)$, taking on $T_x^* M \otimes \mathbb{C}$ the hermitean scalar product induced by the metric on $T_x^* M$ introduced above. Consider the operator

$$d\colon\ C^\infty(M) \to \Lambda^1(M), \tag{22.11}$$

mapping a function $f \in C^\infty(M)$ to its differential $df \in \Lambda^1(M)$, defined by the fact that if $v \in T_x M$, then $\langle df, v\rangle = (vf)(x)$, where $(vf)(x)$ denotes the derivative of f in the direction v. In coordinates:

$$df = \sum_{i=1}^{n} \frac{\partial f}{\partial x^i}\, dx^i.$$

The operator $\delta\colon \Lambda^1(M) \to C^\infty(M)$ is defined as the adjoint of d, i.e. $\delta = d^*$.

Definition 22.1. The *Laplace (or Laplace-Beltrami) operator* on functions

$$\Delta\colon\ C^\infty(M) \to C^\infty(M)$$

on a Riemannian manifold M, is defined by the formula

$$\Delta = -\delta \cdot d. \tag{22.12}$$

Analogously, the Laplace operator is defined on p-forms $\Lambda^p(M)$ as

$$\Delta = -(d\delta + \delta d)\colon\ \Lambda^p(M) \to \Lambda^p(M),$$

but we will not consider this case.

It is immediately clear that the Laplace operator has the following properties:

a) $\Delta^* = \Delta$;

b) if $T\colon M \to M$ preserves the metric on the tangent spaces and $\hat{T}\colon C^\infty(M) \to C^\infty(M)$ is defined by $\hat{T}f = f \circ T$, then

$$\Delta\hat{T} = \hat{T}\Delta,$$

i.e. Δ commutes with isometries;

c) if M is closed, then $(\Delta f, f) \leq 0$ and from $\Delta f = 0$ it follows that $f = $ const.

4) Let us compute δ and Δ in local coordinates. We have

$$\left(\delta\left(\sum_{i=1}^{n} a_i\,dx^i\right), f\right) = \left(\sum_{i=1}^{n} a_i\,dx^i, \sum_{j=1}^{n} \frac{\partial f}{\partial x^j}\,dx^j\right)$$

$$= \sum_{i,\,j=1}^{n} \int g^{ij} a_i\,\frac{\partial \overline{f}}{\partial x^j}\,\sqrt{g}\,dx = \sum_{i,\,j=1}^{n} \int \overline{f}\left[-\frac{\partial}{\partial x^j}\,(\sqrt{g}\,g^{ij} a_i)\right]dx$$

$$= \int \overline{f} \cdot \sum_{i,\,j} \frac{1}{\sqrt{g}}\left[-\frac{\partial}{\partial x^j}\,(\sqrt{g}\,g^{ij} a_i)\right]\sqrt{g}\,dx,$$

from which

$$\delta\left(\sum_{i=1}^{n} a_i\,dx^i\right) = \frac{1}{\sqrt{g}}\sum_{i,\,j=1}^{n} \frac{\partial}{\partial x_j}\,(\sqrt{g}\,g^{ij} a_i) \tag{22.13}$$

and

$$\Delta f = \delta\,df = \frac{1}{\sqrt{g}}\sum_{i,\,j=1}^{n} \frac{\partial}{\partial x^j}\left(\sqrt{g}\,g^{ij}\,\frac{\partial f}{\partial x^i}\right). \tag{22.14}$$

Example 22.1. In the Euclidean space $\mathbb{R}^n$ with its standard metric, we obtain

$$\Delta f = \sum_{i=1}^{n} \frac{\partial^2 f}{\partial x^{i\,2}}.$$

Note that all invariants of Δ are also invariants of the Riemannian manifold, in particular, the residues and the values of the ζ-function.

Exercise 22.1. Compute in local coordinates the principal symbol of the Laplace operator on a Riemannian manifold.

22.2 The Laplace operator on the sphere S^n. The n-sphere S^n is the following submanifold of $\mathbb{R}^{n+1}$:

$$S^n = \left\{(x^0, \ldots, x^n)\colon \sum_{i=0}^{n} (x^i)^2 = 1\right\}.$$

There is a metric on S^n, induced by the standard metric on $\mathbb{R}^{n+1}$ and correspondingly a Laplace operator, which we denote by Δ_S. There is the following method for computing Δ_S, using the operator Δ on $\mathbb{R}^{n+1}$.

Proposition 22.1. *Let $f(\omega)$ be a function on S^n and extend it to $\mathbb{R}^{n+1}$, putting*

$$\hat{f}(x) = f\left(\frac{x}{|x|}\right)$$

i.e. by homogeneity of degree 0. *Then*

$$\Lambda_S f = \Lambda \hat{f}\big|_{S^\cdot}\,, \tag{22.15}$$

or

$$(\Delta \hat{f})(x) = r^{-2}\,\Delta_S f\,, \tag{22.16}$$

where $r = |x|$.

Proof. The equivalence of (22.15) and (22.16) is obvious, since $\Delta\hat{f}$ has degree of homogeneity -2.

Let us prove (22.15). Denote temporarily by Δ_S' the operator on S^n, defined by the right-hand side of (22.15).

The group of isometries $SO\,(n+1)$ acts on S^n by restricting to S^n the rotations of $\mathbb{R}^{n+1}$. Using this group, any unit tangent vector may be mapped into any other such vector. Therefore, the principal symbol of an operator commuting with all isometries, is constant on all unit cotangent vectors (and, consequently, is uniquely defined up to a scalar multiplier). Clearly the operators Δ_S and Δ_S' commute with the action of $SO\,(n+1)$ on functions. Therefore, there exists a constant λ such that the operator $\Delta_S' - \lambda\Delta_S$ has order 1. But this operator also commutes with $SO\,(n+1)$ and since there are no linear functions which are invariant with respect to rotations and since the multiplication term of $\Delta_S' - \lambda\Delta_S$ is constant because of rotational symmetry, then the operator $\Delta_S' - \lambda\Delta_S$ is a multiple of the identity, and, consequently, is zero, because $\Delta_S'1 = \Delta_S 1 = 0$. Hence $\Delta_S' = \lambda\Delta_S$. From what follows, it will be clear that $\lambda = 1$, on the other hand this can be verified by computing Δ_S and Δ_S' on any non-harmonic function on the sphere. $\square$

22.3 Eigenvalues of the operator Δ_S. Let us compute Δ in polar coordinates. Let $r = |x|$ and $\omega = x/|x|$, then

$$\Delta\,(f(r)\,g\,(\omega)) = (\Delta f)\,g + f(\Delta g) + 2\,(\nabla f)\cdot(\nabla g) = (\Delta f)g + f(\Delta g), \tag{22.17}$$

since $(\nabla f)\cdot(\nabla g) = 0$ (∇f is directed along the radius vector and ∇g is directed along the tangent). Let us compute $\Delta f(r)$. We have

$$r'_{x^i} = \frac{x^i}{r}\,,\qquad r''_{x^i x^i} = \frac{1}{r} - \frac{(x^i)^2}{r^3}\,.$$

From this we obtain

$$\Delta f(r) = \sum_{i=0}^{n} \frac{\partial}{\partial x^i}\left(\frac{x^i}{r}\,f'(r)\right) = \sum_{i=0}^{n} f''(r)\,\frac{(x^i)^2}{r^2}$$

$$+ \sum_{i=0}^{n}\left(\frac{1}{r} - \frac{(x^i)^2}{r^3}\right) f'(r) = f''(r) + \frac{n}{r}\,f'(r)\,.$$

Therefore

$$\Delta \equiv \frac{\partial^2}{\partial r^2} + \frac{n}{r}\frac{\partial}{\partial r} + \frac{1}{r^2}\Delta_S, \tag{22.18}$$

where Δ_S is applied on the unit sphere with subsequent extension by homogeneity of degree zero. Formula (22.18) is true also on arbitrary functions (linear combinations of functions of the type $f(r)g(\omega)$ are dense in the space $C^\infty(\mathbb{R}^{n+1}\setminus 0)$).

In particular, for $f(r) = r^\mu$, we obtain

$$\begin{aligned}
\Delta(r^\mu g(\omega)) &= r^{-2+\mu}[\Delta_S g + [\mu(\mu-1)+n\mu]g] \\
&= r^{-2+\mu}[\Delta_S g + \mu(\mu+n-1)g].
\end{aligned} \tag{22.19}$$

It follows from (22.19), that the following proposition holds

Proposition 22.2. *The equality $\Delta(r^\mu g(\omega)) = 0$ (for $r \neq 0$) is equivalent to the fact that $g(\omega)$ is an eigenfunction of the operator $-\Delta_S$ with eigenvalue $\lambda = \mu(\mu+n-1)$.*

Since all eigenvalues of the operator $-\Delta_S$ are non-negative, we may assume that $\mu \geq 0$ or $\mu \leq 1-n$. Note that the quadratic function $\lambda(\mu) = \mu(\mu+n-1)$ takes all values $\lambda \geq 0$ for $\mu \geq 0$ and each of them exactly once. It is therefore clear that the eigenvalues $\lambda \geq 0$ are in a one-to-one correspondence with those $\mu \geq 0$, for which there exists a non-trivial function $g(\omega)$ such that $r^\mu g(\omega)$ is a harmonic function on $R^{n+1}\setminus 0$. But then, by the removable singularity theorem, the function $r^\mu g(\omega)$ is harmonic everywhere on $\mathbb{R}^{n+1}$ and, consequently, is a harmonic polynomial by the Liouville theorem. In particular, μ is an integer. Clearly the converse is also true, i.e. the restrictions to S^n of homogeneous, harmonic polynomials are eigenfunctions of the operator $-\Delta_S$ with eigenvalues $\lambda = k(k+n-1)$, where $k = 0, 1, 2, \ldots$ and by the maximum principle we see that a harmonic polynomial is uniquely defined by its restriction to S^n. Hence, we have proven

Theorem 22.1. *The eigenvalues of the operator $-\Delta_S$ are $\lambda = k(k+n-1)$, where $k = 0, 1, 2, \ldots$ and the multiplicity of the eigenvalue $k(k+n-1)$ equals the dimension of the space of homogenous, harmonic polynomials of degree k.*

22.4 Computing the multiplicity. Let M_k be the space of homogeneous polynomials of degree k. Let us compute $N_k = \dim M_k$. A basis in M_k is given by the monomials $x_0^{k_0} \ldots x_n^{k_n}$, where $k_0 + k_1 + \ldots + k_n = k$. The number of ordered partitions of k into a sum of $n+1$ non-negative numbers equals the number of ordered partitions of $k+n+1$ into a sum of $n+1$ positive numbers, which in turn equals the number of ways of choosing n from $n+k$, i.e. equals

$$N_k = \binom{n+k}{n} = \frac{(n+k)!}{n!\,k!} = \frac{1}{n!}(k+n)(k+n-1)\ldots(k+1). \tag{22.20}$$

Note that Δ determines a map

$$\Delta\colon M_k \to M_{k-2} \tag{22.21}$$

and one has the exact sequence

$$0 \to H_k \to M_k \to M_{k-2} \tag{22.22}$$

where $H_k = \mathrm{Ker}\,\Delta|_{M_k}$ is the space of homogeneous, harmonic polynomials of degree k, the dimension of which we would like to compute.

Theorem 22.2. 1) *The operator $\Delta\colon M_k \to M_{k-2}$ is surjective.*

2) *There is a direct sum decomposition*

$$M_k = \bigoplus_{k-2l \geq 0} r^{2l}\,H_{k-2l}, \tag{22.23}$$

where $r^2 = x_0^2 + \ldots + x_n^2$.

Corollary 22.1. a)

$$\dim H_k = N_k - N_{k-2}. \tag{22.24}$$

b) *If f and φ are two polynomials, then there exists a unique polynomial u such that*

$$\Delta u = f, \quad u|_{|x|=1} = \varphi|_{|x|=1} \tag{22.25}$$

(i.e. the Dirichlet problem for the Poisson equation in the unit ball is solvable in polynomials).

c) *The harmonic polynomials cannot be divided by r^2.*

Derivation of Corollary 22.1 from Theorem 22.2. a) The relation (22.24) follows from the fact that the sequence (22.22) can be rewritten in the form

$$0 \to H_k \to M_k \to M_{k-2} \to 0. \tag{22.26}$$

b) In view of part 1) of the theorem, the solvability of (22.25) reduces to the case $f = 0$, where it is obvious in view of (22.23).

c) Also obvious in view of (22.23). $\square$

Proof of Theorem 22.2. Both statements are proved at the same time by induction in k.

For $k = 0, 1$ the statements of the theorem are true. Let them be true also for all $l < k$.

1°. Show that $H_k \cap r^2 M_{k-2} = 0$. Since by the inductive hypothesis we have

$$M_{k-2} = \bigoplus_{k-2l-2 \geq 0} r^{2l}H_{k-2l-2}$$

then

$$r^2 M_{k-2} = \bigoplus_{\substack{k-2l \geq 0 \\ l > 0}} r^{2l} H_{k-2l}$$

and if $h_k \in H_k \cap r^2 M_{k-2}$ then

$$h_k = \sum_{\substack{l > 0 \\ k-2l \geq 0}} c_l r^{2l} h_{k-2l}. \tag{22.27}$$

Now consider the harmonic polynomial $h = h_k - \sum_{l>0} c_l h_{k-2l}$. We have $\deg h \leq k$ and h_k is the homogeneous component of h of degree k. Since $h\big|_{|x|=1} = 0$, then $h \equiv 0$ from which $h_k \equiv 0$, as required.

2°. From the condition $M_k \supset H_k \oplus r^2 M_{k-2}$, we obtain

$$\dim H_k \leq N_k - N_{k-2} \tag{22.28}$$

where the equality is equivalent to the decomposition

$$M_k = H_k \oplus r^2 M_{k-2} \tag{22.29}$$

3°. From the exact sequence (22.22) it follows that

$$\dim H_k = N_k - \dim \Delta(M_k) \geq N_k - N_{k-2}, \tag{22.30}$$

where the equality is equivalent to the surjectivity of the operator

$$\Delta: M_k \to M_{k-2}$$

4°. From (22.28) and (22.30) it follows that $\dim H_k = N_k - N_{k-2}$ from which follows the surjectivity and the decomposition (22.29) and hence, by the inductive hypothesis the decomposition (22.23) also follows. $\square$

Corollary 22.2. *The multiplicity of the eigenvalue* $\lambda = k(k+n-1)$ *of the operator* $-\Delta_S$ *is equal to* $N_k - N_{k-2}$, *where* N_k *is given by the formula* (22.20).

22.5 The function $N(\lambda)$ for the operator $-\Delta_S$. It is clear that

$$N(k(k+n-1)+0) = \sum_{l \leq k} \dim H_l = \sum_{l \leq k} (N_l - N_{l-2}) = N_k + N_{k-1}. \tag{22.31}$$

But from (22.20) it is also clear that N_k is a polynomial of degree n in k with leading coefficient $1/n!$. It therefore follows from (22.31) that

$$N(k(k+n-1)+0) \sim \frac{2}{n!} k^n \sim \frac{2}{n!} [k(k+n-1)]^{n/2}. \tag{22.32}$$

From (22.31) it also follows that

$$N(k(k+n-1)+0) - N(k(k+n-1)-0) = N_k - N_{k-2} = P_{n-1}(k),$$

where $P_{n-1}(k)$ is a polynomial in k of degree $n-1$. Therefore

$$N(k(k+n-1)+0) - N(k(k+n-1)-0) \geq ck^{n-1} \geq c[k(k+n-1)]^{\frac{n-1}{2}}, \quad (22.33)$$

where $c > 0$. From (22.32) and (22.33) it is clear that

$$N(\lambda) = \frac{2}{n!} \lambda^{n/2}(1 + O(\lambda^{-1/2})), \quad (22.34)$$

where the term $O(\lambda^{-1/2})$ is sometimes greater than $c\lambda^{-1/2}$, i.e. the remainder estimate is the best possible (in any case as far as the exponent is concerned).

Let us verify that (22.34) is exactly the asymptotic formula from Theorem 21.2 (which, in particular, also shows that the multiplier λ from the proof of Proposition 22.1 equals 1). We have to verify that

$$2/n! = (2\pi)^{-n} V_n V_n' \quad (22.35)$$

where V_n is the volume of the unit n-ball and V_n' is the area of S^n (Riemannian volume).

Clearly $V_n = \dfrac{V_{n-1}'}{n}$ and from this everything reduces to the identity

$$V_{n-1}' V_n' = \frac{2(2\pi)^n}{(n-1)!} \quad (22.36)$$

Let us compute V_{n-1}' and prove this relation. Put $I = \int\limits_0^\infty e^{-x^2} dx$; then

$$I^n = \int\limits_{x_i \geq 0} e^{-(x_1^2 + \ldots + x_n^2)} dx_1 \ldots dx_n$$

and in particular

$$I^2 = \int\limits_{x_i \geq 0} e^{-(x_1^2 + x_2^2)} dx_1 dx_2 = \int\limits_0^{\pi/2} d\varphi \int\limits_0^\infty re^{-r^2} dr = \frac{\pi}{4} \int\limits_0^\infty e^{-z} dz = \frac{\pi}{4},$$

from which $I = \dfrac{\sqrt{\pi}}{2}$, $I^n = \dfrac{\pi^{n/2}}{2^n}$. Now note that

$$I^n = \frac{V_{n-1}'}{2^n} \int\limits_0^\infty r^{n-1} e^{-r^2} dr = \frac{V_{n-1}'}{2^n} \cdot \frac{1}{2} \int\limits_0^\infty z^{\frac{n}{2}-1} e^{-z} dz = \frac{V_{n-1}'}{2^{n+1}} \Gamma\left(\frac{n}{2}\right),$$

from which we obtain $V'_{n-1} = \dfrac{2\pi^{n/2}}{\Gamma(n/2)}$.

As is well-known, integration by parts yields $\Gamma(\alpha+1) = \alpha\Gamma(\alpha)$ for $\alpha > 0$, from which for $n = 2k$ we obtain

$$\Gamma\left(\frac{n}{2}\right) = \Gamma(k) = (k-1)! = \left(\frac{n}{2}-1\right)!.$$

If now $n = 2k+1$, then

$$\Gamma\left(\frac{n}{2}\right) = \Gamma\left(k+\frac{1}{2}\right) = \left(k-\frac{1}{2}\right)\left(k-\frac{3}{2}\right)\cdots\frac{5}{2}\cdot\frac{3}{2}\cdot\frac{1}{2}\,\Gamma\left(\frac{1}{2}\right),$$

but $\Gamma\left(\dfrac{1}{2}\right) = \dfrac{2\sqrt{\pi}}{V'_0} = \sqrt{\pi}$, from which

$$\Gamma\left(\frac{n}{2}\right) = \frac{n-2}{2}\cdot\frac{n-4}{2}\cdots\frac{1}{2}\sqrt{\pi} = 2^{-\frac{n-1}{2}}(n-2)!!\,\sqrt{\pi}.$$

Now let n be even. Then

$$\Gamma\left(\frac{n}{2}\right)\Gamma\left(\frac{n+1}{2}\right) = \left(\frac{n}{2}-1\right)!\,2^{-n/2}(n-1)!!\,\sqrt{\pi}$$

$$= 2^{-\left(\frac{n}{2}-1\right)}(n-2)!!\,2^{-n/2}(n-1)!!\,\sqrt{\pi} = 2\cdot 2^{-n}(n-1)!\,\sqrt{\pi},$$

from which

$$V'_{n-1}V'_n = \frac{2\pi^{n/2}\,2\cdot\pi^{(n+1)/2}}{2\cdot 2^{-n}(n-1)!\,\sqrt{\pi}} = \frac{2(2\pi)^n}{(n-1)!},$$

as required. The case of odd n is considered in the same way.

Problem 22.1. Write down the expression for the Laplace operator in the Lobacevskii plane (hyperbolic plane).

Problem 22.2. Compute the eigenvalues for the Laplace operator on the n-dimensional real projective space $\mathbb{R}P^n$ with the natural metric (induced by the metric on the sphere S^n under the two-fold covering $S^n \to \mathbb{R}P^n$).

Problem 22.3. Compute the eigenvalues of the Laplace operator on the n-dimensional complex projective space $\mathbb{C}P^n$ with the natural Kähler metric (cf. S.S. Chern [1]).

Chapter IV
Pseudodifferential Operators in $\mathbb{R}^n$

§23. An Algebra of Pseudodifferential Operators in $\mathbb{R}^n$

The aim of the study of pseudodifferential operators on $\mathbb{R}^n$ is to describe various effects connected with the behaviour of functions as $|x| \to +\infty$. A fundamental role is played here by non-local effects, so we have to give up the requirement of properly supportedness of pseudodifferential operators, unposed, without loss of generality, in the local theory (Chapter I).

23.1 The classes of symbols and amplitudes

Definition 23.1. The symbol class $\Gamma_\varrho^m(\mathbb{R}^N)$, where $m \in \mathbb{R}$, $0 < \varrho \leqq 1$, consists of the functions $a(z) \in C^\infty(\mathbb{R}^N)$, which satisfy the estimates

$$|\partial_z^\alpha a(z)| \leqq C_\alpha \langle z \rangle^{m - \varrho |\alpha|}, \qquad z \in \mathbb{R}^N. \tag{23.1}$$

Example 23.1. Any polynomial $a(z)$ of degree m belongs to $\Gamma_1^m(\mathbb{R}^N)$.

Let us note immediately that if $a \in \Gamma_\varrho^{m_1}(\mathbb{R}^N)$ and $b \in \Gamma_\varrho^{m_2}(\mathbb{R}^N)$, then $ab \in \Gamma_\varrho^{m_1 + m_2}(\mathbb{R}^N)$ and $\partial_z^\alpha a \in \Gamma_\varrho^{m_1 - \varrho |\alpha|}(\mathbb{R}^N)$. Further, given a linear monomorphism $j \colon \mathbb{R}^l \to \mathbb{R}^N$, $a \in \Gamma_\varrho^m(\mathbb{R}^N)$ and $j^*a = a \circ j$, then $j^*a \in \Gamma_\varrho^m(\mathbb{R}^l)$. Note that

$$\bigcap_m \Gamma_\varrho^m(\mathbb{R}^N) = S(\mathbb{R}^N). \tag{23.2}$$

Definition 23.2. Let $a_j \in \Gamma_\varrho^{m_j}(\mathbb{R}^N)$, $j = 1, 2, \ldots$, $m_j \to -\infty$ as $j \to +\infty$ and $a \in C^\infty(\mathbb{R}^N)$. We will write

$$a \sim \sum_{j=1}^\infty a_j, \tag{23.3}$$

if for any integer $r \geqq 2$

$$a - \sum_{j=1}^{r-1} a_j \in \Gamma_\varrho^{\bar{m}_r}(\mathbb{R}^N), \tag{23.4}$$

where $\bar{m}_r = \max_{j \geqq r} m_j$.

The following propositions are similar to Propositions 3.5 and 3.6.

Proposition 23.1. *Let $a_j \in \Gamma_\varrho^{m_j}(\mathbb{R}^N), j = 1, 2, \ldots,$ where $m_j \to -\infty$ as $j \to +\infty$. Then there exists a function a, such that $a \sim \sum_{j=1}^{\infty} a_j$. If another function a' has the same property, then $a - a' \in S(\mathbb{R}^N)$.*

Proposition 23.2. *Let $a_j \in \Gamma_\varrho^{m_j}(\mathbb{R}^N)$, $j = 1, 2, \ldots,$ where $m_j \to -\infty$ as $j \to +\infty$. Let $a \in C^\infty(\mathbb{R}^N)$ and for any multiindex α let the following estimate holds for some constants μ_α and C_α:*

$$|\partial_z^\alpha a(z)| \leq C_\alpha \langle z \rangle^{\mu_\alpha}. \tag{23.5}$$

Finally, let there exist l_j and C_j such that $l_j \to -\infty$ as $j \to +\infty$ and the following estimates hold

$$\left| a(z) - \sum_{j=1}^{r-1} a_j(z) \right| \leq C_r \langle z \rangle^{l_r}. \tag{23.6}$$

Then $a \sim \sum_{j=1}^{\infty} a_j$.

Exercise 23.1. Prove Propositions 23.1 and 23.2.

We now would like to consider operators of the form

$$Au(x) = \int\int e^{i(x-y)\cdot\xi}\, a(x,\xi)\, u(y)\, dy\, d\xi, \tag{23.7}$$

where $a(x, \xi) \in \Gamma_\varrho^m(\mathbb{R}^{2n})$. However, as is evident from the considerations in Chap. I, it is convenient to consider at once a more general formula for the action of the operator

$$Au(x) = \int\int e^{i(x-y)\cdot\xi}\, a(x,y,\xi)\, u(y)\, dy\, d\xi, \tag{23.8}$$

where the function $a(x, y, \xi)$ is called the *amplitude*.

We will describe the class of amplitudes, which will be useful in what follows

Definition 23.3. Let $\Pi_\varrho^m(\mathbb{R}^{3n})$ denote the set of functions $a(x, y, \xi) \in C^\infty(\mathbb{R}^{3n})$, which for some m' satisfy

$$|\partial_\xi^\alpha \partial_x^\beta \partial_y^\gamma a(x, y, \xi)| \leq C_{\alpha\beta\gamma} \langle z \rangle^{m - \varrho|\alpha + \beta + \gamma|} \langle x - y \rangle^{m' + \varrho|\alpha + \beta + \gamma|}, \tag{23.9}$$

where $z = (x, y, \xi) \in \mathbb{R}^{3n}$.

It is clear that if $a \in \Pi_\varrho^m(\mathbb{R}^{3n})$, then $\partial_z^\alpha a(z) \in \Pi_\varrho^{m-\varrho|\alpha|}(\mathbb{R}^{3n})$ and if $b \in \Pi_\varrho^{m_1}(\mathbb{R}^{3n})$, then $ab \in \Pi_\varrho^{m+m_1}(\mathbb{R}^{3n})$. We have $\Gamma_\varrho^m(\mathbb{R}^{3n}) \subset \Pi_\varrho^m(\mathbb{R}^{3n})$. Since $\langle z \rangle / \langle x - y \rangle \geq 1$, then between the classes $\Pi_\varrho^m(\mathbb{R}^{3n})$ there are the inclusions

$$\Pi_{\varrho'}^{m'}(\mathbb{R}^{3n}) \subset \Pi_{\varrho''}^{m''}(\mathbb{R}^{3n}) \quad \text{for} \quad m' \leq m'', \; \varrho' \geq \varrho''.$$

If $a(x, y, \xi) \in \Pi_\varrho^m(\mathbb{R}^{3n})$, then $a(x, x, \xi) \in \Gamma_\varrho^m(\mathbb{R}^{2n})$.

The most important example of an amplitude of the class $\Pi_\varrho^m(\mathbb{R}^{3n})$ is provided by the following

Proposition 23.3. *Let a linear map* $p: \mathbb{R}^{2n} \to \mathbb{R}^n$ *be such that the linear map* $\mathbb{R}^{2n} \to \mathbb{R}^{2n}$, *mapping* (x, y) *into* $(p(x, y), x - y)$, *is an isomorphism. Let* $b(x, \xi) \in \Gamma_\varrho^m(\mathbb{R}^{2n})$. *Define the amplitude* $a(x, y, \xi) \in C^\infty(\mathbb{R}^{3n})$ *by the formula*

$$a(x, y, \xi) = b(p(x, y), \xi). \tag{23.10}$$

Then $a \in \Pi_\varrho^m(\mathbb{R}^{3n})$.

Proof. The functions $|x| + |y|$ and $|p(x, y)| + |x - y|$ give equivalent norms on $\mathbb{R}^{2n}$. Therefore, for the proof of the proposition it remains to use the easily verified inequality

$$\frac{(1 + |p(x, y)| + |\xi|)^s}{(1 + |p(x, y)| + |x - y| + |\xi|)^s} \leq C(1 + |x - y|)^{|s|}, \qquad s \in \mathbb{R},$$

from which the estimates (23.9) follow for $a(x, y, \xi)$ with $m' = |m|$. $\quad\square$

Corollary 23.1. If $b \in \Gamma_\varrho^m(\mathbb{R}^{2n})$, then $a(x, y, \xi) = b(x, \xi)$ and $a(x, y, \xi) = b(y, \xi)$ belong to $\Pi_\varrho^m(\mathbb{R}^{3n})$.

23.2 Function spaces and the action of the operator. Now we introduce the space $C_b^\infty(\mathbb{R}^n)$ consisting of functions $u \in C^\infty(\mathbb{R}^n)$ such that

$$|\partial^\alpha u(x)| \leq C_\alpha \tag{23.11}$$

for any multiindex $|\alpha|$. The best constants C_α in (23.11) constitute a family of semi-norms for a given function, defining a Fréchet space structure on $C_b^\infty(\mathbb{R}^n)$.

The operator A of (23.8) is conveniently studied in the space $C_b^\infty(\mathbb{R}^n)$. In order to give the correct definition of the oscillatory integral appearing in (23.8), we shall have to proceed as in §1. For this purpose, let initially $a(x, y, \xi) \in C_0^\infty(\mathbb{R}^{3n})$. Then the integration in (23.8) in reality is performed over a compact set and we may carry out an integration by parts, using the identities

$$\langle x - y \rangle^{-M} \langle D_\xi \rangle^M e^{i(x-y)\cdot\xi} = e^{i(x-y)\cdot\xi}, \tag{23.12}$$

$$\langle \xi \rangle^{-N} \langle D_y \rangle^N e^{i(x-y)\cdot\xi} = e^{i(x-y)\cdot\xi}, \tag{23.13}$$

where M, N are even non-negative integers. From (23.8) one obtains

$$Au(x) = \iint e^{i(x-y)\cdot\xi} \langle x - y \rangle^{-M} \langle D_\xi \rangle^M \langle D_y \rangle^N$$

$$\times [\langle \xi \rangle^{-N} a(x, y, \xi) u(y)] \, dy \, d\xi \tag{23.14}$$

If the amplitude $a \in \Pi_\varrho^m(\mathbb{R}^{3n})$ satisfies (23.9) and if $u \in C_b^\infty(\mathbb{R}^n)$, then clearly for $m - N < -n$, $m' + m - M < -n$, the integral (23.14) becomes absolutely convergent, defining a continuous function of $x \in \mathbb{R}^n$. Increasing M and N, we will obtain integrals which are convergent also after differentiation with respect to x. Hence the operator defined by (23.14), is a continuous map

$$A: \ C_b^\infty(\mathbb{R}^n) \to C^\infty(\mathbb{R}^n) \tag{23.15}$$

For $m < 0$, we will also obtain a map $A: C_b^\infty(\mathbb{R}^n) \to C_b^\infty(\mathbb{R}^n)$. This same map could also be defined using a cut-off function $\chi(x, y, \xi) \in C_0^\infty(\mathbb{R}^{3n})$, $\chi(0, 0, 0) = 1$ and the formula

$$Au(x) = \lim_{\varepsilon \to +0} \int\int e^{i(x-y)\cdot\xi} \, \chi(\varepsilon x, \varepsilon y, \varepsilon \xi) \, a(x, y, \xi) \, u(y) \, dy \, d\xi. \tag{23.16}$$

The identity of the two definitions (23.14) and (23.16) is verified in the same way as in §1 and we leave this verification as an exercise to the reader.

In particular, the operator A is defined on the space $S(\mathbb{R}^n)$. Let us show that it gives a continuous map

$$A: \ S(\mathbb{R}^n) \to S(\mathbb{R}^n). \tag{23.17}$$

Indeed, using the inequality

$$(1 + |x|)^k \leqq (1 + |y|)^k \, (1 + |x - y|)^k, \quad k > 0,$$

we see from (23.14) that

$$(1 + |x|)^k \, |Au(x)| \leqq C_k$$

for any k and a similar estimate holds if we replace $Au(x)$ by $\partial_x^\alpha(Au(x))$. From this we also have $Au \in S(\mathbb{R}^n)$ for $u \in S(\mathbb{R}^n)$ with an estimate of seminorms guaranteeing the continuity of the map (23.17) (which also could have been obtained from the closed graph theorem).

Finally note, that since the transposed operator

$$^tAu(y) = \int\int e^{i(x-y)\cdot\xi} \, a(x, y, \xi) \, v(x) \, dx \, d\xi \tag{23.18}$$

by similar reasoning, maps $S(\mathbb{R}^n)$ into $S(\mathbb{R}^n)$, then A can be extended by duality to a continuous map

$$A: \ S'(\mathbb{R}^n) \to S'(\mathbb{R}^n).$$

Definition 23.4. The class of operators A of the form (23.8) with amplitudes $a \in \Pi_\varrho^m(\mathbb{R}^{3n})$ will be denoted by $G_\varrho^m(\mathbb{R}^n)$ or simply by G_ϱ^m (if the dimension n is clear or unimportant).

It is useful to have a description of the operators belonging to the intersection $G^{-\infty} = \bigcap_m G_\varrho^m$. We shall show that this intersection is independent of ϱ and consists of operators with kernels $K_A(x, y) \in S(\mathbb{R}^{2n})$. Clearly it suffices to consider the case $\varrho < 1$. Note that the operators with amplitudes $a(x, y, \xi)$ and $\langle x-y \rangle^{-N} \langle D_\xi \rangle^N a(x, y, \xi)$ coincide, from which we see that if $A \in G^{-\infty}$, then A can be determined by an amplitude $a(x, y, \xi)$ satisfying (23.9) with arbitrarily small (arbitrarily close to $-\infty$) numbers m and m'. But then A has the kernel

$$K_A(x, y) = \iint e^{i(x-y)\cdot\xi} a(x, y, \xi)\, d\xi, \tag{23.19}$$

belonging to $S(\mathbb{R}^{2n})$. From this, it follows in particular that A defines a continuous map

$$A\colon S'(\mathbb{R}^n) \to S(\mathbb{R}^n), \tag{23.20}$$

given by the formula

$$Au(x) = \langle K_A(x, \cdot), u(\cdot) \rangle. \tag{23.21}$$

In the general case the kernel $K_A(x, y)$ is defined by the formula

$$\langle K_A, \varphi \rangle = \iint e^{i(x-y)\cdot\xi} a(x, y, \xi)\, \varphi(x, y)\, dx\, dy\, d\xi, \qquad \varphi \in S(\mathbb{R}^{2n}), \tag{23.22}$$

and is a distribution $K_A \in S'(\mathbb{R}^{2n})$.

Exercise 23.2. Denote by $C_t^\infty(\mathbb{R}^n)$ the space of functions $u \in C^\infty(\mathbb{R}^n)$, with the property that for any multi-index α one can find constants C_α and μ_α, such that

$$|\partial^\alpha u(x)| \leqq C_\alpha \langle x \rangle^{\mu_\alpha}. \tag{23.23}$$

Show that an operator $A \in G_\varrho^m$ defines a map

$$A\colon C_t^\infty(\mathbb{R}^n) \to C_t^\infty(\mathbb{R}^n). \tag{23.24}$$

Exercise 23.3. Let $A \in G_\varrho^m(\mathbb{R}^n)$ and K_A the kernel of A. Show that $K_A \in C^\infty(\mathbb{R}^{2n} \setminus \Delta)$, where Δ is the diagonal in $\mathbb{R}^n \times \mathbb{R}^n$.

23.3 Left, right and Weyl symbols

Theorem 23.1. *An operator $A \in G_\varrho^m$ of the form* (23.8) *can be written in any of the following three forms*

$$Au(x) = \iint e^{i(x-y)\cdot\xi} \sigma_{A,l}(x, \xi)\, u(y)\, dy\, d\xi, \tag{23.25}$$

$$Au(x) = \iint e^{i(x-y)\cdot\xi} \sigma_{A,r}(y, \xi)\, u(y)\, dy\, d\xi, \tag{23.26}$$

$$Au(x) = \iint e^{i(x-y)\cdot\xi} \sigma_{A,w}\left(\frac{x+y}{2}, \xi\right) u(y)\, dy\, d\xi. \tag{23.27}$$

Here $\sigma_{A,l}$, $\sigma_{A,r}$ and $\sigma_{A,w}$ belong to $\Gamma_\varrho^m(\mathbb{R}^{2n})$, are uniquely defined and have the following asymptotic expansions:

$$\sigma_{A,l}(x,\xi) \sim \sum_\alpha \frac{1}{\alpha!}\, \partial_\xi^\alpha D_y^\alpha a(x,y,\xi)\,|_{y=x}, \tag{23.28}$$

$$\sigma_{A,r}(y,\xi) \sim \sum_\alpha \frac{1}{\alpha!}\, \partial_\xi^\alpha (-D_x)^\alpha\, a(x,y,\xi)\,|_{x=y}, \tag{23.29}$$

$$\sigma_{A,w}(x,\xi) \sim \sum_{\beta,\gamma} \frac{1}{\beta!\gamma!}\, \left(\frac{1}{2}\right)^{|\beta+\gamma|} \partial_\xi^{\beta+\gamma}(-D_x)^\beta\, D_y^\gamma a(x,y,\xi)\,|_{y=x}. \tag{23.30}$$

This theorem allows the introduction of

Definition 23.5. The functions $\sigma_{A,l}$, $\sigma_{A,r}$ and $\sigma_{A,w}$ from the formulae (23.25)–(23.27) are called, respectively, the *left, right and Weyl symbols* of the operator A.

Although we shall not use any other symbols, let us show the following generalization of Theorem 23.1, containing a parameter $\tau \in \mathbb{R}$ and also allowing us to avoid repetitions in the proof of Theorem 23.1.

Theorem 23.2. *Let $A \in G_\varrho^m$ of the form (23.8) be given. Then for any $\tau \in \mathbb{R}$ A may be uniquely written as*

$$Au(x) = \iint e^{i(x-y)\cdot\xi}\, b_\tau((1-\tau)x+\tau y,\xi)\, u(y)\, dy\, d\!\!\!{}^-\xi, \tag{23.31}$$

where $b_\tau \in \Gamma_\varrho^m(\mathbb{R}^{2n})$. Here b_τ has the following asymptotic expansion

$$b_\tau(x,\xi) \sim \sum_{\beta,\gamma} \frac{1}{\beta!\gamma!}\, \tau^{|\beta|}(1-\tau)^{|\gamma|}\, \partial_\xi^{\beta+\gamma}(-D_x)^\beta\, D_y^\gamma a(x,y,\xi)\,|_{y=x}. \tag{23.32}$$

Definition 23.6. The function $b_\tau(x,\xi)$ will be called the *τ-symbol* of A.

Proof of Theorem 23.2. Putting

$$\begin{cases} v = (1-\tau)x+\tau y, \\ w = x - y, \end{cases} \tag{23.33}$$

we obtain

$$\begin{cases} x = v + \tau w, \\ y = v - (1-\tau)w, \end{cases} \tag{23.34}$$

from which

$$a(x,y,\xi) = a(v+\tau w,\, v-(1-\tau)w,\, \xi). \tag{23.35}$$

Let us now expand the right-hand side of (23.35) at $w=0$ in a Taylor series:

$$a(x,y,\xi) = \sum_{|\beta+\gamma|\le N-1} \frac{(-1)^{|\gamma|}}{\beta!\gamma!}\, \tau^{|\beta|}(1-\tau)^{|\gamma|}(x-y)^{\beta+\gamma}(\partial_x^\beta\partial_y^\gamma a)(v,v,\xi) + r_N, \tag{23.36}$$

where

$$r_N(x, y, \xi) = \sum_{|\beta + \gamma| = N} c_{\beta\gamma} (x - y)^{\beta + \gamma} \int_0^1 (1 - t)^{N-1}$$

$$\times (\partial_x^\beta \partial_y^\gamma a)(v + t\tau w, v - t(1 - \tau) w, \xi) \, dt, \tag{23.37}$$

and $c_{\beta\gamma}$ are constants.

In (23.36) the expression $(\partial_x^\beta \partial_y^\gamma a)(v, v, \xi)$ signifies that in the function $\partial_x^\beta \partial_y^\gamma a(x, y, \xi)$ it is necessary to take $v = (1 - \tau) x + \tau y$ instead of x and y. The expression $(\partial_x^\beta \partial_y^\gamma a)(v + t\tau w, v - t(1 - \tau) w, \xi)$ in formula (23.37) has a similar meaning.

Now note, that the operator with amplitude $(x - y)^{\beta + \gamma} (\partial_x^\beta \partial_y^\gamma a)(v, v, \xi)$ coincides with the one given via the amplitude

$$(-D_\xi)^{\beta + \gamma} (\partial_x^\beta \partial_y^\gamma a)(v, v, \xi) = (-1)^{|\beta| + |\gamma|} (\partial_\xi^{\beta + \gamma} D_x^\beta D_y^\gamma a)(v, v, \xi).$$

Therefore it follows from (23.36) that A can be represented in the form of a sum $A = A_N + R_N$, where A_N is an operator with τ-symbol

$$b_N(x, \xi) = \sum_{|\beta + \gamma| \leq N - 1} \frac{1}{\beta! \gamma!} \tau^{|\beta|} (1 - \tau)^{|\gamma|} \partial_\xi^{\beta + \gamma} (-D_x)^\beta D_y^\gamma a(x, y, \xi)|_{y = x},$$

and R_N is an operator with amplitude $r_N(x, y, \xi)$. Note that R_N is a linear combination of a finite number of terms having amplitudes of the form

$$\int_0^1 (\partial_\xi^{\beta + \gamma} \partial_x^\beta \partial_y^\gamma a)(v + t\tau w, v - t(1 - \tau) w, \xi)(1 - t)^{N-1} dt,$$

$$|\beta + \gamma| = N.$$

Let us show that this amplitude belongs to the class $\Pi_\varrho^{m - 2N\varrho}(\mathbb{R}^{3n})$. For this it suffices to show that this is true for the integrand, with all estimates uniform in t (note that this is obvious for each fixed $t \neq 0$ and true for $t = 0$ by Proposition 23.3). In view of the relations

$$v = (1 - \tau)(v + t\tau w) + \tau(v - t(1 - \tau) w),$$

$$tw = (v + t\tau w) - (v - t(1 - \tau) w)$$

it is obvious that

$$C^{-1} \leqq \frac{|v + t\tau w| + |v - t(1 - \tau) w|}{|v| + |tw|} \leqq C,$$

where $C > 0$ and C does not depend on $t \in [0, 1]$. Therefore

$$|(\partial_\xi^{\beta + \gamma} \partial_x^\beta \partial_y^\gamma a)(v + t\tau w, v - t(1 - \tau) w, \xi)|$$

$$\leqq C(1 + |v| + |tw| + |\xi|)^{m - 2\varrho N}(1 + |tw|)^{m' + 2\varrho N}.$$

Since for $m' + 2 \varrho N \geqq 0$ we have

$$(1+|tw|)^{m'+2\varrho N} \leqq (1+|v|+|tw|+|\xi|)^{m'+2\varrho N} \, (1+|v|+|\xi|)^{-(m'+2\varrho N)},$$

it is clear that if, in addition, $m' + m \geq 0$ and $m - 2\varrho N \leq 0$, then

$$\begin{aligned}
|(\partial_\xi^{\beta+\gamma} & \partial_x^\beta \partial_y^\gamma a)\,(v+t\tau w, v - t(1-\tau)\,w, \xi)| \\
& \leqq C(1+|v|+|\xi|)^{-m'-2\varrho N}\,(1+|v|+|tw|+|\xi|)^{m'+m} \\
& \leqq C(1+|v|+|\xi|)^{m-2\varrho N}\,(1+|w|)^{m'+m} \\
& \leqq C(1+|v|+|w|+|\xi|)^{m-2\varrho N}\,(1+|w|)^{m'+2m+2\varrho N},
\end{aligned}$$

where C does not depend on t. One obtains the estimates for derivatives in an analogous way.

Now let the symbol $b'(x, \xi) \in \Gamma_\varrho^m(\mathbb{R}^{2n})$ be such that

$$b'(x, \xi) \sim \sum_{N=0}^\infty \left(b_N(x, \xi) - b_{N-1}(x, \xi)\right).$$

Then, if A' has τ-symbol $b'(x, \xi)$ it is clear that $A - A' \in G^{-\infty}$, i.e. the operator $A - A'$ has a kernel belonging to $S(\mathbb{R}^{2n})$.

Let us now verify that if A has a kernel $K_A \in S(\mathbb{R}^{2n})$, then it has a τ-symbol $b_\tau(x, \xi) \in S(\mathbb{R}^{2n})$ and the correspondence between kernel and symbol is a one-to-one correspondence. From formula (23.31) it is clear that this correspondence is of the form

$$K_A(x, y) = F_{\xi \to x-y}^{-1} b_\tau((1-\tau)x + \tau y, \xi), \tag{23.38}$$

$$b_\tau(v, \xi) = F_{w \to \xi} K_A(v + \tau w, v - (1-\tau)w) \tag{23.39}$$

((23.39) is obtained from (23.38) by a change of coordinates and the Fourier inversion formula). In particular, for any $K_A \in S(\mathbb{R}^{2n})$, we can find $b_\tau(v, \xi) \in S(\mathbb{R}^{2n})$ by formula (23.39).

We next show the uniqueness of the τ-symbol in the general case. For this we note that (23.38) is always true when A is given via a τ-symbol $b_\tau(x, \xi)$ and if the partial Fourier transform, which appears in this formula, is understood in the same sense as the Fourier transform of distributions (cf. §1). Thereby, the inversion formula is also true, leading to (23.39) after the linear change of coordinates (23.34). Also, from formula (23.39), the uniqueness of the τ-symbol is obvious, taking into account the uniqueness of the kernel K_A. $\square$

Corollary 23.2. *The class of operators G_ϱ^m coincides with the class of operators of the form* (23.25) *with the left symbol $\sigma_{A,l}(x, \xi) \in \Gamma_\varrho^m(\mathbb{R}^{2n})$. The same is also true if we replace* (23.25) *by* (23.26) *or* (23.27) *and $\sigma_{A,l}$ by $\sigma_{A,r}$ or $\sigma_{A,w}$.*

23.4 Relations between the different symbols. The symbols of the transposed and adjoint operators. The expression for the τ-symbol in terms of the τ_1-symbol for a different τ, can be easily obtained from Theorem 23.2 in the form of an asymptotic series. Indeed, if an operator A has the τ_1-symbol $b_{\tau_1}(x, \xi)$, this signifies that it may be determined via the amplitude

$$a(x, y, \xi) = b_{\tau_1}((1 - \tau_1)\, x + \tau_1\, y, \xi).$$

But then, by Theorem 23.2 its τ-symbol has the asymptotic expansion

$$b_\tau(x, \xi) \sim \sum_{\beta, \gamma} \frac{(-1)^{|\beta|}}{\beta!\,\gamma!} \, \tau^{|\beta|} (1 - \tau)^{|\gamma|} (1 - \tau_1)^{|\beta|} \, \tau_1^{|\gamma|} \, \partial_\xi^{\beta + \gamma} D_x^{\beta + \gamma} b_{\tau_1}$$

or

$$b_\tau(x, \xi) \sim \sum_\alpha c_\alpha \partial_\xi^\alpha D_x^\alpha b_{\tau_1}(x, \xi), \tag{23.40}$$

where

$$c_\alpha = \sum_{\beta + \gamma = \alpha} \frac{(-1)^{|\beta|}}{\beta!\,\gamma!} \, [\tau (1 - \tau_1)]^{|\beta|} \, [(1 - \tau)\,\tau_1]^{|\gamma|} \tag{23.41}$$

and, in particular, we have $c_0 = 1$. Now, transforming (23.41) using the Newton binomial formula (Lemma 3.4), we obtain

$$c_\alpha = \frac{1}{\alpha!} \, [(1 - \tau)\,\tau_1 e - \tau(1 - \tau_1)\, e]^\alpha = \frac{1}{\alpha!} \, (\tau_1 - \tau)^{|\alpha|}.$$

where $e = (1, 1, \ldots, 1)$.

Thus, we have proved

Theorem 23.3. *Symbols $b_\tau(x, \xi)$ and $b_{\tau_1}(x, \xi)$ of the same operator $A \in G_\varrho^m$ are related via*

$$b_\tau(x, \xi) \sim \sum_\alpha \frac{1}{\alpha!} \, (\tau_1 - \tau)^{|\alpha|} \, \partial_\xi^\alpha D_x^\alpha b_{\tau_1}(x, \xi). \tag{23.42}$$

In particular, $b_\tau(x, \xi) - b_{\tau_1}(x, \xi) \in \Gamma_\varrho^{m - 2\varrho}(\mathbb{R}^{2n})$.

Let us now consider the transposed operator $\,{}^t A$, defined by the formula

$$\langle Au, v \rangle = \langle u, {}^t\! Av \rangle, \qquad u, v \in S(\mathbb{R}^n), \tag{23.43}$$

where

$$\langle u, v \rangle = \int\limits_{\mathbb{R}^n} u(x)\, v(x)\, dx.$$

From the formula

$$\langle Au, v \rangle = \int\!\!\int\!\!\int e^{i(x - y)\cdot\xi} b_\tau((1 - \tau)\, x + \tau y, \xi)\, u(y)\, v(x)\, dy\, dx\, d\xi$$
$$\equiv \int\!\!\int e^{i(y - x)\cdot\xi} b_\tau((1 - \tau)\, x + \tau y, -\xi)\, u(y)\, v(x)\, dy\, dx\, d\xi$$

it follows that if A has a τ-symbol $b_\tau(x, \xi)$, then ${}^tA$ has the $(1-\tau)$-symbol ${}^tb_{1-\tau}(x, \xi)$, given by the formula

$$ {}^tb_{1-\tau}(x, \xi) = b_\tau(x, -\xi). \tag{23.44} $$

From Theorem 23.3 it now follows

Theorem 23.4. *If $A \in G_\varrho^m$, then ${}^tA \in G_\varrho^m$ and the τ-symbol ${}^tb_\tau(x, \xi)$ of ${}^tA$ can be expressed in terms of the τ-symbol $b_\tau(x, \xi)$ of A by the formula*

$$ {}^tb_\tau(x, \xi) \sim \sum_\alpha \frac{1}{\alpha!} (1-2\tau)^{|\alpha|} \, \partial_\xi^\alpha D_x^\alpha b_\tau(x, -\xi). \tag{23.45} $$

Now let A^* be the adjoint of A, defined by

$$ (Au, v) = (u, A^*v), \qquad u,\ v \in S(\mathbb{R}^n), \tag{23.46} $$

where

$$ (u, v) = \int_{\mathbb{R}^n} u(x) \, \overline{v(x)} \, dx. \tag{23.47} $$

By analogy with Theorem 23.4 one can show

Theorem 23.5. *If $A \in G_\varrho^m$, then $A^* \in G_\varrho^m$ and the τ-symbol $b_\tau^*(x, \xi)$ of A^* is related to the $(1-\tau)$-symbol $b_{1-\tau}(x, \xi)$ of A via the relation*

$$ b_\tau^*(x, \xi) = \overline{b_{1-\tau}(x, \xi)}, \tag{23.48} $$

and can be expressed in terms of the τ-symbol $b_\tau(x, \xi)$ of A via the asymptotic series

$$ b_\tau^*(x, \xi) \sim \sum_\alpha \frac{1}{\alpha!} (1-2\tau)^{|\alpha|} \, \partial_\xi^\alpha D_x^\alpha \overline{b_\tau(x, \xi)}. \tag{23.49} $$

Corollary 23.3. *If $A \in G_\varrho^m$, then*

$$ \sigma_{A^*, w}(x, \xi) = \overline{\sigma_{A, w}(x, \xi)}. \tag{23.50} $$

In particular, the condition $A = A^*$ is equivalent to the real-valuedness of the Weyl symbol $\sigma_{A, w}(x, \xi)$.

23.5 The composition formula

Theorem 23.6. *Let $A' \in G_\varrho^{m_1}$, $A'' \in G_\varrho^{m_2}$. Then $A' \circ A'' \in G_\varrho^{m_1 + m_2}$ and if $b'_{\tau_1}(x, \xi)$ is the τ_1-symbol of A' and $b''_{\tau_2}(x, \xi)$ the τ_2-symbol of A'', then the τ-symbol $b_\tau(x, \xi)$ of $A' \circ A''$ has the asymptotic expansion*

$$ b_\tau(x, \xi) \sim \sum_{\alpha, \beta, \gamma, \delta} c_{\alpha\beta\gamma\delta} (\partial_\xi^\alpha D_x^\beta b'_{\tau_1}(x, \xi))(\partial_\xi^\gamma D_x^\delta b''_{\tau_2}(x, \xi)), \tag{23.51} $$

where $c_{\alpha\beta\gamma\delta}$ are constants (depending on τ, τ_1 and τ_2) such that $c_{0000} = 1$ and the sum runs over sets of multi-indices α, β, γ and δ such that $\alpha + \gamma = \beta + \delta$.

In particular, we have $b_\tau(x, \xi) - b'_{\tau_1}(x, \xi)\, b''_{\tau_2}(x, \xi) \in \Gamma_\varrho^{m_1 + m_2 - 2\varrho}(\mathbb{R}^{2n})$.

Proof. Taking Theorem 23.3 into account, we see that it suffices to consider only one arbitrary triple of the numbers τ, τ_1 and τ_2. Let us take for simplicity $\tau_1 = 0$, $\tau_2 = 1$. A'' can be written, using the symbol $b''_1(y, \xi)$, as

$$A'' u(x) = \iint e^{i(x-y)\cdot\xi}\, b''_1(y, \xi)\, u(y)\, dy\, d\xi\,,$$

or

$$\widehat{A'' u}(\xi) = \int e^{-iy\cdot\xi}\, b''_1(y, \xi)\, u(y)\, dy\,. \tag{23.52}$$

A' has the form

$$A' v(x) = \iint e^{i(x-y)\cdot\xi}\, b'_0(x, \xi)\, u(y)\, dy\, d\xi \;=\; \int e^{ix\cdot\xi}\, b'_0(x, \xi)\, \hat{u}(\xi)\, d\xi\,. \tag{23.53}$$

From (23.52) and (23.53) it follows that

$$A' \circ A'' u(x) = \iint e^{i(x-y)\cdot\xi}\, b'_0(x, \xi)\, b''_1(y, \xi)\, u(y)\, dy\, d\xi\,, \tag{23.54}$$

i.e. $A' \circ A''$ is determined via the amplitude

$$a(x, y, \xi) = b'_0(x, \xi)\, b''_1(y, \xi) \in \Pi_\varrho^{m_1 + m_2}(\mathbb{R}^{3n})\,.$$

From this we have $A' \circ A'' \in G_\varrho^{m_1 + m_2}$. Applying Theorem 23.2 we obtain for $b_\tau(x, \xi)$

$$b_\tau(x, \xi) \sim \sum_{\beta, \gamma} \frac{(-1)^{|\beta|}}{\beta!\gamma!}\, \tau^{|\beta|}(1-\tau)^{|\gamma|}\, \partial_\xi^{\beta+\gamma}\, [(D_x^\beta b'_0(x, \xi))\,(D_x^\gamma b''_1(x, \xi))]$$

and by the Leibniz rule (Lemma 3.3)

$$b_\tau(x, \xi) \sim \sum_{\substack{\beta, \gamma, \delta, \varepsilon \\ \delta + \varepsilon = \beta + \gamma}} \frac{(-1)^{|\beta|}(\beta+\gamma)!}{\beta!\gamma!\delta!\varepsilon!}\, \tau^{|\beta|}(1-\tau)^{|\gamma|}\, (\partial_\xi^\delta D_x^\beta b'_0)\,(\partial_\xi^\varepsilon D_x^\gamma b''_1)\,, \tag{23.55}$$

as required. $\square$

Inserting into (23.55) the expressions for b'_0, b''_1 in terms of b'_{τ_1}, b''_{τ_2} we obtain formulae for the coefficients $c_{\alpha\beta\gamma\delta}$ in (23.51), which may sometimes be simplified. For instance, one can show by analogy with Theorem 3.4 the following

Theorem 23.7. *Under the assumptions of Theorem 23.6 one has*

$$b_0(x, \xi) \sim \sum_\alpha \frac{1}{\alpha!}\, (\partial_\xi^\alpha b'_0(x, \xi))\,(D_x^\alpha b''_0(x, \xi))\,. \tag{23.56}$$

Problem 23.1. Show that the left symbol $\sigma_{A,l}(x,\xi)$ of an operator $A \in G_\varrho^m$ can be expressed in terms of A by the formula

$$\sigma_{A,l}(x,\xi) = e^{-ix \cdot \xi} A(e^{ix \cdot \xi}), \tag{23.57}$$

where A acts on the variable x.

Problem 23.2. Show that if $A' \in G_\varrho^{m_1}$, $A'' \in G_\varrho^{m_2}$, then

$$\sigma_{A' \circ A'', w}(x,\xi) \sim \sum_{\alpha, \beta} \frac{(-1)^{|\beta|}}{\alpha! \beta!} 2^{-|\alpha+\beta|} (\partial_\xi^\alpha D_x^\beta \sigma_{A',w}(x,\xi))(\partial_\xi^\beta D_x^\alpha \sigma_{A'',w}(x,\xi)). \tag{23.58}$$

Problem 23.3. Consider the polynomial

$$(t_1 x_1 + \ldots + t_n x_n + \tau_1 D_{x_1} + \ldots + \tau_n D_{x_n})^N$$

in the variables t, $\tau \in \mathbb{R}^n$ with operator coefficients (x_j is viewed as the multiplication operator by x_j) and write it in the form

$$\sum_{|\alpha+\beta|=N} \frac{N!}{\alpha! \beta!} t^\alpha \tau^\beta A_{\alpha\beta}.$$

Show that $A_{\alpha\beta}$ is an operator with the Weyl symbol $x^\alpha \xi^\beta$.

§24. The Anti-Wick Symbol.
Theorems on Boundedness and Compactness

24.1 Definition and basic properties of the anti-Wick symbol. Put

$$\Phi_0(x) = \pi^{-n/4} \exp[-x^2/2], \qquad x \in \mathbb{R}^n. \tag{24.1}$$

Then $\Phi_0 \in S(\mathbb{R}^n)$ and $\|\Phi_0\| = 1$, where $\|\cdot\|$ denotes the usual norm in $L^2(\mathbb{R}^n)$, generated by the scalar product (23.47). Denote by P_0 the orthogonal projection in $L^2(\mathbb{R}^n)$ onto the vector Φ_0. Clearly, the Schwartz kernel of this projection has the form

$$K(x,y) = \pi^{-n/2} \exp[-(x^2+y^2)/2]. \tag{24.2}$$

Computing the Weyl symbol σ_0 of P_0, we get, using formula (23.39) with $\tau = \tfrac{1}{2}$;

$$\sigma_0(x,\xi) = F_{v \to \xi} K\left(x + \frac{1}{2} v, \ x - \frac{1}{2} v\right). \tag{24.3}$$

Now, taking into account that

$$F_{v \to \xi} \exp(-\alpha v^2) = \left(\frac{\pi}{\alpha}\right)^{n/2} \exp\left(-\frac{\xi^2}{4\alpha}\right), \qquad \alpha > 0, \qquad (24.4)$$

we obtain from (24.3) that

$$\sigma_0(x, \xi) = F_{v \to \xi}\, \pi^{-n/2} \exp\left(-x^2 - \frac{v^2}{4}\right) = 2^n \exp\left[-(x^2 + \xi^2)\right]. \qquad (24.5)$$

Now let $z = (x, \xi)$, $z_0 = (x_0, \xi_0) \in \mathbb{R}^{2n}$. Consider the operator P_{z_0} with Weyl symbol $\sigma_{z_0}(z)$ of the form

$$\sigma_{z_0}(z) = \sigma_0(z - z_0). \qquad (24.6)$$

It is easily verified that

$$P_{z_0} = M_{\xi_0} T_{x_0} P_0 T_{x_0}^{-1} M_{\xi_0}^{-1}, \qquad (24.7)$$

where M_{ξ_0} is the multiplication operator by $e^{ix \cdot \xi_0}$ and T_{x_0} is the shift operator by x_0 in $L^2(\mathbb{R}^n)$, i.e. $T_{x_0} u(x) = u(x - x_0)$. Setting $U_{z_0} = M_{\xi_0} T_{x_0}$, we see that P_{z_0} can be written in the form

$$P_{z_0} = U_{z_0} P_0 U_{z_0}^{-1} \qquad (24.8)$$

from which it is obvious, since U_{z_0} is unitary that P_{z_0} is the orthogonal projection onto the vector $\Phi_{z_0} = U_{z_0} \Phi_0$.

We wish to consider the following operator, being a linear combination of operators P_z, $z \in \mathbb{R}^{2n}$:

$$A = \int a(x, \xi)\, P_{x, \xi}\, dx\, d\xi, \qquad (24.9)$$

where $a(x, \xi) \in \Gamma_\varrho^m(\mathbb{R}^{2n})$. As for the meaning of this formula, let us note that the following can be immediately verified: if $u(x) \in S(\mathbb{R}^n)$, then $(P_{x, \xi} u)(x_0) \in S(\mathbb{R}^{2n}_{x, \xi})$. Due to this, (24.9) makes sence if we consider it on functions $u(x) \in S(\mathbb{R}^n)$, and one can easily ensure that A maps $S(\mathbb{R}^n)$ into $S(\mathbb{R}^n)$.

Definition 24.1. An operator A of the form (24.9) is called *operator with anti-Wick symbol* $a(x, \xi)$.

The convenience of anti-Wick symbols is demonstrated by

Proposition 24.1. *If* $a(z) \geq 0$, *then* $A \geq 0$, *i.e.* $(Au, u) \geq 0$ *for* $u \in S(\mathbb{R}^n)$.

Proof. This follows from the fact that $P_{z_0} \geq 0$ for any $z_0 \in \mathbb{R}^{2n}$. $\quad \square$

Corollary 24.1. *If* $a(z)$ *is real-valued, then*

$$\|A\| \leq \sup_{z \in \mathbb{R}^{2n}} |a(z)|, \qquad (24.10)$$

where

$$\|A\| = \sup_{u \in S(\mathbb{R}^n),\, u \neq 0} \|Au\|/\|u\|.$$

Proof. The statement $\|A\| \leq M$ is equivalent, for self-adjoint A, to the non-negativity of the operators $M - A$ and $A + M$, which is obvious for $M = \sup_{z \in \mathbb{R}^{2n}} |a(z)|$ from Proposition 24.1. $\square$

Corollary 24.2. *For a complex-valued function $a(z)$ one has the following estimate*

$$\|A\| \leq 2 \sup_{z \in \mathbb{R}^{2n}} |a(z)|. \tag{24.11}$$

Proof. It suffices to use Corollary 24.1 for $\operatorname{Re} a$ and $\operatorname{Im} a$. $\square$

Remark 24.1. In fact, (24.10) holds also for complex-valued functions $a(z)$ (cf. Problem 24.4), although we shall not use this.

24.2 Connection between the anti-Wick symbol and the other symbols.

Theorem 24.1. *Let A be an operator with anti-Wick symbol $a(z) = a(x, \xi) \in \Gamma_\varrho^m(\mathbb{R}^{2n})$. Then $A \in G_\varrho^m$ and the Weyl symbol of A can be expressed as*

$$b(z) = \pi^{-n} \int e^{-|z-z'|^2} a(z')\, dz'. \tag{24.12}$$

Further, any τ-symbol $b_\tau(z)$ of A has the asymptotic expansion

$$b_\tau(z) \sim \sum_\alpha c_\alpha \partial^\alpha a(z), \tag{24.13}$$

where c_α are constants (depending on τ), such that $c_0 = 1$ and $c_\alpha = 0$ for odd $|\alpha|$. In particular

$$b_\tau(z) - a(z) \in \Gamma_\varrho^{m-2\varrho}(\mathbb{R}^{2n}). \tag{24.14}$$

Proof. The relation (24.12) is obvious from (24.9), (24.5) and (24.6). In view of Theorem 23.3 it suffices to show (24.13) for $\tau = \frac{1}{2}$, i.e. for the Weyl symbol. Let us expand $a(z')$ in the Taylor series at z:

$$a(z') = \sum_{|\alpha| < N} \frac{1}{\alpha!} (\partial^\alpha a(z))(z' - z)^\alpha + r_N(z', z), \tag{24.15}$$

where

$$r_N(z', z) = \sum_{|\alpha| = N} c'_\alpha (z' - z)^\alpha \int_0^1 \partial^\alpha a(z + t(z' - z))(1 - t)^{N-1}\, dt, \tag{24.16}$$

and where c'_α are constants.

Inserting (24.15) into (24.12) we obtain

$$b(z) = \sum_{|\alpha| < N} c_\alpha \partial^\alpha a(z) + R_N, \qquad (24.17)$$

where

$$c_\alpha = \frac{1}{\alpha! \pi^n} \int z^\alpha e^{-|z|^2} dz \qquad (24.18)$$

(from which obtain in particular, $c_0 = 1$ and $c_\alpha = 0$ for odd $|\alpha|$), and

$$R_N(z) = \pi^{-n} \int e^{-|z - z'|^2} r_N(z', z) \, dz'.$$

Let us show that $R_N(z) \in \Gamma_\varrho^{m - \varrho N}(\mathbb{R}^{2N})$ (from which we have, in an obvious manner, the required expansion (24.13)). It is convenient to rewrite $R_N(z)$ in the form

$$R_N(z) = \sum_{|\alpha| = N} c_\alpha'' \int_0^1 dt \int dw \cdot w^\alpha e^{-|w|^2} (\partial^\alpha a)(z + tw)(1 - t)^{N-1},$$

implying that $\partial_z^\gamma R_N(z)$ has the form of a sum of terms like

$$\int_0^1 dt \int dw \cdot w^\alpha e^{-|w|^2} (\partial^\beta a)(z + tw)(1 - t)^{N-1},$$

where $|\beta| = N + |\gamma|$. Clearly it suffices to estimate the expression

$$I_{\alpha\beta}(z) = \int dw \cdot w^\alpha e^{-|w|^2} (\partial^\beta a)(z + tw). \qquad (24.19)$$

uniformly in $t \in [0, 1]$. To estimate $I_{\alpha\beta}(z)$ we decompose it into the sum of two integrals:

$$I_{\alpha\beta}'(z) \quad \text{over the domain } |w| < |z|/2,$$
$$I_{\alpha\beta}''(z) \quad \text{over the domain } |w| > |z|/2.$$

For $I_{\alpha\beta}'(z)$ one has the estimate

$$|I_{\alpha\beta}'(z)| \leq C_\beta \langle z \rangle^{m - \varrho(N + |\gamma|)} \int \langle w \rangle^N e^{-|w|^2} dw = C_\beta' \langle z \rangle^{m - \varrho(N + |\gamma|)}, \quad (24.20)$$

and $I_{\alpha\beta}''(z)$ can be estimated as follows:

$$|I_{\alpha\beta}''(z)| \leq C_{\alpha\beta} \int_{|w| > |z|/2} e^{-|w|^2} \langle w \rangle^{N + m} \langle z \rangle^m \, dw \leq C_k \langle z \rangle^{-k} \qquad (24.21)$$

for an arbitrary k. From (24.20) and (24.21) we immediately have the required estimate for $I_{\alpha\beta}(z)$. $\square$

Theorem 24.2. *Let $A' \in G_\varrho^m$. Then there exists an operator $A \in G_\varrho^m$, such that A is given by the anti-Wick symbol*

$$a(z) \in \Gamma_\varrho^m(\mathbb{R}^{2n}) \qquad \text{and} \qquad A - A' \in G^{-\infty}, \tag{24.22}$$

i.e. the operator $A - A'$ has the Schwartz kernel $K_{A-A'}(x, y) \in S(\mathbb{R}^{2n})$.

Remark 24.2. Not every operator $A \in G_\varrho^m$ has an anti-Wick symbol $a \in \Gamma_\varrho^m$ (this is clear for instance, from the fact that the Weyl symbol $b(z)$ as defined by formula (24.12), must be a real-analytic function in $z \in \mathbb{R}^{2n}$). The actual finding of the anti-Wick symbol from a given Weyl symbol requires the solution of the inverse heat equation. Theorem 24.2 shows that if one disregards symbols in $S(\mathbb{R}^n)$, this process becomes possible.

Proof of Theorem 24.2. Let A' have the Weyl symbol $b'(x, \xi)$. Consider the operator A_0 with anti-Wick symbol $a_0(x, \xi) = b'(x, \xi)$ and put $A_1' = A' - A_0$. Then $A_1' \in G_\varrho^{m-2\varrho}$ by Theorem 24.1. Denote by A_1 the operator with anti-Wick symbol $a_1(x, \xi)$, equal to the Weyl symbol of A_1'. We have

$$A' = A_0 + A_1 + A_2', \qquad A_2' \in G_\varrho^{m-4\varrho}.$$

Continuing by induction, we may construct a sequence of operators $A_j, j = 0, 1, 2, \ldots$, with anti-Wick symbols $a_j(z) \in \Gamma_\varrho^{m-2j\varrho}(\mathbb{R}^{2n})$, such that

$$A' - \sum_{j=0}^{N-1} A_j \in G_\varrho^{m-2\varrho N}. \tag{24.23}$$

Let $a(z) \in \Gamma_\varrho^m(\mathbb{R}^{2n})$ be such that

$$a \sim \sum_{j=0}^{\infty} a_j.$$

Then if A is the operator with anti-Wick symbol $a(z)$, (24.22) follows from (24.23) as required. $\square$

24.3 Theorems on boundedness and compactness

Theorem 24.3. *The operator $A \in G_\varrho^0$ can be extended to a bounded operator on $L^2(\mathbb{R}^n)$.*

Proof. In view of Theorem 24.2 and Corollary 24.2 the statement reduces to the case $A \in G^{-\infty}$, where it is obtained from the obvious estimate

$$\|A\|^2 \leq \iint |K_A(x, y)|^2 \, dx \, dy, \tag{24.24}$$

resulting from the Cauchy-Bunjakovskij-Schwarz inequality (cf. also Appendix 3). $\square$

Theorem 24.4. *The operator $A \in G_\varrho^m$ for $m < 0$ can be extended to a compact operator on $L^2(\mathbb{R}^n)$.*

Proof. First note, that the operators in $G^{-\infty}$ are Hilbert-Schmidt operators (cf. Appendix 3), hence they are compact.

Now let $A \in G_\varrho^m$, $m < 0$. Using Theorem 24.2, we may assume that A has the anti-Wick symbol $a(z) \in \Gamma_\varrho^m(\mathbb{R}^{2n})$. Let $\chi(z) \in C_0^\infty(\mathbb{R}^{2n})$, $\chi(z) = 1$ for $|z| \leq 1$. Put $a_L(z) = \chi(z/L) a(z)$ and let A_L be the operator with the anti-Wick symbol $a_L(z)$. Then $\sup\limits_{z \in \mathbb{R}^{2n}} |a(z) - a_L(z)| \to 0$ as $L \to +\infty$ and in view of Corollary 24.2 we therefore have $\|A - A_L\| \to 0$ as $L \to +\infty$.

But A_L is compact for any L since $A_L \in G^{-\infty}$. From this the compactness of A also follows. $\square$

24.4 Problems

Problem 24.1. Prove the boundedness theorem in the same way as Theorem 6.1 was proved.

Problem 24.2. Prove the compactness theorem 24.4 without using the theory of the anti-Wick symbol.

Hint: Verify with the help of polar decomposition of A, that compactness of A is equivalent to compactness of A^*A; show that compactness of A^*A is equivalent to compactness of $B_N = (A^*A)^N$ and obtain the latter for large N from the fact that $K_{B_N}(x, y) \in L^2(\mathbb{R}^{2n})$.

Problem 24.3. Consider the system of vectors $\{\Phi_z\}_{z \in \mathbb{R}^{2n}}$ defined in 24.1. Show that the map $I: f \to (f, \Phi_z)$ defines an isometric embedding of $L^2(\mathbb{R}^n)$ into $L^2(\mathbb{R}^{2n})$, i.e.

$$\|f\|^2 = (2\pi)^{-n} \int\limits_{\mathbb{R}^{2n}} |(f, \Phi_z)|^2 \, dz \tag{24.25}$$

(one says in this case that the vectors φ_z constitutes an *overcomplete* system with respect to the measure $(2\pi)^{-n} dz$ in the space $\mathbb{R}^{2n}$).

Problem 24.4. Show that an operator A with an anti-Wick symbol $a(z)$ can be written in the form

$$A = I^* M_a I, \tag{24.26}$$

where M_a is the multiplication operator by $a(z)$ in $L^2(\mathbb{R}^{2n})$ and the operator I: $L^2(\mathbb{R}^n) \to L^2(\mathbb{R}^{2n})$ was introduced in Problem 24.3. Derive from this the validity of (24.10) for complex-valued functions $a(z)$.

Problem 24.5. Introduce the Wick symbol $c(z)$ of an operator $A \in G_\varrho^m$ with a Weyl symbol $b(z)$ by the formula

$$c(z) = \pi^{-n} \int e^{-|z - z'|^2} b(z') \, dz'. \tag{24.27}$$

Show that $c(z)$ can be expressed in terms of $b(z)$ via the asymptotic series

$$c(z) \sim \sum_{\alpha} c_\alpha \partial^\alpha b(z) , \qquad (24.28)$$

where $c_0 = 1$ and $c_\alpha = 0$ for odd $|\alpha|$.

Verify that the Wick symbol $c(z)$ of A can be expressed by the formula

$$c(z) = (A\Phi_z, \Phi_z). \qquad (24.29)$$

Derive from this that

$$\sup_{z \in \mathrm{I\!R}^{2n}} |c(z)| \leqq \|A\|.$$

Problem 24.6. Let $a(z) \to 0$ as $|z| \to +\infty$. Show that the operator A, defined by the formula (24.26) is compact.

Problem 24.7. Let A be a compact operator in $L^2(\mathrm{I\!R}^n)$ and let $c(z)$ be defined by the formula (24.29). Show that $c(z) \to 0$ as $|z| \to +\infty$.

Problem 24.8. Let $A \in G_\varrho^m$ for some $m \in \mathrm{I\!R}$ and let $b(z)$ be the Weyl symbol of A. Show that the boundedness of A is equivalent to $\sup_{z \in \mathrm{I\!R}^{2n}} |b(z)| < \infty$.

Hint: Use Corollary 24.2, Problem 24.5 and the construction from the proof of Theorem 24.2.

Problem 24.9. Let A and $b(z)$ be as in the foregoing problem. Show that the compactness of A is equivalent to the condition $b(z) \to 0$ as $|z| \to +\infty$.

Hint: Use Problems 24.6, 24.7 and the construction in the proof of Theorem 24.2.

Problem 24.10. Show that a differential operator with polynomial coefficients in $\mathrm{I\!R}^n$ has a polynomial in $z = (x, \xi)$ as its anti-Wick symbol.

Hint: On polynomials one can solve the inverse heat equation, since the Laplacian is nilpotent on the polynomials of a given degree; if $b(z)$ is a polynomial Weyl symbol of A, then its anti-Wick symbol has the form

$$a(z) = e^{-\frac{1}{4}\Delta} b(z) = \sum_{k=0}^{\infty} \frac{1}{k!} \left(-\frac{\Delta}{4} \right)^k b(z) . \qquad (24.30)$$

Problem 24.11. Compute the coefficients c_α in formula (24.13) for $\tau = \frac{1}{2}$, i.e. in expressing the Weyl symbol through the anti-Wick symbol. Show that this series can be written in the form

$$b(z) \sim e^{\frac{\Delta}{4}} a(z) = \sum_{k=0}^{\infty} \frac{1}{k!} \left(\frac{\Delta}{4} \right)^k a(z) . \qquad (24.31)$$

Hint: In computing the coefficients c_α one may assume that $a(z)$ is a polynomial.

Problem 24.12. Let A be a differential operator with polynomial coefficients on $\mathbb{R}^n$. Show that A may be uniquely expressed in any of the two forms

$$A = \sum_{\alpha,\,\beta} c_{\alpha\beta}\, a^\alpha\, (a^+)^\beta\,, \tag{24.32}$$

$$A = \sum_{\alpha,\,\beta} c'_{\alpha\beta}\, (a^+)^\beta\, a^\alpha\,, \tag{24.33}$$

where $a^+ = \left(x_1 - \dfrac{\partial}{\partial x_1}, \ldots, x_n - \dfrac{\partial}{\partial x_n} \right)$, $a = \left(x_1 + \dfrac{\partial}{\partial x_1}, \ldots, x_n + \dfrac{\partial}{\partial x_n} \right)$, and α, β are n-dimensional multi-indices, the sums (24.32) and (24.33) are finite and $c_{\alpha\beta}$, $c'_{\alpha\beta}$ are complex constants. Show that in this case the anti-Wick symbol $a(x,\xi)$ and the Wick symbol $c(x,\xi)$ of A are given by the formulae

$$a(x,\xi) = \sum_{\alpha,\,\beta} c_{\alpha\beta}\,(x+i\xi)^\alpha\,(x-i\xi)^\beta\,, \tag{24.34}$$

$$c(x,\xi) = \sum_{\alpha,\,\beta} c'_{\alpha\beta}\,(x-i\xi)^\beta\,(x+i\xi)^\alpha\,. \tag{24.35}$$

§25. Hypoellipticity and Parametrix. Sobolev Spaces. The Fredholm Property

25.1 The class of hypoelliptic symbols and operators

Definition 25.1. We shall write $a(z) \in H\Gamma^{m,\,m_0}_\varrho(\mathbb{R}^N)$, if $a(z) \in C^\infty(\mathbb{R}^N)$ and there is an R such that the following estimates hold for $|z| \geq R$

$$C\,|z|^{m_0} \leq |a(z)| \leq C_1\,|z|^m\,, \tag{25.1}$$

$$|\partial^\alpha a(z)| \leq C_\alpha\,|a(z)|\,|z|^{-\varrho|\alpha|}\,, \tag{25.2}$$

where C, C_1, C_α are positive constants.

From this it follows that $a(z) \in \Gamma^m_\varrho(\mathbb{R}^N)$.

This class of symbols has properties close to those of the symbols in §5.

Definition 25.2. By $HG^{m,\,m_0}_\varrho(\mathbb{R}^n)$ or $HG^{m,\,m_0}_\varrho$ we denote the class of operators A, given by the τ-symbols

$$b_\tau \in H\Gamma^{m,\,m_0}_\varrho(\mathbb{R}^{2n})\,.$$

Clearly, $HG^{m,\,m_0}_\varrho \subset G^m_\varrho$.

For the justification of Definition 25.2 we need

Proposition 25.1. *If it is true for some $\tau \in \mathbb{R}$ that $b_\tau \in H\Gamma^{m,\,m_0}_\varrho$, then it is true also for all $\tau \in \mathbb{R}$.*

Proof. The proof is based upon Theorem 23.3 and the following Lemma proved in the same way as Lemmas 5.1–5.3 and Propositions 5.2–5.4.

Lemma 25.1. 1) *The classes* $H\Gamma_\varrho^{m,\,m_0}(\mathbb{R}^N)$ *satisfy:*

a) *if* $a(z) \in H\Gamma_\varrho^{m,\,m_0}$, *then* $a^{-1}(z) \in H\Gamma_\varrho^{-m_0,\,-m}$ $(|z| \geq R)$ *and* $(\partial^\alpha a)/a \in \Gamma_\varrho^{-\varrho|\alpha|}$ $(|z| \geq R)$;

b) *if* $a \in H\Gamma_\varrho^{m,\,m_0}$, $a' \in H\Gamma_\varrho^{m',\,m_0'}$, *then* $aa' \in H\Gamma_\varrho^{m+m',\,m_0+m_0'}$;

c) *if* $a \in H\Gamma_\varrho^{m,\,m_0}$ *and* $r \in \Gamma_\varrho^{m_1}$ *where* $m_1 < m_0$, *then* $a + r \in H\Gamma_\varrho^{m,\,m_0}$.

2) *The classes* $HG^{m,\,m_0}$ *satisfy:*

d) *if* $A \in HG_\varrho^{m,\,m_0}$, $A' \in HG_\varrho^{m',\,m_0'}$, *then* $A \cdot A' \in HG_\varrho^{m+m',\,m_0+m_0'}$;

e) *if* $A \in HG_\varrho^{m,\,m_0}$, *then* ${}^tA \in HG_\varrho^{m,\,m_0}$ *and* $A^* \in HG_\varrho^{m,\,m_0}$;

f) *if* $A \in HG_\varrho^{m,\,m_0}$ *and* $R \in G_\varrho^{m_1}$ *where* $m_1 < m_0$, *then* $A + R \in HG_\varrho^{m,\,m_0}$.

Exercise 25.1. Prove Lemma 25.1.

Exercise 25.2. Prove Proposition 25.1.

25.2 The Parametrix and regularity. By analogy with Theorem 5.1 we can prove

Theorem 25.1. *If* $A \in HG_\varrho^{m,\,m_0}$ *then there is an operator* $B \in HG_\varrho^{-m_0,\,-m}$ *such that*

$$BA = I + R_1, \qquad AB = I + R_2 \tag{25.3}$$

where $R_j \in G^{-\infty}$, $j = 1, 2$. *If* B' *is another operator in* $G_\varrho^{m'}$ *for which either* $B'A - I \in G^{-\infty}$ *or* $AB' - I \in G^{-\infty}$, *then* $B' - B \in G^{-\infty}$.

Exercise 25.3. Prove Theorem 25.1.

Corollary 25.1. *Let* $A \in HG_\varrho^{m,\,m_0}$. *Then the following statements hold*:

a) *If* $u \in S'(\mathbb{R}^n)$ *and* $Au \in S(\mathbb{R}^n)$, *then* $u \in S(\mathbb{R}^n)$.

b) *If* $u \in S'(\mathbb{R}^n)$ *and* $Au \in C_t^\infty(\mathbb{R}^n)$, *then* $u \in C_t^\infty(\mathbb{R}^n)$.

Proof. Follows in an obvious way from Theorem 25.1 and the results in 23.2. $\square$

25.3 Sobolev spaces. Consider for arbitrary $s \in \mathbb{R}$ the operator L_s, differing from the operator with the left symbol $b(x, \xi) = (1 + |\xi|^2 + |x|^2)^{s/2}$ by an operator in $G_1^{s'}$, $s' < s$. It is easily seen that $L_s \in HG_1^{s,\,s}$.

Definition 25.3. The space $Q^s = Q^s(\mathbb{R}^n)$ consists of the distributions $u \in S'(\mathbb{R}^n)$ for which $L_s u \in L^2(\mathbb{R}^n)$.

By analogy with Theorems 7.1 and 7.2 we can prove

Theorem 25.2. 1) *If* $A \in G_\varrho^m$ *then* A *maps*

$$A: \; Q^s \to Q^{s-m}. \tag{25.4}$$

2) *If $A \in HG_\varrho^{m,\,m_0}$, $u \in S'(\mathbb{R}^n)$ and $Au \in Q^s$, then $u \in Q^{s+m_0}$.*

Exercise 25.4. Prove Theorem 25.2.

Since the class $\bigcup_m G_1^m$ contains all differential operators with polynomial coefficients, it is clear that Theorem 25.2 implies

Corollary 25.2.

$$\bigcap_s Q^s = S(\mathbb{R}^n), \qquad \bigcup_s Q^s = S'(\mathbb{R}^n).$$

Now we introduce a Hilbert space structure on Q^s. Note that we may assume that L_{-s} is a parametrix of L_s in the sense of Theorem 25.1. Then

$$L_{-s}L_s = I + R_s, \qquad R_s \in G^{-\infty}.$$

Let $p > s$, p an even integer. Put

$$(u,v)_s = (L_s u, L_s v) + \sum_{|\alpha|+|\beta| \leq p} (x^\alpha D^\beta R_s u,\, x^\alpha D^\beta R_s v). \tag{25.5}$$

From the representation

$$u = L_{-s}L_s u - R_s u \tag{25.6}$$

it is clear that (25.5) defines a pre-Hilbert structure on Q^s. By analogy with Proposition 7.2′ one verifies that

Proposition 25.2. *The scalar product (25.5) defines a Hilbert space structure on Q^s.*

Exercise 25.5. Prove Proposition 25.2.

Finally, by analogy with the argument in §7, one proves the following statements.

Proposition 25.3. *The scalar product $(\cdot,\cdot)$ in $L^2(\mathbb{R}^n)$ induces a duality between Q^s and Q^{-s} (the exact formulation as in Theorem 7.7).*

Proposition 25.4. *The operator $A \in G_\varrho^m$ can be extended to a continuous operator $A: Q^s \to Q^{s-m}$ and to a compact operator $A: Q^s \to Q^{s-m-\varepsilon}$ for $\varepsilon > 0$. The embedding operator $Q^s \hookrightarrow Q^{s-\varepsilon}$, $\varepsilon > 0$, is compact for any $s \in \mathbb{R}$.*

Exercise 25.6. Prove Propositions 25.3 and 25.4.

25.4 The Fredholm property. By analogy with Theorem 8.1 is proved

Proposition 25.5. *If $A \in HG_\varrho^{m,\,m}$, then $A \in \mathrm{Fred}\,(Q^s, Q^{s-m})$ for any $s \in \mathbb{R}$. The space $\mathrm{Im}\,(A\,|_{Q^s})$ in Q^{s-m} is the orthogonal complement to $\mathrm{Ker}\,A^*$ with respect to the scalar product $(\cdot,\cdot)$ in $L^2(\mathbb{R}^n)$.*

Note that

$$\mathrm{Ker}\,(A\,|_{Q^s}) = \mathrm{Ker}\,(A\,|_{S'(\mathbb{R}^n)}) = \mathrm{Ker}\,(A\,|_{S(\mathbb{R}^n)}) \tag{25.7}$$

for any $A \in HG_\varrho^{m,\,m_0}$.

To extend Proposition 25.5 to operators $A \in HG_\varrho^{m,\,m_0}$ (with $m_0 < m$), it is necessary to regard A not as an operator from Q^s into Q^{s-m}, but as an operator in the topological vector spaces $S(\mathbb{R}^n)$, $S'(\mathbb{R}^n)$ and similar spaces or, as an unbounded operator

$$A_{s,\,s'}\colon\ Q^s \to Q^{s'}, \qquad\qquad (25.8)$$

where $s' \geqq s - m_0$, with the domain $D_{A_{s,\,s'}}$ consisting of those $u \in Q^s$ such that $Au \in Q^{s'}$.

Definition 25.4. Let E_1 and E_2 be two topological vector spaces, A an unbounded operator from E_1 into E_2 with the domain D_A. The operator A is called *Fredholm operator* if the following conditions are fulfilled:

a) $\dim \operatorname{Ker} A < +\infty$;

b) $\operatorname{Im} A$ in a closed subspace in E_2;

c) $\dim \operatorname{Coker} A < +\infty$.

Theorem 25.3. 1) *The operator* $A \in HG_\varrho^{m,\,m_0}$ *defines a Fredholm operator from* $S(\mathbb{R}^n)$ *into* $S(\mathbb{R}^n)$ *and from* $S'(\mathbb{R}^n)$ *into* $S'(\mathbb{R}^n)$.

2) *The operators* $A_{s,\,s'}$ *of the form* (25.8) *defined by* A *are, for* $s' \geqq s - m_0$, *also Fredholm operators.*

Remark 25.1. We consider the weak topology in $S'(\mathbb{R}^n)$. Since in Definition 25.4 the topology appears only in b), it is clear that the Fredholm property also holds in all stronger topologies.

Proof of Theorem 25.3. Let the duality between $S(\mathbb{R}^n)$ and $S'(\mathbb{R}^n)$ be given by the extension of the scalar product $(\cdot,\cdot)$ from $L^2(\mathbb{R}^n)$. Note that the finite-dimensionality of $\operatorname{Ker} A$ and $\operatorname{Ker} A^*$ follows from Theorem 25.1 since due to the inclusion

$$\operatorname{Ker} A \subset \operatorname{Ker} BA \subset \operatorname{Ker}(I + R_1)$$

the question reduces to the case $A = I + R_1$ for which everything is obvious. We shall now consider the inclusion

$$A(S'(\mathbb{R}^n)) \supset AB(S'(\mathbb{R}^n)) = (I + R_2)(S'(\mathbb{R}^n)).$$

For the operator $I + R_2$ the Fredholm property on $S'(\mathbb{R}^n)$ follows from Proposition 25.5. Therefore the subspace $A(S'(\mathbb{R}^n))$ is closed in $S'(\mathbb{R}^n)$ and $\operatorname{codim} A(S'(\mathbb{R}^n)) < +\infty$ which proves the Fredholm property of A in $S'(\mathbb{R}^n)$.

Let us prove the Fredholm property of A in $S(\mathbb{R}^n)$. It suffices to verify only conditions b) and c) in Definition 25.4. We shall show that

$$A(S(\mathbb{R}^n)) = \{u\colon u \in S(\mathbb{R}^n),\ u \perp \operatorname{Ker} A^*\}, \qquad (25.9)$$

where orthogonality is in the sense of $L^2(\mathbb{R}^n)$. First, note that

$$A(S'(\mathbb{R}^n)) = \{u\colon u \in S'(\mathbb{R}^n),\ u \perp \operatorname{Ker} A^*\}, \qquad (25.10)$$

since $A(S'(\mathbb{R}^n))$ is closed in $S'(\mathbb{R}^n)$, and $S(\mathbb{R}^n)$ is the dual of $S'(\mathbb{R}^n)$. But now (25.9) follows from (25.10) since

$$A(S(\mathbb{R}^n)) = A(S'(\mathbb{R}^n)) \cap S(\mathbb{R}^n)$$

in view of Corollary 25.1. (25.9) shows the Fredholm property of A on $S(\mathbb{R}^n)$.

Finally let us verify the Fredholm property of $A_{s,s'}$ for $s' \geq s - m_0$. Once again, it only remains to verify that

$$\text{Im } A_{s,s'} = \{u: u \in Q^{s'}, u \perp \text{Ker } A^*\}. \tag{25.11}$$

Let $u \in Q^{s'}$, $u \in (\text{Ker } A^*)^\perp$. Then $u = Av$, where $v \in S'(\mathbb{R}^n)$ in view of the already proven relation (25.10). But from Corollary 25.1 we then obtain that $v \in Q^{s'+m_0}$, i.e. $v \in Q^s$, since $s \leq s' + m_0$. This shows (25.11). $\square$

By analogy with Theorem 8.2 one proves

Theorem 25.4. *Let* $A \in HG_\varrho^{m,m_0}$ *and* $\text{Ker } A = \text{Ker } A^* = \{0\}$. *Then there is an operator* $A^{-1} \in HG_\varrho^{-m_0,-m}$, *which is the inverse to* A.

Exercise 25.7. Prove Theorem 25.4.

Problem 25.1. Show that the operator $A \in HG_\varrho^{m,m_0}$ is Fredholm in the space $C_t^\infty(\mathbb{R}^n)$.

Problem 25.2. Show that if a differential operator A with polynomial coefficients has a τ-symbol $a(z)$ elliptic in $z = (x, \xi)$, then the symbol of its parametrix B has an asymptotic expansion in terms of homogeneous functions in z for $|z| > 1$.

§26. Essential Self-Adjointness. Discreteness of the Spectrum

26.1 Symmetric and self-adjoint operators. Let H_1 and H_2 be Hilbert spaces and suppose we are given an, in general unbounded, operator

$$A: H_1 \rightarrow H_2. \tag{26.1}$$

As usual, D_A denotes the domain of A (it is understood that this domain is given with A, which is then a linear map from the linear subspace D_A into H_2; note that writing (26.1) does not imply that A is defined on all of H_1). The *adjoint operator*

$$A^*: H_2 \rightarrow H_1 \tag{26.2}$$

is *defined* if D_A is dense in H_1 and, in this case, D_{A^*} consists of all $v \in H_2$, for which there exists a vector $g \in H_1$ with

$$(Au, v) = (u, g), \quad u \in D_A, \tag{26.3}$$

(on the left-hand side of (26.3) is the scalar product in H_2 and on the right-hand side that in H_1). It is clear that g is uniquely defined and by definition $A^*v = g$. In particular, we have the identity

$$(Au, v) = (u, A^*v), \quad u \in D_A, \quad v \in D_{A^*}. \tag{26.4}$$

Definition 26.1. Let H be a Hilbert space. An operator $A: H \to H$ is called *symmetric* if

$$(Au, v) = (u, Av), \quad u, v \in D_A \tag{26.5}$$

Definition 26.2. An operator $A: H \to H$ is called *self-adjoint* if $A = A^*$.

It is obvious that a self-adjoint operator is symmetric. The converse is in general not true.

Definition 26.3. An operator $A: H_1 \to H_2$ is called *closed*, if the graph G_A, consisting of all pairs $\{u, Au\}$, where $u \in D_A$, is a closed subspace in $H_1 \oplus H_2$.

Exercise 26.1. Show that if A^* is defined, then it is closed.

Exercise 26.2. Let an operator A be bounded, i.e. there exists a constant $C > 0$, such that $\|Au\| \le C\|u\|$, $u \in D_A$. Show that A is closed if and only if D_A is a closed subspace of H_1.

The well-known closed-graph theorem (cf. Rudin [1]) states that if $D_A = H_1$ and A is closed, then A is bounded. Obviously the same holds if D_A is a closed subspace in H_1.

Let an operator $A: H_1 \to H_2$ be given. We say that A has a *closure* $\overline{A}$, if the closure $\overline{G}_A$ of the graph G_A is again the graph of (closed) operator, which we denote by $\overline{A}$. In particular, any symmetric operator $A: H \to H$ has a closure if D_A is dense. Indeed, it is enough to verify, that if u_n is a sequence of vectors in D_A, such that $\lim_{n \to \infty} u_n = 0$ *and* $\lim_{n \to \infty} Au_n = f$, then $f = 0$. But for $v \in D_A$ we obtain

$$(f, v) = \lim_{n \to \infty} (Au_n, v) = \lim_{n \to \infty} (u_n, Av) = 0,$$

from which we have $f = 0$. Note that if A is a symmetric operator, then so is $\overline{A}$.

Definition 26.4. An operator $A: H \to H$ is called *essentially self-adjoint* if D_A is dense in H and $\overline{A} = A^*$.

In particular, A^* is then an extension of A and, hence, A is symmetric.
A criterion for essential self-adjointness is given by

Theorem 26.1. *A symmetric operator $A: H \to H$ with dense domain is essentially self-adjoint if and only if the following inclusions hold*

$$\operatorname{Ker}(A^* - iI) \subset D_{\overline{A}}, \tag{26.6}$$

$$\operatorname{Ker}(A^* + iI) \subset D_{\overline{A}}. \tag{26.6'}$$

Proof. 1. The necessity of (26.6) and (26.6′) is obvious. To verify their sufficiency, let us first note that since A^* is an extension of $\overline{A}$, it follows from (26.6) that $\operatorname{Ker}(A^* - iI) = \operatorname{Ker}(\overline{A} - iI)$. But $\operatorname{Ker}(\overline{A} - iI) = 0$ since $\overline{A}$ is symmetric. Therefore, from (26.6) it follows that

$$\operatorname{Ker}(A^* - iI) = 0. \tag{26.7}$$

Similarly, from (26.6′) we find that

$$\operatorname{Ker}(A^* + iI) = 0 \tag{26.7′}$$

2. Let us now verify that $(\overline{A} - iI)^{-1}$ (defined on $(\overline{A} - iI)(H)$) is bounded. We have

$$\|(\overline{A} - iI)f\|^2 = ((\overline{A} - iI)f, (\overline{A} - iI)f) = \|Af\|^2 + \|f\|^2, \tag{26.8}$$

since $(\overline{A}f, f)$ is a real number in view of the fact that $\overline{A}$ is symmetric. It follows from (26.8) that $\|f\|^2 \leq \|(\overline{A} - iI)f\|^2$, i.e.

$$\|(\overline{A} - iI)^{-1}g\| \leq \|g\|, \qquad g \in (\overline{A} - iI)(H).$$

3. It is clear that $\overline{A} - iI$ is closed. Therefore $(\overline{A} - iI)^{-1}$ is also closed and since $(\overline{A} - iI)^{-1}$ is bounded, its domain $(\overline{A} - iI)(H)$ is closed in H. However the orthogonal complement of $(\overline{A} - iI)(H)$ is obviously equal to $\operatorname{Ker}(\overline{A} - iI)^* = \operatorname{Ker}(A^* + iI) = 0$. Therefore $(\overline{A} - iI)^{-1}$ is everywhere defined. By similar reasoning, $(\overline{A} + iI)^{-1}$ is also everywhere defined.

4. Let us verify that $(\overline{A} - iI)^{-1}$ and $(\overline{A} + iI)^{-1}$ are adjoint to each other. We obviously have

$$((\overline{A} - iI)u, v) = (u, (\overline{A} + iI)v), \qquad u, v \in D_{\overline{A}}.$$

Denoting $(\overline{A} - iI)u = f$ and $(\overline{A} + iI)v = g$, we obtain the required relation

$$(f, (A + iI)^{-1}g) = ((A - iI)^{-1}f, g), \qquad f, g \in H.$$

5. Let us finally verify that $\overline{A} = A^*$. We will use the following easily verified fact: if B is an operator in H, such that $(B^{-1})^*$ and $(B^*)^{-1}$ are defined, then $(B^{-1})^* = (B^*)^{-1}$. We have

$$A^* = \overline{A}^* = (\overline{A} + iI)^* + iI = \{[(\overline{A} + iI)^{-1}]^{-1}\}^* + iI$$
$$= \{[(\overline{A} + iI)^{-1}]^*\}^{-1} + iI = [(\overline{A} - iI)^{-1}]^{-1} + iI = \overline{A} - iI + iI = \overline{A},$$

as required. $\square$

26.2 Essential self-adjointness of hypoelliptic symmetric operators. In this section we shall denote by A^+ the operator which is formally adjoint to an operator $A \in G_\varrho^m$, i.e. the operator $A^+ \in G_\varrho^m$, such that

$$(Au, v) = (u, A^+v), \qquad u, \ v \in C_0^\infty(\mathbb{R}^n).$$

In the preceding sections we have written A^* instead of A^+, but here the notation A^* will be reserved for the adjoint operator in the sense of section 26.1.

Theorem 26.2. *Let* $A \in HG_\varrho^{m, m_0}$, *where* $m_0 > 0$ *and* $A^+ = A$. *In* $L^2(\mathbb{R}^n)$ *consider the unbounded operator* A_0, *defined as the operator* A *on the domain* $C_0^\infty(\mathbb{R}^n)$. *Then* A_0 *is essentially self-adjoint and its closure coincides with the restriction of the operator* A *(defined on* $S'(\mathbb{R}^n)$*) to the set*

$$D_{\bar{A}_0} = \{u : u \in L^2(\mathbb{R}^n), \ Au \in L^2(\mathbb{R}^n)\}. \tag{26.9}$$

Proof. 1. Denote by D the right-hand side of (26.9). Since

$$(Au, v) = (u, Av), \qquad u \in S(\mathbb{R}^n), \qquad v \in S'(\mathbb{R}^n), \tag{26.10}$$

it is clear that $D \subset D_{A_0^*}$ and in addition

$$A|_D = A_0^*|_D.$$

Let us verify that indeed $D = D_{A_0^*}$. Let $v \in D_{A_0^*}$, i.e. $v \in L^2(\mathbb{R}^n)$ and for some $f \in L^2(\mathbb{R}^n)$ the identity

$$(Au, v) = (u, f), \quad u \in C_0^\infty(\mathbb{R}^n), \tag{26.11}$$

holds. But it follows from (26.10) that the same identity holds if we replace f by Av. Therefore $Av = f$, i.e. $v \in D$ as required. Thus we have demonstrated that the right-hand side of (26.9) equals $D_{A_0^*}$.

2. In order to now use Theorem 26.1, we will verify the inclusion

$$\mathrm{Ker}\,(A_0^* - iI) \subset D_{\bar{A}_0}. \tag{26.12}$$

From what we have already shown, it is clear that

$$\mathrm{Ker}\,(A_0^* - iI) = \{u : u \in L^2(\mathbb{R}^n), \ (A - iI)u = 0\}.$$

Taking into account that $A - iI \in HG_\varrho^{m, m_0}$, it follows from Corollary 25.1 that $\mathrm{Ker}\,(A_0^* - iI) \subset S(\mathbb{R}^n)$, from which (26.12) follows, since A maps $S(\mathbb{R}^n)$ into $S(\mathbb{R}^n)$ continuously and $C_0^\infty(\mathbb{R}^n)$ is dense in $S(\mathbb{R}^n)$. Similarly one proves the inclusion $\mathrm{Ker}\,(A_0^* + iI) \subset D_{\bar{A}_0}$, which concludes the proof of Theorem 26.2. $\quad\square$

26.3 Discreteness of the spectrum

Theorem 26.3. *Let $A \in HG_\varrho^{m, m_0}$, where $m_0 > 0$ and $A^+ = A$. Then A has discrete spectrum in $L^2(\mathbb{R}^n)$. More precisely, there exists an orthonormal basis of eigenfunctions $\varphi_j(x) \in S(\mathbb{R}^n)$, $j = 1, 2, \ldots$, with eigenvalues $\lambda_j \in \mathbb{R}$, such that $|\lambda_j| \to +\infty$ as $j \to +\infty$. The spectrum $\sigma(A)$ of $\overline{A} = A^*$ in $L^2(\mathbb{R}^n)$ coincides with the set of all eigenvalues $\{\lambda_j\}$.*

Proof. The proof is similar to that of Theorem 8.3. In view of the separability of $L^2(\mathbb{R}^n)$, there exists a number $\lambda_0 \in \mathbb{R} \setminus \sigma(A)$. But then Theorem 25.4 implies that $(A - \lambda_0 I)^{-1} \in HG_\varrho^{-m_0, -m}$ and, in particular we see that $(A - \lambda_0 I)^{-1}$ is compact and self-adjoint in $L^2(\mathbb{R}^n)$. The remainder of the proof is a verbatim repetition of the proof of Theorem 8.3. $\square$

Problem 26.1. Let $A \in G_\varrho^m$ be such that there are numbers $\lambda_\pm \in \mathbb{C}$, such that $\operatorname{Im} \lambda_+ > 0$, $\operatorname{Im} \lambda_- < 0$ and

$$(A - \lambda_+ I) \in HG_\varrho^{m, m_0} \quad \text{and} \quad (A - \lambda_- I) \in HG_\varrho^{m, m_0}$$

for some $m_0 \in \mathbb{R}$. Show that if $A^+ = A$, then A is essentially self-adjoint.

Problem 26.2. Let H_1, H_2 be Hilbert spaces and let $A: H_1 \to H_2$, and $A^+: H_2 \to H_1$, be such that

$$(Au, v) = (u, A^+ v), \quad u \in D_A, \quad v \in D_{A^+}.$$

Show that if $A^+ A$ is essentially self-adjoint, then $\overline{A}^+ = A^*$ and $\overline{A} = (A^+)^*$.

Hint: Consider in $H_1 \oplus H_2$ the operator defined as the matrix

$$\mathfrak{A} = \begin{pmatrix} 0 & A^+ \\ A & 0 \end{pmatrix}.$$

Then the conditions $\overline{A}^+ = A^*$ and $\overline{A} = (A^+)^*$ are equivalent to the essential self-adjointness of the operator $\mathfrak{A}$. Compute $\operatorname{Ker}(\mathfrak{A} + iI)$.

Problem 26.3. Let $A \in HG_\varrho^{m, m_0}$, $m_0 > 0$. Denote by A_0^+ the operator A^+, restricted to $C_0^\infty(\mathbb{R}^n)$. Show that $\overline{A}_0^+ = A_0^*$ and $\overline{A}_0 = (A_0^+)^*$.

Hint: Use the result of Problem 26.2, after extending the operators A_0 and A_0^+ to $S(\mathbb{R}^n)$.

Remark 26.1. The result in Problem 26.3 means that the "strong and weak extensions coincide" for an operator $A \in HG_\varrho^{m, m_0}$ for $m_0 > 0$: if $u \in L^2(\mathbb{R}^n)$ and $Au \in L^2(\mathbb{R}^n)$, then there exists a sequence $u_j \in C_0^\infty(\mathbb{R}^n)$ such that $u_j \to u$ and $Au_j \to Au$ as $j \to +\infty$ in the $L^2(\mathbb{R}^n)$-norm.

Problem 26.4. Prove analogue of Theorem 8.4 on the structure of the spectrum, eigenfunctions and associated functions for operators $A \in HG_\varrho^{m, m_0}$, $m_0 > 0$.

Problem 26.5. Let an operator A have the anti-Wick symbol $a(z) \in \Gamma_\varrho^m(\mathbb{R}^{2n})$ and let $\overline{A}_0$ be self-adjoint in $L^2(\mathbb{R}^n)$. Let $a(z) \to +\infty$ as $|z| \to \infty$. Show that $\overline{A}_0$ has discrete spectrum in the sense of Theorem 26.3 and that $\lambda_j \to +\infty$ as $j \to +\infty$.

Problem 26.6. Let $A \in G_\varrho^m$ be such that $\overline{A}_0$ is self-adjoint in $L^2(\mathbb{R}^n)$ and has discrete spectrum such that $\lambda_j \to +\infty$ as $j \to +\infty$. Let $c(z)$ be the Wick symbol of the operator A. Show that $c(z) \to +\infty$ as $|z| \to +\infty$.

§27. Trace and Trace Class Norm

27.1 The trace and the Hilbert-Schmidt norm expressed in terms of the symbol. Here we make use of notations and facts concerning Hilbert-Schmidt and trace class operators which are presented in Appendix 3.

Let us begin with the formal expression for the trace in terms of the τ-symbol. Let $A \in G_\varrho^m$, let $b_\tau(x, \xi)$ be the τ-symbol of A and K_A its kernel. We have formally

$$K_A(x, y) = \int e^{i(x-y)\cdot\xi} \, b_\tau((1-\tau)x + \tau y, \xi) \, d\xi, \tag{27.1}$$

from which

$$K_A(x, x) = \int b_\tau(x, \xi) \, d\xi$$

and

$$\operatorname{Sp} A = \int b_\tau(x, \xi) \, dx \, d\xi. \tag{27.2}$$

Note that (27.2) means in particular, that its right-hand side is independent of τ.

Proposition A.3.2 yields

$$\|A\|_2^2 = \int |K_A(x, y)|^2 \, dx \, dy = \int |K_A(x, x+z)|^2 \, dx \, dz. \tag{27.3}$$

But by (27.1)

$$K_A(x, x+z) = \int e^{-iz\cdot\xi} \, b_\tau(x + \tau z, \xi) \, d\xi. \tag{27.4}$$

Therefore we have formally

$$\int |K_A(x, x+z)|^2 \, dx \, dz = \int K_A(x, x+z) \, \overline{K_A(x, x+z)} \, dx \, dz$$

$$= \int e^{iz\cdot(\eta-\xi)} b_\tau(x + \tau z, \xi) \, \overline{b_\tau(x + \tau z, \eta)} \, d\xi \, d\eta \, dx \, dz$$

$$= \int e^{iz\cdot(\eta-\xi)} b_\tau(x, \xi) \, \overline{b_\tau(x, \eta)} \, d\xi \, d\eta \, dx \, dz$$

$$= \int |\int e^{-iz\cdot\xi} b_\tau(x, \xi) \, d\xi|^2 \, dx \, dz = \int |b_\tau(x, \xi)|^2 \, d\xi \, dx \tag{27.5}$$

(we have here used the shift invariance of the integral and the Parceval identity for the Fourier transform). As a result we obtain

$$\|A\|_2^2 = \int |b_\tau(x,\xi)|^2 \, dx \, d\xi, \tag{27.6}$$

where again the right-hand side is independent of τ

Proposition 27.1. *The correspondence between operators $A \in G^{-\infty}$ and τ-symbols $b_\tau(x,\xi) \in S(\mathbb{R}^{2n})$ extends by continuity to an isometry between $S_2(L^2(\mathbb{R}^n))$ and $L^2(\mathbb{R}^n)$ such that (27.6) holds. If $A \in G_\varrho^m$, then the condition $A \in S_2(L^2(\mathbb{R}^n))$ is equivalent to $b_\tau \in L^2(\mathbb{R}^{2n})$ for some τ and this then holds for all τ and the formula (27.6) also holds in this situation.*

Proof. The computations (27.3)–(27.6) are justified for $A \in G^{-\infty}$ or, what is the same, for $K_A \in S(\mathbb{R}^{2n})$. Since $G^{-\infty}$ is dense in $S_2(L^2(\mathbb{R}^n))$ and $S(\mathbb{R}^{2n})$ is dense in $L^2(\mathbb{R}^{2n})$, the existence and uniqueness of the required isometry is obvious. Finally, the last statement is obvious from the uniqueness of the τ-symbol. $\square$

Corollary 27.1. *If $A \in G_\varrho^m$ and $m < -n$, then $A \in S_2(L^2(\mathbb{R}^n))$.*

Proposition 27.2. *1) If $A \in G_\varrho^m$ and $m < -2n$, then $A \in S_1(L^2(\mathbb{R}^n))$ and for any fixed $m < -2n$ and $\tau \in \mathbb{R}$ there exist constants C and N, such that the following estimate holds*

$$\|A\|_1 \leqq C \sum_{|\gamma| \leqq N} \sup_z \left\{ |\partial_z^\gamma b_\tau(z)| \langle z \rangle^{-m + \varrho |\gamma|} \right\}. \tag{27.7}$$

2) For $A \in G_\varrho^m$, $m < -2n$, formula (27.2) for the trace $\mathrm{Sp}\, A$ holds for any $\tau \in \mathbb{R}$.

Proof. 1) Choose an operator $P \in HG_\varrho^{m/2, m/2}$ for $\mathrm{Ker}\, P = \mathrm{Ker}\, P^* = 0$, so that $P^{-1} \in HG_\varrho^{-m/2, -m/2}$ exists (the existence of an operator P of this type follows, for instance, from Theorem 26.3). In view of Corollary 27.1, we have $P^2 \in S_1(L^2(\mathbb{R}^n))$. But from the obvious representation $A = P^2(P^{-2}A)$ and the fact that $P^{-2}A \in G_\varrho^0 \subset \mathscr{L}(L^2(\mathbb{R}^n))$, it follows that $A \in S_1(L^2(\mathbb{R}^n))$. Therefore the inclusion

$$G_\varrho^m \subset S_1(L^2(\mathbb{R}^n)), \qquad m < -2n. \tag{27.8}$$

is proved.

Let us now prove (27.7). It can be obtained in two ways: either by a direct sharpening of the arguments carried out so far (from similar estimates in the composition formula and the boundedness theorem) or from the closed graph theorem. The latter route is shorter and is commonplace for many argument of this type, although it is also rougher. We will carry out carefully the corresponding arguments.

Introduce in G_ϱ^m a Fréchet topology, defined by semi-norms of the form

$$\|A\|_{(N)} = \sum_{|\gamma| \leqq N} \sup_z \left\{ |\partial_z^\gamma b_\tau(z)| \langle z \rangle^{-m + \varrho |\gamma|} \right\}. \tag{27.9}$$

We have to show that the embedding (27.8) is continuous in the natural Banach space topology on $S_1(L^2(\mathbb{R}^n))$. In view of the closed graph theorem (cf. e.g. Rudin [1]) it is only necessary to show that this embedding has a closed graph. This is most conveniently proved constructing a Hausdorff space M such that

$$G_\varrho^m \subset S_1(L^2(\mathbb{R}^n)) \subset M \tag{27.10}$$

where both embeddings $G_\varrho^m \subset M$ and $S_1(L^2(\mathbb{R}^n)) \subset M$ are continuous. Now as M we may, for instance, take $S_2(L^2(\mathbb{R}^n))$, since the continuity of the embedding of G_ϱ^m and $S_1(L^2(\mathbb{R}^n))$ in $S_2(L^2(\mathbb{R}^n))$ follows immediately from Propositions 27.1 (formula (27.6)) and A.3.7 (estimate (A.3.29)).

2) Now we will prove (27.2) for $A \in G_\varrho^m$, $m < -2n$. Note that both its parts are continuous on G_ϱ^m. But for any $m' > m$, $G^{-\infty}$ is dense in G_ϱ^m in the topology of $G_\varrho^{m'}$. Therefore, it suffices to prove (27.2) for $A \in G^{-\infty}$.

We would like to carry out carefully the argument from A.3.5. This is trivial, if we present A in the form $A = L_1 \circ L_2$, where the operators L_1 and L_2 have kernels with enough continuous and rapidly decreasing derivatives. But the latter representation can be constructed by an argument similar to the one used in 1) of this proof. $\square$

27.2 A more precise estimate of the trace class norm in terms of the τ-symbol. The estimate (27.7) is not very convenient, since it contains a weight-function increasing in z. At the same time, we see that $\|A\|_1$ does not change if we shift the τ-symbol by some vector $z_0 = (x_0, \xi_0) \in \mathbb{R}^{2n}$. Indeed, if $b_\tau(x, \xi)$ is the τ-symbol of A and if we denote by A_{z_0} the operator with the τ-symbol $b_\tau(x - x_0, \xi - \xi_0)$, then we obtain

$$
\begin{aligned}
A_{z_0} u(x) &= \int e^{i(x-y)\cdot\xi} b_\tau((1-\tau)x + \tau y - x_0, \xi - \xi_0) u(y)\, dy\, d\xi \\
&= \int e^{i[(x-x_0)-(y-x_0)]\cdot\xi} b_\tau((1-\tau)(x-x_0) \\
&\quad + \tau(y-x_0), \xi - \xi_0)\, u((y-x_0) + x_0)\, dy\, d\xi \\
&= \int e^{i\xi_0\cdot(x'-y')} e^{i(x'-y')\cdot\xi'} b_\tau((1-\tau)x' + \tau y', \xi')\, u(y' + x_0)\, dy'\, d\xi',
\end{aligned}
$$

where $x' = x - x_0$, $y' = y - y_0$, $\xi' = \xi - \xi_0$. Denote by U the unitary operator, mapping $u(x)$ into $(Uu)(x) = e^{-i\xi_0\cdot x} u(x + x_0)$, then we see that $A_{z_0} = U^{-1} A U$, from which

$$\|A_{z_0}\|_1 = \|A\|_1. \tag{27.11}$$

An estimate of the trace class norm which is invariant relatively to the shifts of the τ-symbol is given by the following

Proposition 27.3. *There exist constants C and N, such that for $A \in G_\varrho^m$, $m < -2n$, the following estimate holds*

$$\|A\|_1 \leqq C \sum_{|\gamma| \leqq N} \int |\partial_z^\gamma b_\tau(z)|\, dz. \tag{27.12}$$

Proof. It suffices to show the estimate (27.12) in the case where $b_\tau(z) \in C_0^\infty(\mathbb{R}^{2n})$. First let,

$$\operatorname{supp} b_\tau \subset \{z: |z| \leq R_0\}, \tag{27.13}$$

where R_0 is some fixed constant. Then it follows from Proposition 27.2 that there are constants C_1 and M (depending on R_0) such that

$$\|A\|_1 \leq C_1 \sum_{|\gamma| \leq M} \sup_z |\partial_z^\gamma b_\tau(z)|. \tag{27.14}$$

But since for $b(z) \in C_0^\infty(\mathbb{R}^{2n})$ we have

$$b(z) = \int_{\substack{t_j \leq z_j \\ j=1, \ldots, 2n}} \frac{\partial^{2n} b}{\partial z_1 \ldots \partial z_{2n}}(t_1, \ldots, t_{2n}) \, dt_1 \ldots dt_{2n}$$

and consequently

$$\sup_z |b(z)| \leq \int \left| \frac{\partial^{2n} b(z)}{\partial z_1 \ldots \partial z_{2n}} \right| dz,$$

it follows from (27.14) that we have (27.12) (with $N = M + 2n$), provided that (27.13) is satisfied.

Now, using the invariance of the trace class norm (formula (27.11)) and the invariance of the right-hand side of (27.12) with respect to shift in the argument of $b(z)$, we see that (27.12) always holds, with the same constants C and N, provided

$$\operatorname{diam} \operatorname{supp} b_\tau \leq R_0 \tag{27.15}$$

Let us finally get rid of the condition (27.15). Take a partition of unity

$$1 \equiv \sum_{j=1}^\infty \varphi_j(z)$$

such that $\operatorname{diam} \operatorname{supp} \varphi_j \leq R_0$, there is a number l such that any ball of unit radius does not intersect more than l sets $\operatorname{supp} \varphi_j$, and, in addition, $|\partial^\gamma \varphi_j(z)| \leq C_\gamma$, $j = 1, 2, \ldots$, with constants C_γ not depending on j. Introducing the operators A_j with the τ-symbols $\varphi_j(z) b_\tau(z)$, we obtain

$$\|A\|_1 \leq \sum_{j=1}^\infty \|A_j\|_1 \leq C \sum_{j=1}^\infty \sum_{|\gamma| \leq N} \int |\partial_z^\gamma(\varphi_j b_\tau)(z)| \, dz$$

$$\leq C_1 \sum_{|\gamma| \leq N} \int |\partial_z^\gamma b_\tau(z)| \, dz,$$

as required. $\quad\square$

Problem 27.1. Let A be determined by the anti-Wick symbol $a(z) \in \Gamma_\varrho^m$. Show that

$$\|A\|_1 \leq \int |a(x, \xi)| \, dx \, d\xi .$$

Problem 27.2. Let $c(z)$ be the Wick symbol of $A \in G_\varrho^m$. Show that

$$\int |c(x, \xi)| \, dx \, d\xi \leq \|A\|_1 .$$

Hint: Use the polar decomposition of A and the result of Problem 24.5.

Problem 27.3. Let A have the anti-Wick symbol $a(z) \in \Gamma_\varrho^m$, $m < -2n$. Show that

$$\mathrm{Sp}\, A = \int a(x, \xi) \, dx \, d\xi .$$

Problem 27.4. Let $A \in G_\varrho^m$, $m < -2n$, and let $c(z)$ be the Wick symbol of A. Show that

$$\mathrm{Sp}\, A = \int c(x, \xi) \, dx \, d\xi .$$

Problem 27.5. Let M be a closed n-manifold and let $A \in L_{\varrho, \delta}^m(M)$, $1 - \varrho \leq \delta < \varrho$. Show that if $m < -n/2$, then $A \in S_2(L^2(M))$ and if $m < -n$, then $A \in S_1(L^2(M))$.

§28. The Approximate Spectral Projection

28.1 The Glazman lemma. In this paragraph we shall describe an abstract scheme of yet another method to obtain the asymptotic behaviour of eigenvalues which is based on the construction of an approximate spectral projection. At the basis on this method lies the following well-known variational lemma of Glazman.

Lemma 28.1. *Let A be a self-adjoint, semi-bounded from below operator with discrete spectrum in a Hilbert space H, i.e. A has an orthonormal basis of eigenvectors $e_1, e_2, \ldots$ with eigenvalues $\lambda_1, \lambda_2, \ldots$, such that $\lambda_k \to +\infty$ as $k \to +\infty$.*

Let $N(\lambda)$ be the number of eigenvalues of A not exceeding λ (multiplicities counted). Then

$$N(\lambda) = \sup_{\substack{L \subset D_A \\ (Au, u) \leq \lambda(u, u), u \in L}} \dim L \tag{28.1}$$

(L is a linear subspace of D_A).

Proof. Let E_λ be the spectral projection of A. Since

$$\dim(E_\lambda H) = \operatorname{Sp} E_\lambda = N(\lambda), \tag{28.2}$$

then, putting $L = E_\lambda H$, we see that the right-hand side of (28.1) cannot be smaller than the left-hand side.

Next we show that the right-hand side of (28.1) cannot exceed the left-hand side.

Let the linear subspace $L \subset D_A$ be such that

$$(Au, u) \leqq \lambda(u, u), \quad u \in L. \tag{28.3}$$

Since

$$(Au, u) > \lambda(u, u), \quad u \in (I - E_\lambda) H \setminus \{0\}, \tag{28.4}$$

it follows from (28.3) that

$$L \cap (I - E_\lambda) H = 0.$$

But then E_λ is injective as a mapping of L into $E_\lambda H$, from which it follows that

$$\dim L \leqq N(\lambda),$$

as required. $\square$

28.2 Properties of the spectral projections. The spectral projections operators enjoy the following properties:

1) $E_\lambda^* = E_\lambda$;
2) $E_\lambda^2 = E_\lambda$;
3) $E_\lambda(A - \lambda I) E_\lambda \leqq 0$;
4) $(I - E_\lambda)(A - \lambda I)(I - E_\lambda) > 0$

(meaning that the corresponding quadratic form is strictly greater than 0 on the non-zero vectors in D_A);

5) $\operatorname{Sp} E_\lambda = N(\lambda)$.

Basic in what follows is

Proposition 28.1. *Let E_λ' be a family of operators, for which $E_\lambda' H \subset D_A$ and which satisfies conditions* 1)–4). *Then* 5) *is also fulfils, i.e.*

$$\operatorname{Sp} E_\lambda' = N(\lambda) = \operatorname{Sp} E_\lambda. \tag{28.5}$$

Proof. It follows from 1) and 2) that E_λ' is an orthogonal projection. Putting $L_\lambda = E_\lambda' H$, $M_\lambda = (I - E_\lambda') H$, we have, in view of 3), that $(Au, u) \leqq \lambda(u, u)$, $u \in L_\lambda$, from which, by Lemma 28.1, it follows that $\operatorname{Sp} E_\lambda' = \dim L_\lambda \leqq N(\lambda)$. Further, from 4), we have $(E_\lambda H) \cap M_\lambda = 0$, which implies $\dim(E_\lambda H) = N(\lambda) \leqq \dim L_\lambda = \operatorname{Sp} E_\lambda'$, proving (28.5). $\square$

Remark 28.1. Note that under the conditions of Proposition 28.1 we do not necessarily have $E'_\lambda = E_\lambda$ (cf. Problem 28.1).

28.3 Approximate spectral projection operator

Theorem 28.1. *Let A be an operator as in Lemma 28.1 and $\{\mathscr{E}_\lambda\}_{\lambda \in \mathbb{R}}$ a family of operators such that $\mathscr{E}_\lambda H \subset D_A$ and that for some $\varepsilon > 0$, $\delta > 0$, we have:*

$1°$. $\mathscr{E}_\lambda^* = \mathscr{E}_\lambda$;

$2°$. *$\mathscr{E}_\lambda$ is a trace class operator and*

$$\| \mathscr{E}_\lambda^2 - \mathscr{E}_\lambda \|_1 = O(V(\lambda) \cdot \lambda^{-\delta}) \qquad as \qquad \lambda \to +\infty, \tag{28.6}$$

where $V(\lambda)$ is some positive, non-decreasing function, defined for $\lambda \geq \lambda_0$;

$3°$. $\mathscr{E}_\lambda (A - \lambda I) \mathscr{E}_\lambda \leq C\lambda^{1-\varepsilon}$;

$4°$. $(I - \mathscr{E}_\lambda)(A - \lambda I)(I - \mathscr{E}_\lambda) \geq -C\lambda^{1-\varepsilon}$;

$5°$. $\operatorname{Sp} \mathscr{E}_\lambda = V(\lambda)(1 + 0(\lambda^{-\delta}))$ *as* $\lambda \to +\infty$.

Let us also assume that the function $V(\lambda)$ appearing in $2°$ and $5°$, is such that

$$[V(\lambda + C\lambda^{1-\varepsilon}) - V(\lambda)]/V(\lambda) = O(\lambda^{-\delta}) \qquad as \qquad \lambda \to +\infty \tag{28.7}$$

for some $C > 0$. Then we have

$$N(\lambda) = V(\lambda)(1 + O(\lambda^{-\delta})) \qquad as \qquad \lambda \to +\infty. \tag{28.8}$$

Proof. The idea is to apply Lemma 28.1 to the linear subspace L, spanned by the eigenvectors of $\mathscr{E}_\lambda$, having eigenvalues close to 1 (they are all close to either 1 or 0, as we shall see later).

Let α_j be eigenvalues of $\mathscr{E}$. They are real by $1°$ and by $2°$ and $5°$ satisfy the conditions

$$\sum_j |\alpha_j^2 - \alpha_j| = O(\lambda^{-\delta} V(\lambda)),$$

$$\sum_j \alpha_j = V(\lambda)(1 + O(\lambda^{-\delta})). \tag{28.9}$$

Lemma 28.2.

$$\sum_{|\alpha_j - 1| \leq 1/2} \alpha_j = V(\lambda)(1 + O(\lambda^{-\delta})). \tag{28.10}$$

Proof. For $|\alpha_j - 1| > \frac{1}{2}$ we have $|\alpha_j^2 - \alpha_j| = |\alpha_j| \, |\alpha_j - 1| \geq \frac{1}{2} |\alpha_j|$. Therefore

$$\sum_{|\alpha_j - 1| > 1/2} |\alpha_j| \leq 2 \sum_{|\alpha_j - 1| > 1/2} |\alpha_j^2 - \alpha_j| \leq 2 \sum_j |\alpha_j^2 - \alpha_j| = O(\lambda^{-\delta} V(\lambda)),$$

which together with (28.9) implies (28.10). $\square$

Lemma 28.3. *Let $\tilde{N}(\lambda)$ be the number of eigenvalues of $\mathscr{E}_\lambda$ in the interval $[\frac{1}{2}, 3/2]$. Then*

$$\tilde{N}(\lambda) = V(\lambda)\,(1 + O(\lambda^{-\delta})). \tag{28.11}$$

Proof. Put $\varepsilon_j = 1 - \alpha_j$. Then $2°$ can be rewritten as

$$\sum_j |\varepsilon_j^2 - \varepsilon_j| = O(\lambda^{-\delta} V(\lambda)), \tag{28.12}$$

and the statement of Lemma 28.2 gives that

$$\sum_{|\varepsilon_j| \leq 1/2} (1 - \varepsilon_j) = V(\lambda)\,(1 + O(\lambda^{-\delta})),$$

or

$$\tilde{N}(\lambda) = V(\lambda)\,(1 + O(\lambda^{-\delta})) + \sum_{|\varepsilon_j| \leq 1/2} \varepsilon_j.$$

But, as in Lemma 28.2, it follows from (28.12) that

$$\sum_{|\varepsilon_j| \leq 1/2} |\varepsilon_j| \leq 2 \sum_{|\varepsilon_j| \leq 1/2} |\varepsilon_j^2 - \varepsilon_j| = O(\lambda^{-\delta} V(\lambda)),$$

giving also (28.11). $\square$

Let us continue the proof of Theorem 28.1.

a) Let L_λ be the linear manifold spanned by the eigenvectors of $\mathscr{E}_\lambda$ with eigenvalues α_j such that $|\alpha_j - 1| \leq \frac{1}{2}$, so that

$$\tilde{N}(\lambda) = \dim L_\lambda = V(\lambda)\,(1 + O(\lambda^{-\delta})) \tag{28.13}$$

by Lemma 28.3. Condition $3°$ implies that

$$(\mathscr{E}_\lambda (A - \lambda I)\,\mathscr{E}_\lambda u, u) \leq C\lambda^{1-\varepsilon} (u, u), \qquad u \in H. \tag{28.14}$$

But since

$$(u, u) \leq 4\,(\mathscr{E}_\lambda u, \mathscr{E}_\lambda u), \qquad u \in L_\lambda, \tag{28.15}$$

it follows from (28.14) that

$$([A - (\lambda + 4C\lambda^{1-\varepsilon})I]\,\mathscr{E}_\lambda u, \mathscr{E}_\lambda u) \leq 0, \qquad u \in L_\lambda.$$

Because $\mathscr{E}_\lambda$ is an isomorphism of L_λ onto itself, it follows that

$$([A - (\lambda + 4C\lambda^{1-\varepsilon})I]v, v) \leq 0, \qquad v \in L_\lambda.$$

But by Lemma 28.1

$$\tilde{N}(\lambda) = V(\lambda)\,(1 + O(\lambda^{-\delta})) \leq N(\lambda + 4C\lambda^{1-\varepsilon}).$$

Putting $\lambda + 4C\lambda^{1-\varepsilon} = t$, we obtain $t = \lambda(1 + O(\lambda^{-\varepsilon}))$, which implies $\lambda = t(1 + O(t^{-\varepsilon}))$, i.e.

$$V(t(1+O(t^{-\varepsilon})))\,(1+O(t^{-\delta})) \leqq N(t).$$

Using condition (28.7) we see that

$$V(t(1+O(t^{-\varepsilon}))) = V(t)\,(1+O(t^{-\delta})),$$

hence

$$V(t)\,(1+O(t^{-\delta})) \leqq N(t). \tag{28.16}$$

b) Let $M_\lambda = (L_\lambda)^\perp$. Obviously we have

$$(u,u) \leqq 4((I-\mathscr{E}_\lambda)\,u,(I-\mathscr{E}_\lambda)u)\,, \qquad u \in M_\lambda. \tag{28.17}$$

It therefore follows from 4° that

$$([A-(\lambda-4C\lambda^{1-\varepsilon})I]v,v) \geqq 0, \qquad v \in D_A \cap M_\lambda. \tag{28.18}$$

From (28.18) we see that

$$(E_{\lambda-4C\lambda^{1-\varepsilon}-\varepsilon'}\,H)\cap M_\lambda = 0,$$

for any $\varepsilon' > 0$ which implies by analogy with the reasoning in the proof of Lemma 28.1, that

$$\tilde{N}(\lambda) \geqq N(\lambda-4C\lambda^{1-\varepsilon}-\varepsilon').$$

Now, arguing as in step a) of this proof, we obtain

$$N(t) \leqq V(t)\,(1+O(t^{-\delta})). \tag{28.19}$$

The estimates (28.16) and (28.19) together give the required asymptotic formula (28.8). $\square$

28.4 Sufficient conditions on $V(\lambda)$. Condition (28.7) looks difficult to verify. We therefore give here a simpler sufficient condition.

Proposition 28.2. *Let $V(\lambda)$ be a positive, non-decreasing function defined and differentiable for $\lambda \geqq \lambda_0$. Assume that*

$$V'(\lambda)/V(\lambda) = O(\lambda^{\varepsilon-\delta-1}). \tag{28.20}$$

Then $V(\lambda)$ satisfies (28.7).

Example 28.1. The function $V(\lambda) = \lambda^\alpha$, where $\alpha > 0$ satisfies (28.20) for $\varepsilon \geqq \delta$.

Proof of Proposition 28.2. Set

$$\varphi(\lambda) = V'(\lambda)/V(\lambda).$$

Then

$$V(\lambda) = V(\lambda_0)\,\exp\left(\int_{\lambda_0}^{\lambda}\varphi(\tau)\,d\tau\right).$$

This gives

$$V(\lambda + C\lambda^{1-\varepsilon}) - V(\lambda) = V(\lambda_0)\left\{\exp\left(\int_{\lambda_0}^{\lambda+C\lambda^{1-\varepsilon}}\varphi(\tau)\,d\tau\right) - \exp\left(\int_{\lambda_0}^{\lambda}\varphi(\tau)\,d\tau\right)\right\}$$

$$= V(\lambda)\left\{\exp\left(\int_{\lambda}^{\lambda+C\lambda^{1-\varepsilon}}\varphi(\tau)\,d\tau\right) - 1\right\}. \tag{28.21}$$

Since $|\varphi(\lambda)| \leqq C\lambda^{\varepsilon-\delta-1}$, then for $\varepsilon \neq \delta$ we obtain

$$\int_{\lambda}^{\lambda+C\lambda^{1-\varepsilon}}\varphi(\tau)\,d\tau \leqq C_1\int_{\lambda}^{\lambda+C\lambda^{1-\varepsilon}}\tau^{\varepsilon-\delta-1}\,d\tau = C_2\,[(\lambda + C\lambda^{1-\varepsilon})^{\varepsilon-\delta} - \lambda^{\varepsilon-\delta}]$$

$$= C_2\,\lambda^{\varepsilon-\delta}\,[(1 + C\lambda^{-\varepsilon})^{\varepsilon-\delta} - 1] \leqq C_3\lambda^{\varepsilon-\delta}\cdot\lambda^{-\varepsilon} = C_3\lambda^{-\delta}.$$

For $\varepsilon = \delta$, we obtain the same estimate

$$\int_{\lambda}^{\lambda+C\lambda^{1-\varepsilon}}\varphi(\tau)\,d\tau \leqq C_1\int_{\lambda}^{\lambda+C\lambda^{1-\varepsilon}}\tau^{-1}\,d\tau = C_1\ln(1 + C\lambda^{-\varepsilon}) \leqq C_2\lambda^{-\varepsilon} = C_2\lambda^{-\delta}.$$

It now follows from (28.21) that

$$V(\lambda + C\lambda^{1-\varepsilon}) - V(\lambda) \leqq CV(\lambda)\cdot\lambda^{-\delta}$$

(we used here that $e^x - 1 \sim x$ as $x \to 0$), but this is the required inequality (28.7).　□

28.5 The idea for applying Theorem 28.1 (an heuristic outline). Let an operator A have the Weyl symbol $b(z)$, $z \in \mathbb{R}^{2n}$. Put

$$V(\lambda) = (2\pi)^{-n}\int_{b(z)<\lambda} dz. \tag{28.22}$$

The spectral projection E_λ of the operator A can be defined as the operator $\chi_\lambda(A)$, when $\chi_\lambda(\cdot)$ is the characteristic function of the ray $(-\infty, \lambda]$. Let us consider the operator $\mathscr{E}_\lambda$ with the Weyl symbol $\chi_\lambda(b(\cdot))$. It is natural to expect that $\mathscr{E}_\lambda$ for large λ should imitate the spectral projection operator E_λ. It then remains to note that

$$\mathrm{Sp}\,\mathscr{E}_\lambda = (2\pi)^{-n}\int\chi_\lambda(b(z))\,dz = V(\lambda).$$

The technical implementation of this idea consists in applying Theorem 28.1. However, in view of the necessity to consider the composition of operators, we have to smooth the characteristic function χ_λ. In general, we arrive at a situation, which show the necessity of dealing with pseudodifferential operators with parameter, giving us the possibility of verifying $1°\text{--}5°$.

28.6 The exact construction. Let us introduce a function $\chi(t, \lambda, \varkappa)$, t, λ, $\varkappa \in \mathbb{R}$, $\lambda \geqq 1$, $\varkappa > 0$, such that

$$\chi(t, \lambda, \varkappa) = \begin{cases} 1 & \text{for } t \leqq \lambda, \\ 0 & \text{for } t \geqq \lambda + 2\varkappa. \end{cases} \tag{28.23}$$

and such that the following estimate for the derivatives holds

$$|(\partial/\partial t)^k \chi(t, \lambda, \varkappa)| \leqq C_k \varkappa^{-k}. \tag{28.24}$$

The existence of $\chi(t, \lambda, \varkappa)$ is easily verified, for instance, in the following manner. Let $\psi(t, \lambda, \varkappa)$ be the characteristic function of the set $\{(t, \lambda, \varkappa) : t \leqq \lambda + \varkappa\}$. Then we may put

$$\chi(t, \lambda, \varkappa) = \frac{1}{\varkappa} \int \psi(\tau, \lambda, \varkappa)\, \chi_0((t-\tau)/\varkappa)\, d\tau,$$

where $\chi_0(v) \in C_0^\infty(\mathbb{R}^1)$, $\chi_0(v) = 0$ for $|v| > 1$ and $\int \chi_0(v)\, dv = 1$.

Let now A have a real Weyl symbol

$$b(z) \in H\Gamma_\varrho^{m, m_0}(\mathbb{R}^{2n}), \qquad m_0 > 0, \tag{28.25}$$

such that for some $C > 0$ and $R_0 > 0$

$$b(z) \geqq C|z|^{m_0}, \qquad |z| \geqq R_0 \tag{28.26}$$

(it follows from (28.25), that (28.26) holds either for $b(z)$ or for $-b(z)$; we fix the sign in such a way that A becomes semi-bounded from below, this fact should be obvious from what follows. Put

$$e(z, \lambda, \varkappa) = \chi(b(z), \lambda, \varkappa) \tag{28.27}$$

and now, choosing $\varkappa = \lambda^{1-v}$, where $v > 0$, define $\mathscr{E}_\lambda$ as the operator with the Weyl symbol

$$e(z, \lambda) = \chi(b(z), \lambda, \lambda^{1-v}), \tag{28.28}$$

where $v > 0$ will be chosen later. Let us note immediately that

$$e(z, \lambda) = \begin{cases} 1 & \text{for } b(z) \leqq \lambda, \\ 0 & \text{for } b(z) \geqq \lambda + 2\lambda^{1-v}. \end{cases} \tag{28.29}$$

Now we try to estimate the derivatives in z of $e(z,\lambda)$.

Note that estimates for the class $H\Gamma_\varrho^{m,m_0}$, with $m_0 > 0$ can be written in the form

$$|\partial^\gamma b(z)| \leqq C_\gamma\, b(z)^{1-\varrho'|\gamma|}, \qquad \varrho' > 0, \quad |z| \geqq R_0, \tag{28.30}$$

where as ϱ' one can take, for instance, $\varrho' = \varrho/m$ or possibly larger values.

Now, differentiate (28.28):

$$\partial_z^\gamma e(z,\lambda) = \sum_{\substack{\gamma_1+\ldots+\gamma_k=\gamma \\ |\gamma_j|>0}} c_{\gamma_1\ldots\gamma_k}(\partial^{\gamma_1} b(z)) \ldots (\partial^{\gamma_k} b(z)) \left(\frac{\partial^k \chi}{\partial t^k}(t,\lambda,\lambda^{1-\nu})\bigg|_{t=b(z)}\right), \tag{28.31}$$

where the sum runs over all possible decompositions of γ into a sum $\gamma_1 + \ldots + \gamma_k$ with an arbitrary number of terms $k \leqq |\gamma|$.

Denote by $T_k(z,\lambda)$ the sum of all terms corresponding to a fixed k in (28.31). It follows from (28.31), (28.30) and (28.24) that

$$|T_k(z,\lambda)| \leq C_\gamma b(z)^{k-\varrho'|\gamma|} \cdot \lambda^{-k(1-\nu)}. \tag{28.32}$$

Due to (28.29) we can replace λ by $b(z)$ and rewrite (28.32) in the form

$$|\partial_z^\gamma T_k(z,\lambda)| \leq C_\gamma b(z)^{(\mu-\varrho')|\gamma|} \cdot \lambda^{k\nu-\mu|\gamma|}, \tag{28.33}$$

where μ is an arbitary real number.

Since $k \leqq |\gamma|$, this implies that

$$|\partial_z^\gamma e(z,\lambda)| \leqq C_\gamma b(z)^{(\mu-\varrho')|\gamma|} \cdot \lambda^{(\nu-\mu)|\gamma|}. \tag{28.34}$$

These estimates are true for $\lambda \geqq 1$ and for $|z| \geqq R_0$. In addition, we have the obvious relations

$$|e(z,\lambda)| \leqq C, \qquad |z| \leqq R_0, \tag{28.35}$$

$$\partial^\gamma e(z,\lambda) = 0, \qquad |z| \leqq R_0, \quad \lambda \geqq \lambda_0, \tag{28.36}$$

if $|\gamma| > 0$ and λ_0 is sufficiently large.

It is obvious from (28.34) that it is advantageous to take μ such that $\nu < \mu < \varrho'$, implying the necessity of selecting ν such that $\nu < \varrho'$, which we will assume in the sequel. Under some complementary conditions this will guarantee the applicability of Theorem 28.1.

28.7 Equipotential surfaces of the symbol and the properties of $V(\lambda)$. In this section, we assume that the equipotential set $\{z: b(z) = \lambda\}$ for large λ, is a smooth hypersurface and moreover that

$$\nabla b(z) \neq 0 \qquad \text{for} \qquad |z| \geqq R_0, \tag{28.37}$$

where $\nabla b(z) = \left(\dfrac{\partial b}{\partial z_1}, \ldots, \dfrac{\partial b}{\partial z_{2n}}\right)$ is the gradient of $b(z)$.

Let $V(\lambda)$ be defined by the formula (28.22). Since the $2n$-form

$$dz = dz_1 \wedge \ldots \wedge dz_{2n}$$

is the differential of the $(2n-1)$-form

$$\omega = \frac{1}{2n}\sum_{j=1}^{2n}(-1)^{j+1}z_j dz_1 \wedge \ldots \wedge \widehat{dz_j} \wedge \ldots \wedge dz_{2n}$$

(the cap on dz_j denotes that dz_j is omitted), we can transform the integral in (28.22) into an integral over the surface $b(z) = \lambda$:

$$\int_{b(z)<\lambda} dz = \int_{b(z)=\lambda} \omega. \tag{28.38}$$

Let n_z be the unit outward normal vector to the surface $b(z) = \lambda$ at z, i.e.

$$n_z = \frac{\nabla b(z)}{|\nabla b(z)|}.$$

Denoting by dS_z the area element of the surface $b(z) = \lambda$, we derive from (28.38) that

$$V(\lambda) = \frac{(2\pi)^{-n}}{2n}\int_{b(z)=\lambda}(z \cdot n_z)\,dS_z = \frac{(2\pi)^{-n}}{2n}\int_{b(z)=\lambda}\frac{dS_z}{|\nabla b(z)|}(z \cdot \nabla b(z)). \tag{28.39}$$

Now, we will calculate $V'(\lambda)$. Note that the distance at z from the surface $b(z) = \lambda$ to the near equipotential surface $b(z) = \lambda + \Delta\lambda$ is equal to $\dfrac{\Delta\lambda(1+o(1))}{|\nabla b(z)|}$. Therefore

$$V'(\lambda) = (2\pi)^{-n}\int_{b(z)=\lambda}\frac{dS_z}{|\nabla b(z)|}. \tag{28.40}$$

Comparing (28.39) and (28.40), we see that

$$\frac{V'(\lambda)}{V(\lambda)} \leq \left[\min_{b(z)=\lambda}(z \cdot \nabla b(z))\right]^{-1}. \tag{28.41}$$

The formula (28.41) implies

Proposition 28.3. *Let*

$$|z \cdot \nabla b(z)| \geq Cb(z)^{1-\varkappa}, \qquad |z| \geq R_0, \quad C > 0. \tag{28.42}$$

Then

$$V'(\lambda)/V(\lambda) = O(\lambda^{\varkappa-1}). \qquad (28.43)$$

Remark 28.2. It follows from (28.42) that $\nabla b(z) \neq 0$ for $|z| \geq R_0$, so it is not necessary to require this in advance. Also, the geometric condition (28.42) guarantees that the surface $b(z) = \lambda$ is star-shaped with respect to the origin, i.e. any ray, starting from 0, intersects this surface in exactly one point and at a non-zero angle.

Exercise 28.1. Prove that (28.42) is satisfied if $b(z)$ is an elliptic polynomial. *Hint*: Use the Euler identity for homogeneous functions.

Problem 28.1. For dim $H = 2$ construct an operator A, such that for some λ there is an operator E'_λ for which the conditions 1) – 4) are satisfied, but $E'_\lambda \neq E_\lambda$.

Problem 28.2. On the torus $\mathbb{T}^n = \mathbb{R}^n/2\pi \mathbb{Z}^n$ consider the operator $A = -\Delta + Q$, where Δ is the Laplace operator and Q any bounded self-adjoint operator on $L^2(\mathbb{T}^n)$, with $\|Q\| < M$. Let $N(\lambda)$ be the number of eigenvalues of A, smaller than λ.

Using the approximate spectral projection operator $\mathscr{E}_\lambda$, which equals the exact spectral projection operator for $-\Delta$, show that

$$N_0(\lambda - M) \leq N(\lambda) \leq N_0(\lambda + M),$$

where $N_0(\lambda)$ is the number of points of the lattice $\mathbb{Z}^n$, belonging to the ball $|x| \leq \sqrt{\lambda}$. Derive from this the asymptotic formula

$$N(\lambda) = c_n \lambda^{n/2} (1 + O(\lambda^{-1/2})).$$

§29. Operators with Parameter

29.1. The class of symbols and operators. The estimates (28.34) obtained for $e(z, \lambda)$ in §28 motivate the following

Definition 29.1. Denote by $\Gamma^{m,\mu}_{\varrho,\sigma}$ the class of functions $a(z, \lambda)$ defined for $z \in \mathbb{R}^{2n}$, $\lambda \geq \lambda_0$, infinitely differentiable in z and satisfying

$$|\partial_z^\gamma a(z, \lambda)| \leq C_\gamma \langle z \rangle^{m - \varrho|\gamma|} \lambda^{\mu - \sigma|\gamma|}. \qquad (29.1)$$

Here m, μ, ϱ, $\sigma \in \mathbb{R}$, $\varrho > 0$, $\sigma \geq 0$.

It is clear that if $a_j \in \Gamma^{m_j, \mu_j}_{\varrho_j, \sigma_j}$, $j = 1$, 2, then $a_1 a_2 \in \Gamma^{m_1 + m_2, \mu_1 + \mu_2}_{\varrho, \sigma}$, where $\varrho = \min(\varrho_1, \varrho_2)$, $\sigma = \min(\sigma_1, \sigma_2)$. Further, if $a \in \Gamma^{m,\mu}_{\varrho,\sigma}$ then $\partial_z^\gamma a \in \Gamma^{m - \varrho|\gamma|, \mu - \sigma|\gamma|}_{\varrho, \sigma}$.

Note that if $a \in \Gamma^{m,\mu}_{\varrho,\sigma}$ then for any fixed $\lambda \geq \lambda_0$, $a(z, \lambda) \in \Gamma^m_\varrho(\mathbb{R}^{2n})$, which allows us to define a class of operators $A(\lambda)$, depending on a parameter and with

Weyl symbols $a(z, \lambda) \in \Gamma_{\varrho, \sigma}^{m, \mu}$. This class of operator-valued functions $A(\lambda)$ acts, for instance, on $S(\mathbb{R}^n)$ and we will denote this class by $G_{\varrho, \sigma}^{m, \mu}$.

29.2 The composition formula

Theorem 29.1. *Let $a_j \in \Gamma_{\varrho_j, \sigma_j}^{m_j, \mu_j}$, $j = 1, 2$; and let $A_j(\lambda)$ be the corresponding operator-functions. Then $A_1(\lambda) \circ A_2(\lambda) \in G_{\varrho, \sigma}^{m_1 + m_2, \mu_1 + \mu_2}$, where $\varrho = \min(\varrho_1, \varrho_2)$, $\sigma = \min(\sigma_1, \sigma_2)$ and where the Weyl symbol $b(z, \lambda)$ of the composition $B(\lambda) = A_1(\lambda) \circ A_2(\lambda)$ is given by*

$$b = \sum_{|\alpha + \beta| \leq N - 1} \frac{(-1)^{|\beta|}}{\alpha! \beta!} \, 2^{-|\alpha + \beta|} \, (\partial_\xi^\alpha D_x^\beta a_1) \, (\partial_\xi^\beta D_x^\alpha a_2) + r_N , \tag{29.2}$$

where

$$r_N \in \Gamma_{\varrho, \sigma}^{m_1 + m_2 - N(\varrho_1 + \varrho_2), \mu_1 + \mu_2 - N(\sigma_1 + \sigma_2)} . \tag{29.3}$$

Proof. The proof could be carried out according to the scheme used for proving Theorem 23.6, introducing first the corresponding class of amplitudes and repeating the arguments from §23. However, for brevity, we will give a direct proof.

To begin with, we obtain a formula for the composition $B = A_1 \circ A_2$ of the operators A_1 and A_2 with Weyl symbols $a_1(z)$, $a_2(z) \in C_0^\infty(\mathbb{R}^{2n})$. Clearly

$$Bu(x) = \int a_1\left(\frac{x + x_1}{2}, \zeta\right) a_2\left(\frac{x_1 + y}{2}, \eta\right)$$

$$\times \, e^{i[(x - x_1) \cdot \zeta + (x_1 - y) \cdot \eta]} \, u(y) \, dy \, dx_1 \, d\eta \, d\zeta .$$

If $K_B(x, y)$ is the kernel of B, we obtain that then

$$K_B(x, y) = \int a_1\left(\frac{x + x_1}{2}, \zeta\right) a_2\left(\frac{x_1 + y}{2}, \eta\right)$$

$$\times \, e^{i[(x - x_1) \cdot \zeta + (x_1 - y) \cdot \eta]} \, dx_1 \, d\eta \, d\zeta . \tag{29.4}$$

Now using formula (23.39) (with $\tau = \tfrac{1}{2}$), that yields an expression for the symbol in terms of the kernel

$$b(x, \xi) = \int e^{-ix_2 \cdot \xi} \, K_B\left(x + \frac{x_2}{2}, x - \frac{x_2}{2}\right) dx_2 . \tag{29.5}$$

Putting $x_2/2 = x_3$, we can also write

$$b(x, \xi) = 2^n \int e^{-2ix_3 \cdot \xi} \, K_B(x + x_3, x - x_3) \, dx_3 . \tag{29.6}$$

From this and (29.4) we find that

$$b(x, \xi) = 2^n \int a_1 \left(\frac{x + x_1 + x_3}{2}, \zeta \right) a_2 \left(\frac{x + x_1 - x_3}{2}, \eta \right)$$

$$\times e^{i[(x + x_3 - x_1) \cdot \zeta + (x_1 + x_3 - x) \cdot \eta - 2x_3 \cdot \xi]} \, dx_1 \, dx_3 \, d\eta \, d\zeta.$$

Instead of x_1 and x_3 we introduce new integration variables

$$x_4 = \frac{x + x_1 + x_3}{2}, \qquad x_5 = \frac{x + x_1 - x_3}{2},$$

so that $x_1 = x_4 + x_5 - x$, $x_3 = x_4 - x_5$. Observing that $\left| \dfrac{\partial(x_1, x_3)}{\partial(x_4, x_5)} \right| = 2^n$, we obtain

$$b(x, \xi) = 2^{2n} \int a_1(x_4, \zeta) \, a_2(x_5, \eta)$$

$$\times e^{2i[(x - x_5) \cdot \zeta + (x_4 - x) \cdot \eta + (x_5 - x_4) \cdot \xi]} \, dx_4 \, dx_5 \, d\zeta \, d\eta$$

or

$$b(x, \xi) = 2^{2n} \int a_1(y, \eta) \, a_2(z, \zeta)$$

$$\times e^{2i[(x - z) \cdot \eta + (y - x) \cdot \zeta + (z - y) \cdot \xi]} \, dy \, dz \, d\eta \, d\zeta. \tag{29.7}$$

Note that the exponent in (29.7) may also be written as

$$2i \begin{vmatrix} 1 & 1 & 1 \\ x & y & z \\ \xi & \eta & \zeta \end{vmatrix} = 2i \sum_{j=1}^{n} \begin{vmatrix} 1 & 1 & 1 \\ x_j & y_j & z_j \\ \xi_j & \eta_j & \zeta_j \end{vmatrix}.$$

From the form of this exponent, the possibility of integrating by parts follows, resulting in the appearance of decreasing factors of the type $\langle x - z \rangle^{-N}$, $\langle y - x \rangle^{-N}, \langle \eta - \xi \rangle^{-N}, \langle \zeta - \xi \rangle^{-N}$. Therefore, the points y, η, z, ζ, where $y = z = x$, $\eta = \zeta = \xi$, play the most important role in the integral (29.7). This leads to the idea of expanding $a_1(y, \eta)$ and $a_2(z, \zeta)$ in a Taylor series at $y = x$ and $z = x$. First, make the change of variables $y' = y - x$, $z' = z - x$, $\eta' = \eta - \xi$, $\zeta' = \zeta - \xi$. Omitting the dashes, we obtain

$$b(x, \xi) = 2^{2n} \int a_1(x + y, \xi + \eta) \, a_2(x + z, \xi + \zeta)$$

$$\times e^{2i(y \cdot \zeta - z \cdot \eta)} \, d\eta \, d\zeta \, dy \, dz. \tag{29.8}$$

We shall view this integral as an iterated integral with the order of the differentials $d\eta \, d\zeta \, dy \, dz$. Write out the expansions

$$a_1(x + y, \xi + \eta) = \sum_{|\alpha| \leq N - 1} \frac{y^\alpha}{\alpha!} \, \partial_x^\alpha a_1(x, \xi + \eta) + r_N^{(1)}(x, y, \xi, \eta), \tag{29.9}$$

$$a_2(x+z,\xi+\zeta) = \sum_{|\beta|\leq N-1} \frac{z^\beta}{\beta!}\, \partial_x^\beta a_2(x,\xi+\zeta) + r_N^{(2)}(x,z,\xi,\zeta)\,, \tag{29.10}$$

where

$$r_N^{(1)}(x,y,\xi,\eta) = \sum_{|\alpha|=N} c_\alpha \int_0^1 (1-\tau)^{N-1}\,d\tau \cdot y^\alpha\,(\partial_x^\alpha a_1)\,(x+\tau y,\xi+\eta)\,, \tag{29.11}$$

$$r_N^{(2)}(x,z,\xi,\zeta) = \sum_{|\beta|=N} c_\beta \int_0^1 (1-\tau)^{N-1}\,d\tau\; z^\beta\,(\partial_x^\beta a_2)\,(x+\tau z,\xi+\zeta)\,. \tag{29.12}$$

Inserting these expressions into (29.8), we obtain

$$b(x,\xi) = 2^{2n} \sum_{|\alpha+\beta|\leq N-1} \int \frac{y^\alpha z^\beta}{\alpha!\beta!}\, [\partial_x^\alpha a_1(x,\xi+\eta)]\,[\partial_x^\beta a_2(x,\xi+\zeta)]$$

$$\times\, e^{2i(y\cdot\zeta-z\cdot\eta)}\,d\eta\,d\zeta\,dy\,dz + r_N^{(3)}(x,\xi)\,, \tag{29.13}$$

where $r_N^{(3)}(x,\xi)$ has the form of a linear combination of terms of four types:
$r_{\alpha,\beta}^{(4)}(x,\xi)$, being the same type of integral as the summands in (29.13), but
with $|\alpha+\beta|\geq N$;

$$r_{\alpha,N}^{(5)}(x,\xi) = \int y^\alpha\,[\partial_x^\alpha a_1(x,\xi+\eta)]\,[r_N^{(2)}(x,z,\xi,\zeta)]$$

$$\times\, e^{2i(y\cdot\zeta-z\cdot\eta)}\,d\eta\,d\zeta\,dy\,dz\,; \tag{29.14}$$

$$r_{N,\beta}^{(6)}(x,\xi) = \int z^\beta\,[r_N^{(1)}(x,y,\xi,\eta)]\,[\partial_x^\beta a_2(x,\xi+\zeta)]$$

$$\times\, e^{2i(y\cdot\zeta-z\cdot\eta)}\,d\eta\,d\zeta\,dy\,dz\,; \tag{29.15}$$

$$r_N^{(7)}(x,\xi) = \int r_N^{(1)}(x,y,\xi,\eta)\, r_N^{(2)}(x,z,\xi,\zeta)\, e^{2i(y\cdot\zeta-z\cdot\eta)}\,d\eta\,d\zeta\,dy\,dz\,. \tag{29.16}$$

Let us calculate one of the integrals in (29.13). For this, note first that

$$y^\alpha e^{2iy\cdot\zeta} = \left(\frac{1}{2}\,D_\zeta\right)^\alpha (e^{2iy\cdot\zeta})\,, \tag{29.17}$$

$$z^\beta e^{-2iz\cdot\eta} = \left(-\frac{1}{2}\,D_\eta\right)^\beta (e^{-2iz\cdot\eta})\,, \tag{29.18}$$

and carry out the integration by parts, using these identities. We obtain

$$2^{2n} \int \frac{y^\alpha z^\beta}{\alpha! \beta!} \, [\partial_x^\alpha a_1(x, \xi+\eta)] \, [\partial_x^\beta a_2(x, \xi+\zeta)] \, e^{2i(y\cdot\zeta - z\cdot\eta)} \, d\eta \, d\zeta \, dy \, dz$$

$$= \frac{2^{2n-|\alpha+\beta|} \cdot (-1)^{|\alpha|}}{a! \beta!} \int [\partial_x^\alpha D_\xi^\beta a_1(x, \xi+\eta)] \, [\partial_x^\beta D_\xi^\alpha a_2(x, \xi+\zeta)]$$

$$\times e^{2i(y\cdot\zeta - z\cdot\eta)} \, d\eta \, d\zeta \, dy \, dz = \frac{2^{-|\beta|}(-2)^{-|\alpha|}}{\alpha! \beta!} \int \left[\partial_x^\alpha D_\xi^\beta a_1\left(x, \xi+\frac{\eta}{2}\right)\right]$$

$$\times \left[\partial_x^\beta D_\xi^\alpha a_2\left(x, \xi+\frac{\zeta}{2}\right)\right] e^{i(y\cdot\zeta - z\cdot\eta)} \, d\eta \, d\zeta \, dy \, dz = \frac{2^{-|\beta|}(-2)^{-|\alpha|}}{\alpha! \beta!}$$

$$\times \int \left[\partial_x^\alpha D_\xi^\beta a_1\left(x, \xi+\frac{\eta}{2}\right)\right] e^{-iz\cdot\eta} \, d\eta \, dz \cdot \int \left[\partial_x^\beta D_\xi^\alpha a_2\left(x, \xi+\frac{\zeta}{2}\right)\right]$$

$$\times e^{iy\cdot\zeta} \, d\zeta \, dy = \frac{2^{-|\beta|}(-2)^{-|\alpha|}}{\alpha! \beta!} \, [\partial_x^\alpha D_\xi^\beta a_1(x, \xi))] \, [\partial_x^\beta D_\xi^\alpha a_2(x, \zeta)],$$

which gives terms with compact support in the formula (29.2).

Now, note that formula (29.8) and all the following computations remain valid for arbitrary symbols a_j in the classes Γ_ϱ^m. This fact can be established by substituting the oscillatory integral (29.8) for a convergent one, but it can also be verified by a standard passing to the limit from compactly supported symbols, as in §1.

Assume now that the symbols a_1 and a_2 are as in the formulation of the theorem. It suffices to prove the inclusion (29.3) for each of the remainders $r_{\alpha,\beta}^{(4)}$, $r_{\alpha,N}^{(5)}$, $r_{N,\beta}^{(6)}$, $r_N^{(7)}$. As far as $r_{\alpha,\beta}^{(4)}$ is concerned, this inclusion is trivial, since it has the same form as the terms of the sum in (29.13).

To estimate $r_{\alpha,N}^{(5)}$, $r_{N,\beta}^{(6)}$ and $r_N^{(7)}$ it is convenient as before to integrate (29.14)–(29.16) by parts (using (29.17) and (29.18)). As a result y^α is replaced by D_ζ^α and z^β by D_η^β and we arrive at the situation of having to estimate, uniformly in τ_1, $\tau_2 \in [0,1]$, symbols of the type

$$\int [\partial_x^\alpha D_\xi^\beta a_1(x+\tau_1 y, \xi+\eta, \lambda)] \, [\partial_x^\beta D_\xi^\alpha a_2(x+\tau_2 z, \xi+\zeta, \lambda)]$$

$$\times e^{2i(y\cdot\zeta - z\cdot\eta)} \, d\eta \, d\zeta \, dy \, dz, \qquad |\alpha+\beta| \geqq N. \tag{29.19}$$

Differentiating (29.19) with respect to x and ξ, we find that the derivative $\partial_x^\gamma \partial_\xi^\delta$ of this expression is a linear combination of terms of the form

$$\int [\partial_x^{\alpha+\gamma'} \partial_\xi^{\beta+\delta'} a_1(x+\tau_1 y, \xi+\eta, \lambda)] \, [\partial_x^{\beta+\gamma''} \partial_\xi^{\alpha+\delta''} a_2(x+\tau_2 z, \xi+\zeta, \lambda)]$$

$$\times e^{2i(y\cdot\zeta - z\cdot\eta)} \, d\eta \, d\zeta \, dy \, dz, \qquad \gamma'+\gamma'' = \gamma, \quad \delta'+\delta'' = \delta. \tag{29.20}$$

Using the identities

$$e^{2i(y\cdot\zeta - z\cdot\eta)} = (1 + |y|^2 + |\eta|^2)^{-M} (1 - \tfrac{1}{4}\Delta_{z,\zeta})^M e^{2i(y\cdot\zeta - z\cdot\eta)},$$

$$e^{2i(y\cdot\zeta - z\cdot\eta)} = (1 + |z|^2 + |\zeta|^2)^{-M} (1 - \tfrac{1}{4}\Delta_{y,\eta})^M e^{2i(y\cdot\zeta - z\cdot\eta)}$$

and introducing the notation

$$\langle z,\zeta\rangle = (1+|z|^2+|\zeta|^2)^{1/2}, \qquad \langle y,\eta\rangle = (1+|y|^2+|\eta|^2)^{1/2},$$

we see that (29.20) reduces to a linear combination of expressions of the form

$$\int [\partial_{x,\xi}^{\varkappa'+v'} a_1(x+\tau_1 y,\xi+\eta,\lambda)]\,[\delta_{x,\xi}^{\varkappa''+v''} a_2(x+\tau_2 z,\xi+\zeta,\lambda)]$$

$$\times \langle y,\eta\rangle^{-2M}\langle z,\zeta\rangle^{-2M} e^{2i(y\cdot\zeta-z\cdot\eta)}\,d\eta\,d\zeta\,dy\,dz, \tag{29.21}$$

where $\varkappa'$, $\varkappa''$, v', v'' are $2n$-dimensional multi-indices, such that

$$|\varkappa'|\geq N, \qquad |\varkappa''|\geq N, \qquad |v'|+|v''|\geq |\gamma|+|\delta|. \tag{29.22}$$

The integral (29.21) can be estimated in absolute value by

$$C\int \langle x+\tau_1 y,\xi+\eta\rangle^{m_1-\varrho_1|\varkappa'+v'|}\cdot \lambda^{\mu_1-\sigma_1|\varkappa'+v'|}$$

$$\times \langle x+\tau_2 z,\xi+\zeta\rangle^{m_2-\varrho_2|\varkappa''+v''|}\cdot \lambda^{\mu_2-\sigma_2|\varkappa''+v''|}$$

$$\times \langle y,\eta\rangle^{-2M}\langle z,\zeta\rangle^{-2M}\,d\eta\,d\zeta\,dy\,dz,$$

which, due to (29.22), does not exceed the expression

$$C\lambda^{\mu_1+\mu_2-N(\sigma_1+\sigma_2)-|\gamma+\delta|\sigma}\cdot \int \langle x+\tau_1 y,\xi+\eta\rangle^{m_1-\varrho_1 N-\varrho_1|v'|}$$

$$\times \langle x+\tau_2 z,\xi+\zeta\rangle^{m_2-\varrho_2 N-\varrho_2|v''|}\langle y,\eta\rangle^{-2M}\langle z,\zeta\rangle^{-2M}\quad d\eta\,d\zeta\,dy\,dz. \tag{29.23}$$

Here the power of λ corresponds exactly to the statement of the theorem, so that it suffices to estimate the integral in (29.23) by the desired powers of $\langle x,\xi\rangle$.

Note, that the integral in (29.23) coincides with the product of the integrals

$$\int \langle x+\tau_1 y,\xi+\eta\rangle^{m_1-\varrho_1 N-\varrho_1|v'|}\langle y,\eta\rangle^{-2M}\,d\eta\,dy, \tag{29.24}$$

$$\int \langle x+\tau_2 z,\xi+\zeta\rangle^{m_2-\varrho_2 N-\varrho_2|v''|}\langle z,\zeta\rangle^{-2M}\,d\zeta\,dz. \tag{29.25}$$

Let us estimate the integral (29.24). Decompose the domain of integration into two parts

$$\Omega_1 = \{y,\eta: |y|+|\eta|\leq \tfrac{1}{2}\,(|x|+|\xi|)\},$$

$$\Omega_2 = \{y,\eta: |y|+|\eta|> \tfrac{1}{2}\,(|x|+|\xi|)\},$$

and denote the integral over Ω_j by I_j, $j=1,2$. Since the Lebesgue volume of Ω_1 does not exceed $C_1\langle x,\xi\rangle^n$ and since the integrand can be estimated over Ω_1 by $C_2\langle x,\xi\rangle^{m_1-\varrho_1 N-\varrho_1|v'|}$, we have for I_1 the estimate

$$|I_1|\leq C_3\langle x,\xi\rangle^{m_1-\varrho_1 N-\varrho_1|v'|+n}. \tag{29.26}$$

Consider now I_2. We may assume that $|x|+|\xi|\geq 1$ (for $|x|+|\xi|<1$ the desired estimates are trivial). Then, using the obvious estimate

$$(1+|x+\tau_1 y|+|\xi+\eta|)^p\leq (1+|x|+|\xi|)^p\,(1+|y|+|\eta|)^{|p|}$$

and taking M sufficiently large, we obtain

$$|I_2| \leq C_4 \langle x, \xi \rangle^{m_1 - \varrho_1 N - \varrho_1 |v'|}. \tag{29.27}$$

From (29.26) and (29.27) it follows that (29.24) can be estimated by

$$C \langle x, \xi \rangle^{m_1 - \varrho_1 N - \varrho_1 |v'| + n}.$$

Taking into account that similar estimate holds for (29.25) we see that (29.23) does not exceed

$$C \langle x, \xi \rangle^{m_1 + m_2 - (\varrho_1 + \varrho_2) N - \varrho_1 |v'| - \varrho_2 |v''| + 2n} \cdot \lambda^{\mu_1 + \mu_2 - N(\sigma_1 + \sigma_2) - \sigma|\gamma + \delta|}$$

$$\leq C \langle x, \xi \rangle^{m_1 + m_2 - N(\varrho_1 + \varrho_2) - \varrho|v + \delta| + 2n} \cdot \lambda^{\mu_1 + \mu_2 - N(\sigma_1 + \sigma_2) - \sigma|\gamma + \delta|},$$

which guarantees the inclusion

$$r_N \in \Gamma_{\varrho, \sigma}^{m_1 + m_2 - N(\varrho_1 + \varrho_2) + 2n, \, \mu_1 + \mu_2 - N(\sigma_1 + \sigma_2)}. \tag{29.28}$$

Now, increasing N and considering the additional terms which appear in the sum (29.2), we see that (29.3) follows from (29.28), thus proving the theorem. $\square$

29.3 Positivity of operators with parameter

Theorem 29.2. *Let* $a(z, \lambda) \in \Gamma_{\varrho, \sigma}^{m, \mu}$, $\sigma > 0$, $a(z, \lambda) \geq \varepsilon > 0$, ε *a constant, and assume that the estimates*

$$|\partial_z^\gamma a(z, \lambda)| \leq C_\gamma a(z, \lambda) \cdot \lambda^{-\sigma_0 |\gamma|}, \qquad \lambda \geq \lambda_0, \quad z \in \mathbb{R}^{2n}, \tag{29.29}$$

hold, where $\sigma_0 > 0$ *and* σ_0 *does not depend on* γ. *If* $A(\lambda)$ *is the operator with the Weyl symbol* $a(z, \lambda)$, *then for sufficiently large* λ *we have* $A(\lambda) \geq 0$ *(i.e.* $(A(\lambda)u, u) \geq 0$ *for* $u \in S(\mathbb{R}^n)$*).*

For the proof, we need the following lemma which allows us to use the anti-Wick symbol.

Lemma 29.1. *Consider an operator* $B(\lambda)$ *with anti-Wick symbol* $a(z, \lambda) \in \Gamma_{\varrho, \sigma}^{m, \mu}$ *and let* $b(z, \lambda)$ *be its Weyl symbol. Then*

$$a - b = \sum_{0 < |\gamma| < N} c_\gamma (\partial_z^\gamma a) + r_N, \tag{29.30}$$

where $c_\gamma = 0$ *for odd* $|\gamma|$ *and* $r_N \in \Gamma_{\varrho, \sigma}^{m - \varrho N, \, \mu - \sigma N}$.

Proof. Similar to the proof of Theorem 24.1.

Exercise 29.1. Prove Lemma 29.1.

Proof of Theorem 29.2. Let $B_0(\lambda)$ be an operator with the anti-Wick symbol $a(z, \lambda)$ and let $b_0(z, \lambda)$ be the Weyl symbol of $B_0(\lambda)$. Consider now the operator $B_1(\lambda)$ with the anti-Wick symbol $a(z, \lambda) - b_0(z, \lambda)$ and denote by $b_1(z, \lambda)$ its Weyl symbol. By induction we may construct a sequence of operators $B_0, B_1, B_2, \ldots$, such that B_j is the operator with the anti-Wick symbol $a - b_0 - b_1 - \ldots - b_{j-1}$, where $b_0, b_1, \ldots, b_{j-1}$ are the Weyl symbols of the operators $B_0, B_1, \ldots, B_{j-1}$.

It follows from Lemma 29.1 that if

$$A_k = A - B_0 - B_1 - \ldots - B_{k-1},$$

then $A_k \in \Gamma_{\varrho, \sigma}^{m - 2\varrho k, \mu - 2\sigma k}$. Put

$$Q_k = B_0 + B_1 + \ldots + B_{k-1}.$$

Thus, $A = A_k + Q_k$.

An induction in k shows that $b_j(z, \lambda)$ for $j > 0$ is of the form

$$b_j = \sum_{2j \leq |\gamma| < N} c_{\gamma, j}(\partial^\gamma a) + r_{N, j}, \tag{29.31}$$

where $c_{\gamma, j}$ are constants and $r_{N, j} \in \Gamma_{\varrho, \sigma}^{m - \varrho N, \mu - \sigma N}$. Therefore the operator Q_k has an anti-Wick symbol of the form

$$q_k(z, \lambda) = a(z, \lambda) + \sum_{2 \leq |\gamma| < N} c_\gamma [\partial_z^\gamma a(z, \lambda)] + r_N'(z, \lambda), \tag{29.32}$$

where $r_N'(z, \lambda) \in \Gamma_{\varrho, \sigma}^{m - \varrho N, \mu - \sigma N}$. Taking (29.29) into account, we obtain

$$q_k(z, \lambda) = a(z, \lambda)(1 + \lambda^{-2\sigma_0} s(z, \lambda)),$$

where $s(z, \lambda)$ is such that

$$\sup_{z \in \mathbb{R}^{2n}} |s(z, \lambda)| \leq C, \qquad \lambda \geq \lambda_0,$$

and C is independent of λ. In particular, it is clear that $q_k(z, \lambda) \geq \varepsilon/2 > 0$ for sufficiently large λ implying

$$Q_k \geq \frac{\varepsilon}{2} I \tag{29.33}$$

in view of Proposition 24.1.

Now note that for large k

$$\|A_k(\lambda)\| \to 0 \qquad \text{as} \qquad \lambda \to +\infty. \tag{29.34}$$

Indeed, it suffices to verify that this holds for the Hilbert-Schmidt norm $\|A(\lambda)\|_2$. But

$$\|A_k(\lambda)\|_2^2 = (2\pi)^{-n} \int |b_k(z, \lambda)|^2 \, dz \to 0 \qquad \text{as} \qquad \lambda \to +\infty,$$

if we choose k so large that

$$m - 2\varrho k < -n, \qquad \mu - 2\sigma k < 0.$$

It follows directly from (29.33) and (29.34) that $A \geq \dfrac{\varepsilon}{4} I$, proving Theorem 29.2. $\square$

§30. Asymptotic Behaviour of the Eigenvalues

Consider an operator A with a real Weyl symbol $b(z) \in H\Gamma_{\varrho}^{m,\,m_0}$, $m_0 > 0$. By Theorem 26.2, A is essentially self-adjoint and by Theorem 26.3 it has discrete spectrum. Since $b(z)$ has no zeros for large $|z|$, then, changing sign if necessary, we may assume that

$$C_1 |z|^{m_0} \leq b(z) \leq C_2 |z|^m, \qquad |z| \geq R_0, \qquad m_0 > 0, \tag{30.1}$$

holds, where C_1, C_2 are positive constants. It is easy to show that in this case the operator is semi-bounded from below. Indeed, repeating the argument of the proof of Theorem 24.2 and using Theorem 24.1, we see that there exists an operator A' with the anti-Wick symbol

$$b(z) + \sum_{0 < |\gamma| < N} c_\gamma \partial^\gamma b(z), \tag{30.2}$$

such that $A - A'$ is bounded. But then it suffices to verify the semi-boundedness from below for A', which follows from the semi-boundedness from below of any function of the form (30.2), which in turn is a consequence of the fact that

$$b(z) + \sum_{0 < |\gamma| < N} c_\gamma \partial^\gamma b(z) = b(z) \left(1 + \sum_{0 < |\gamma| < N} c_\gamma \frac{\partial^\gamma b(z)}{b(z)} \right),$$

where all the terms in the parenthesis, except the first one, tend to 0 as $|z| \to +\infty$. Let ϱ' be a positive number such that the following estimates hold

$$|\partial^\gamma b(z)| \leq C_\gamma b(z)^{1 - \varrho'|\gamma|}, \qquad |z| \geq R_0 \tag{30.3}$$

(as we have already remarked in 28.6, one may take, for instance, $\varrho' = \varrho/m$, although this might not be the best value for ϱ'). Finally assume that

$$|z \cdot \nabla b(z)| \geq c b(z)^{1 - \varkappa}, \qquad |z| \geq R_0, \tag{30.4}$$

where $0 \leq \varkappa < 1$, $c > 0$. Set

$$V(\lambda) = (2\pi)^{-n} \int_{b(z) < \lambda} dz, \tag{30.5}$$

and let $N(\lambda)$ be the number of eigenvalues of A, not exceeding λ.

Theorem 30.1. *Let the operator A have the Weyl symbol $b(z) \in C^\infty(\mathbb{R}^{2n})$, satisfying conditions (30.1), (30.3) and (30.4) with $\varkappa < \varrho'$. Then for any $\varepsilon > 0$, one has the asymptotic formula*

$$N(\lambda) = V(\lambda)\,(1 + O(\lambda^{\varkappa - \varrho' + \varepsilon}))\,. \tag{30.6}$$

Proof. We will make use of Theorem 28.1 with the approximate spectral projection operators $\mathscr{E}_\lambda$ constructed in section 28.6. Recall that $\mathscr{E}_\lambda$ has a real-valued Weyl symbol $e(z, \lambda)$, equal to 1 for $b(z) \leqq \lambda$ and 0 for $b(z) \geqq \lambda + 2\lambda^{1-\nu}$ and if $\nu < \varrho'$ the estimates (28.34)–(28.36) guarantee that

$$e(z, \lambda) \in \Gamma^{0,\,0}_{\tilde{\varrho},\,\tilde{\sigma}}\,, \qquad \tilde{\varrho} > 0\,, \qquad \tilde{\sigma} > 0\,. \tag{30.7}$$

We have to verify that all the conditions of Theorem 28.1 are fulfilled. The condition $\mathscr{E}_\lambda^* = \mathscr{E}_\lambda$ is obvious since $e(z, \lambda)$ is real-valued. The fact that $\mathscr{E}_\lambda$ belongs to the trace class follows from Proposition 27.2.

Denote the Weyl symbol of an arbitrary operator A by $\sigma(A)$. Obviously

$$\sigma(\mathscr{E}_\lambda^2 - \mathscr{E}_\lambda) = \sum_{0 < |\alpha + \beta| < N} c_{\alpha\beta}\,[\partial_\xi^\beta \partial_x^\alpha e(z, \lambda)]\,[\partial_\xi^\alpha \partial_x^\beta e(z, \lambda)] + (e^2 - e) + r_N\,, \tag{30.8}$$

where $r_N \in \Gamma^{-2N\tilde{\varrho},\,-2N\tilde{\sigma}}_{\varrho,\,\sigma}$. Note that all terms in the sum, except for r_N, are supported where $\lambda \leqq a(z) \leqq \lambda(1 + 2\lambda^{-\nu})$, and if we apply Proposition 27.3 to each term, we obtain the estimate

$$\|\mathscr{E}_\lambda^2 - \mathscr{E}_\lambda\|_1 = O(V(\lambda + 2\lambda^{1-\nu}) - V(\lambda))\,.$$

But it follows from Proposition 28.3 that

$$V'(\lambda)/V(\lambda) = O(\lambda^{\varkappa - 1})\,,$$

which, by Proposition 28.2, gives the estimate

$$V(\lambda + 2\lambda^{1-\nu}) - V(\lambda) = O(\lambda^{\varkappa - \nu} V(\lambda))\,.$$

Therefore

$$\|\mathscr{E}_\lambda^2 - \mathscr{E}_\lambda\|_1 = O(\lambda^{\varkappa - \nu} V(\lambda))\,. \tag{30.9}$$

In addition, it follows from Proposition 27.2 that

$$\mathrm{Sp}\,\mathscr{E}_\lambda = (2\pi)^{-n} \int e(z, \lambda)\,dz$$

$$= V(\lambda) + O(V(\lambda + 2\lambda^{1-\nu}) - V(\lambda)) = V(\lambda)\,(1 + O(\lambda^{\varkappa - \nu}))\,. \tag{30.10}$$

Note that we must take $v < \varrho'$. Choosing $v = \varrho' - \varepsilon$, where $\varepsilon > 0$, we may rewrite (30.9) and (30.10) in the form

$$\| \mathscr{E}_\lambda^2 - \mathscr{E}_\lambda \|_1 = O(\lambda^{\varkappa - \varrho' + \varepsilon} V(\lambda)) \tag{30.11}$$

$$\mathrm{Sp}\, \mathscr{E}_\lambda = V(\lambda)\,(1 + O(\lambda^{\varkappa - \varrho' + \varepsilon})). \tag{30.12}$$

2. Let us now verify requirement 3° of Theorem 28.1:

$$\mathscr{E}_\lambda (A - \lambda I)\, \mathscr{E}_\lambda \leqq C\lambda^{1 - v}.$$

We write this inequality in the form

$$\mathscr{E}_\lambda (\lambda I - A)\, \mathscr{E}_\lambda + C\lambda^{1 - v} \geqq 0. \tag{30.13}$$

Next we compute the Weyl symbol of $\mathscr{E}_\lambda (\lambda I - A)\, \mathscr{E}_\lambda$. We have

$$\sigma(\mathscr{E}_\lambda (\lambda I - A)) = \sum_{|\alpha + \beta| < N} c_{\alpha\beta}\,(\partial_x^\alpha \partial_\xi^\beta e)\,(\partial_x^\beta \partial_\xi^\alpha (\lambda - a(z))) + r_N,$$

where $r_N \in \Gamma^{m - N(\varrho + \tilde{\varrho}),\, 1 - N\tilde{\sigma}}_{\min(\varrho,\tilde{\varrho}),\, 0}$.

Applying once again the composition formula, we obtain

$$\sigma(\mathscr{E}_\lambda (\lambda I - A)\, \mathscr{E}_\lambda) = \sum c_{\gamma_1 \gamma_2 \gamma_3}\,(\partial_z^{\gamma_1} e)\,(\partial_z^{\gamma_2} e)\,(\partial_z^{\gamma_3}(\lambda - a(z))) + r_N, \tag{30.14}$$

where $\tilde{r}_N \in \Gamma^{m - N(\varrho + \tilde{\varrho}),\, 1 - N\tilde{\sigma}}_{\min(\varrho,\tilde{\varrho}),\, 0}$, the sum is finite and $c_{000} = 1$.

It is clear that the remainder $\tilde{r}_N$ cannot affect (30.13), and therefore we will estimate the compactly supported terms.

We will show that for $|\gamma_1| + |\gamma_2| \neq 0$ and for some $\tilde{\varrho} > 0$ and $\tilde{\sigma} > 0$, we have the estimates

$$|(\partial^{\gamma_1} e)\,(\partial^{\gamma_2} e)\,(\partial^{\gamma_3}(\lambda - a))|$$

$$\leqq C\langle z \rangle^{-\tilde{\varrho}(|\gamma_1| + |\gamma_2| + |\gamma_3|)}\, \lambda^{1 - v - \tilde{\sigma}(|\gamma_1| + |\gamma_2| + |\gamma_3|)}. \tag{30.15}$$

To begin with let $\gamma_3 = 0$. Then (30.15) holds in view of the fact that $|\lambda - a(z)| \leqq 2\lambda^{1 - v}$ on the support of $(\partial^{\gamma_1} e)\,(\partial^{\gamma_2} e)$. Let $\gamma_3 \neq 0$, then

$$|(\partial^{\gamma_1} e)\,(\partial^{\gamma_2} e)\,(\partial^{\gamma_3} a)| \leqq C\langle z \rangle^{-\tilde{\varrho}(|\gamma_1| + |\gamma_2|)}\, \lambda^{-\tilde{\sigma}(|\gamma_1| + |\gamma_2|)}\, \lambda^{1 - \varrho'|\gamma_3|}$$

$$= C\langle z \rangle^{-\tilde{\varrho}(|\gamma_1| + |\gamma_2|)}\, \lambda^{1 - v}\, \lambda^{-|\gamma_3|(\varrho' - v/|\gamma_3|)}.$$

Note that $\varrho' - v/|\gamma_3| \geqq \varrho' - v > 0$. Therefore

$$|(\partial^{\gamma_1} e)\,(\partial^{\gamma_2} e)\,(\partial^{\gamma_3} a)| \leqq C\lambda^{1 - v}\langle z \rangle^{-\tilde{\varrho}(|\gamma_1| + |\gamma_2|)}\, \lambda^{-\tilde{\sigma}(|\gamma_1| + |\gamma_2|)}\, \lambda^{-\sigma_1|\gamma_3|},$$

where $\sigma_1 = \varrho' - \nu$. Taking into account, that the inequalities $\lambda \leq a \leq \lambda + 2\lambda^{1-\nu}$ and $\langle z \rangle^{m_0} \leq C\lambda$ hold on the support of $(\partial^{\gamma_1} e)(\partial^{\gamma_2} e)$, we obtain the estimate (30.15) (although perhaps with smaller $\tilde{\varrho}$ and $\tilde{\sigma}$ than in (30.7)).

Therefore we obtain

$$(\partial^{\gamma_1} e)(\partial^{\gamma_2} e)(\partial^{\gamma_3}(\lambda - a)) \in \Gamma^{0,\, 1-\nu}_{\tilde{\varrho},\, \tilde{\sigma}}, \qquad |\gamma_1| + |\gamma_2| \neq 0. \tag{30.16}$$

Now, repeating the reasoning, used at the beginning of this section to prove the semi-boundedness of A, we obtain that the $L^2(\mathbb{R}^n)$-norm of the operator with Weyl symbol $(\partial^{\gamma_1} e)(\partial^{\gamma_2} e)(\partial^{\gamma_3}(\lambda - a))$ does not exceed $C\lambda^{1-\nu}$, so this term also does not affect (30.13).

By similar arguments, one verifies that

$$e^2 \partial^{\gamma_3}(\lambda - a) \in \Gamma^{0,\, 1-\nu}_{\tilde{\varrho},\, 0}, \qquad \gamma_3 \neq 0, \tag{30.17}$$

hence this term also cannot affect (30.13).

Finally let us investigate the function $e^2(\lambda - a) = q$. The function $q(z, \lambda)$ has the following properties:

$$q(z, \lambda) \geq -C\lambda^{1-\nu}, \qquad \partial^{\gamma} q \in \Gamma^{0,\, 1-\nu}_{\tilde{\varrho},\, 0}, \qquad |\gamma| > 0.$$

If P is the operator with the anti-Wick symbol $q(z, \lambda)$ and $p(z, \lambda)$ is its Weyl symbol, then it follows from Lemma 29.1 that $q - p \in \Gamma^{0,\, 1-\nu}_{\tilde{\varrho},\, 0}$. But it is clear that $P \geq -C\lambda^{1-\nu}$ and $\|Q - P\| \leq C\lambda^{1-\nu}$. (30.13) follows from this. Putting $\nu = \varrho' - \varepsilon$ in this relation, we obtain

$$\mathscr{E}_\lambda (A - \lambda I) \, \mathscr{E}_\lambda \leq C\lambda^{1 - \varrho' + \varepsilon}. \tag{30.18}$$

3. Let us now verify that

$$(I - \mathscr{E}_\lambda)(A - \lambda I)(I - \mathscr{E}_\lambda) + C\lambda^{1-\nu} \geq 0 \tag{30.19}$$

for sufficiently large $C > 0$. Applying Theorem 29.1, we see that the symbol of the left-hand side of (30.19) has the form

$$C\lambda^{1-\nu} + \sum c_{\gamma_1 \gamma_2 \gamma_3} \partial^{\gamma_1}(1 - e) \cdot \partial^{\gamma_2}(1 - e) \cdot \partial^{\gamma_3}(a - \lambda) + r_N, \tag{30.20}$$

where $r_N \in \Gamma^{m - N(\varrho + \tilde{\varrho}),\, 1 - N\tilde{\sigma}}_{\min(\varrho,\, \tilde{\varrho}),\, 0}$ and $c_{000} = 1$.

The operator R_N with the Weyl symbol r_N can be estimated in norm by $\|R_N\| \leq C\lambda^{1 - N\tilde{\sigma}}$ and so cannot affect (30.19).

Let us estimate the principal part of (30.20), the symbol

$$q(z, \lambda) = (1 - e(z, \lambda))^2 \, (a(z) - \lambda) + C\lambda^{1-\nu}. \tag{30.21}$$

First note that

$$q(z,\lambda) \geqq \lambda^{1-\nu}. \tag{30.22}$$

for sufficiently large C. Now we show that

$$|\partial_z^\gamma q(z,\lambda)| \leqq C_\gamma q(z,\lambda) \langle z \rangle^{-\tilde{\varrho}|\gamma|} \lambda^{-\tilde{\sigma}|\gamma|} \tag{30.23}$$

for some $\tilde{\varrho} > 0$ and $\tilde{\sigma} > 0$.

We have

$$\partial_z^\gamma q = \sum c_{\gamma'\gamma''} \partial_z^{\gamma'} (1-e)^2 \partial_z^{\gamma''}(a-\lambda).$$

If $\gamma' \neq 0$ then the corresponding term may be estimated as in step 2. of this proof. If $\gamma' = 0$, then for $\gamma \neq 0$ we have the estimate

$$|(1-e)^2 \partial^\gamma(a-\lambda)| \leqq Ca(z)^{1-\varrho'|\gamma|} \leqq (q(z,\lambda)+2\lambda)(a(z))^{-\varrho'|\gamma|}, \tag{30.24}$$

on the support of $1-e$, since $a(z) \leqq q(z,\lambda) + 2\lambda$. Furthermore

$$(q(z,\lambda)+2\lambda)(a(z))^{-\varrho'|\gamma|} \leqq q(z,\lambda)(a(z))^{-\varrho'|\gamma|} + 2\lambda^{1-\nu} \lambda^\nu a(z)^{-\varrho'|\gamma|}$$

$$\leqq q(z,\lambda)\lambda^{-\varrho'|\gamma|/2}\langle z\rangle^{-\varrho'|\gamma|/2m_0} + q(z,\lambda)(a(z))^{\nu-\varrho'|\gamma|} \tag{30.25}$$

on $\operatorname{supp}(1-e)^2$, since $\lambda^{1-\nu} \leqq q(z,\lambda)$ and $\lambda^\nu \leqq a(z)^\nu$. Finally

$$q(z,\lambda)(a(z))^{\nu-\varrho'|\gamma|} \leqq q(z,\lambda)(a(z))^{-|\gamma|(\varrho'-\nu/|\gamma|)}$$

$$\leqq q(z,\lambda)(a(z))^{-|\gamma|\cdot(\varrho'-\nu)/2} \langle z\rangle^{-|\gamma|(\varrho-\nu)/2m_0}. \tag{30.26}$$

Now, (30.23) follows from (30.24)–(30.26). Note also that in view of the obvious estimate $|q(z,\lambda)| \leqq C\langle z\rangle^m \lambda^{1-\nu}$ (30.23) implies that $q(z,\lambda) \in \Gamma_{\tilde{\varrho},\tilde{\sigma}}^{m,1-\nu}$.

Estimating the remaining terms in the sum (30.20) in a similar fashion, we see that if we denote this whole sum by $q(z,\lambda)$, then the estimates (30.22) and (30.23) still hold and so $q(z,\lambda) \in \Gamma_{\tilde{\varrho},\tilde{\sigma}}^{m,1-\nu}$. As we have seen in the preceeding sections, we may take $\nu = \varrho' - \varepsilon$, with $\varepsilon > 0$. Applying Theorem 29.2, we see that (30.19) holds, or

$$(I-\mathscr{E}_\lambda)(A-\lambda I)(I-\mathscr{E}_\lambda) + C\lambda^{1-\varrho'+\varepsilon} \geqq 0. \tag{30.27}$$

4. To complete the proof of Theorem 30.1, it remains to note that the requirements 1°–5° of Theorem 28.1 for the construction of the "almost-projection operator" $\mathscr{E}_\lambda$ have already been verified ((30.11), (30.12), (30.18) and (30.27)) and property (28.7) for $V(\lambda)$ follows from (30.4) and Proposition 28.3. An Application of Theorem 28.1 then completes the proof of the asymptotic formula (30.6). $\square$

Problem 30.1. Compute the eigenvalues of the operator $A = -\Delta + |x|^2$ and verify directly the asymptotic formula (30.6).

Hint: The operator $A = -\Delta + |x|^2$ is the quantum mechanical energy operator for the harmonic oscillator, and its eigenvalues may be found in any text-book on quantum mechanics.

Problem 30.2. Show that if the Weyl symbol $b(z)$ of A is an elliptic polynomial whose principal homogeneous part does not take values in the ray $\arg \lambda = \varphi_0$, then the complex powers A^z and the ζ-function $\zeta(z) = \operatorname{Sp} A^z$ can be defined and the ζ-function admits a meromorphic continuation to the whole complex plane $\mathbb{C}$.

Find its poles, its residues and the values of $\zeta(z)$ at the points $0, 1, 2, \ldots$. Obtain here the asymptotic formula for $N(\lambda)$ (without estimating the remainder term) using the Tauberian theorem of Ikehara.

Appendix 1
Wave Fronts and Propagation of Singularities

In this appendix we present the definition and the simplest properties of the wave front of a distribution as introduced by Hörmander [6]. The concept of wave front is important in that it allows a microlocal (localized at a point of the cotangent bundle) formulation of the theorems on regularity of solutions of differential equations and also clarifies questions connected with the propagation of singularities. The wave fronts also play an important role on spectral theory, and are naturally connected with pseudodifferential operators. While leaving out many important questions of the theory of wave fronts, *I* nevertheless thought it useful to add this short appendix.

A.1.1 Wave front of a distribution

Definition A.1.1. Let X be an open set in $\mathbb{R}^n$, let $(x_0, \xi_0) \in X \times (\mathbb{R}^n \setminus \{0\})$ and $u \in \mathcal{D}'(X)$. We shall write $(x_0, \xi_0) \notin WF(u)$ if there exists $v \in \mathcal{E}'(X)$ such that $u = v$ in a neighborhood of x_0 and

$$|\tilde{v}(\xi)| \leq C_N \langle \xi \rangle^{-N} \quad \text{if} \quad \left| \frac{\xi}{|\xi|} - \frac{\xi_0}{|\xi_0|} \right| < \varepsilon, \tag{A.1.1}$$

for sufficiently small $\varepsilon > 0$ and arbitrary $N > 0$, i.e. $\tilde{v}(\xi)$ is rapidly decreasing in a conic neighbourhood of ξ_0.

Thus $WF(u)$ is a closed conic set in $X \times (\mathbb{R}^n \setminus \{0\})$ which we call *the wave front of the distribution u*.

Lemma A.1.1. *If $\varphi(x) \in C_0^\infty(X)$ and $(x_0, \xi_0) \notin WF(u)$ then $(x_0, \xi_0) \notin WF(\varphi u)$.*

Proof. We have to show that if $\tilde{v}$ is rapidly decreasing in an open cone Γ, then so is $\widetilde{\varphi v}$. Now

$$\widetilde{\varphi v}(\xi) = \int \tilde{v}(\xi - \eta)\, \tilde{\varphi}(\eta)\, d\eta$$

$$= \int_{|\eta| \leq R} \tilde{v}(\xi - \eta)\, \tilde{\varphi}(\eta)\, d\eta + \int_{|\eta| \geq R} \tilde{v}(\xi - \eta)\, \tilde{\varphi}(\eta)\, d\eta$$

and

$$|\widetilde{\varphi v}(\xi)| \leq C \sup_{|\eta| \leq R} |\tilde{v}(\xi-\eta)| + C_L \int_{|\eta| \geq R} (1+|\xi-\eta|)^p (1+|\eta|)^{-L} d\eta$$

$$\leq C \sup_{|\eta| \leq R} |\tilde{v}(\xi-\eta)| + C_L (1+|\xi|)^p \int_{|\eta| \geq R} (1+|\eta|)^{p-L} d\eta$$

$$\leq C \sup_{|\eta| \leq R} |\tilde{v}(\xi-\eta)| + C_L (1+|\xi|)^p R^{n+p-L}.$$

Putting $R = |\xi|^{1/2}$, we see that if ξ belongs to a cone slightly smaller than Γ, then $\xi - \eta \in \Gamma$ for large $|\xi|$ and $|\eta| \leq R$. In addition, $|\xi-\eta| \sim |\xi|$ and $R^{n+p-L} \sim |\xi|^{(n+p-L)/2}$. Picking a large L we see that $\widetilde{\varphi v}(\xi)$ is rapidly decreasing in ξ as $|\xi| \to \infty$, $\xi \in \Gamma$. $\square$

Corollary A.1.1. *In Definition A.1.1 we may put* $v = \varphi u$, *for* $\varphi \in C_0^\infty(X)$.

Proof. We may choose first v as in the definition and then φ such $\varphi = 1$ in a neighborhood of x_0 and $\varphi u = \varphi v$. It remains only to apply Lemma A.1.1. $\square$

Lemma A.1.2. *Let* $\pi\colon X \times (\mathbb{R}^n \setminus 0) \to X$ *be the natural projection and* $u \in \mathcal{D}'(X)$. *Then*

$$\pi WF(u) = \operatorname{sing\,supp} u.$$

Proof. a) If $x_0 \notin \operatorname{sing\,supp} u$ pick $\varphi \in C_0^\infty(X)$ such that $\varphi = 1$ in a neighbourhood of x_0, $\varphi = 0$ in a neighbourhood of $\operatorname{sing\,supp} u$. Then we see that $\varphi u \in C_0^\infty(X)$, from which $\widetilde{\varphi u} \in S(\mathbb{R}^n)$ i.e. $x_0 \notin \pi WF(u)$.

b) Let $x_0 \notin \pi WF(u)$. Then for any $\xi_0 \in \mathbb{R}^n \setminus \{0\}$ there exists a function $\varphi_{\xi_0}(x) \in C_0^\infty(X)$ and a conical neighbourhood Γ_{ξ_0} of ξ_0 such that $\varphi_{\xi_0}(x) = 1$ near x_0 and $(\widetilde{\varphi_{\xi_0} u})(\xi)$ decreases rapidly in Γ_{ξ_0}. Let $\Gamma_{\xi_1}, \ldots, \Gamma_{\xi_N}$ be a covering of $\mathbb{R}^n \setminus \{0\}$. Putting $\varphi = \prod_{j=1}^N \varphi_{\xi_j}$ we see that $\widetilde{\varphi u}(\xi)$ decreases rapidly everywhere so that $\varphi u \in C_0^\infty(X)$ i.e. $u \in C^\infty$ in a neighbourhood of x_0 so $x_0 \notin \operatorname{sing\,supp} u$.

Proposition A.1.1. *Let* $u \in \mathcal{D}'(X)$ *and* $(x_0, \xi_0) \notin WF(u)$. *Then there exists a classical properly supported* ΨDO $A \in CL^0(X)$ *such that* $\sigma_A \equiv 1 \pmod{S^{-\infty}}$ *in a conic neighbourhood of* (x_0, ξ_0) *and* $Au \in C_0^\infty(X)$.

Proof. Let $\varphi \in C_0^\infty(X)$, $\varphi = 1$ in a neighbourhood of x_0 and suppose $\widetilde{\varphi u}(\xi)$ decreases rapidly in a conic neighbourhood of ξ_0. Let $\chi(\xi)$ be supported in this neighbourhood with $\chi(t\xi) = \chi(\xi)$ for $t \geq 1$, $|\xi| \geq 1$ and $\chi(\xi) \in C^\infty(X)$ with $\chi(\xi) = 1$ in some smaller conic neighbourhood of ξ_0. Then $\chi(\xi)\,\widetilde{\varphi u}(\xi)$ decreases rapidly so that $\chi(D)(\varphi(x) u(x)) \in C^\infty$. But then $\psi(x) \chi(D)(\varphi(x) u(x)) \in C_0^\infty(X)$ if $\psi \in C_0^\infty(X)$. We may pick ψ so that $\psi(x) = 1$ in a neighbourhood of x_0 and then the ΨDO $A = \psi(x)\chi(D)\varphi(x)$ satisfies all the required conditions. $\square$

Proposition A.1.2. *Suppose we are given* $u \in \mathcal{D}'(X)$, $(x_0, \xi_0) \in X \times (\mathbb{R}^n \setminus \{0\})$ *and an operator* $A \in CL^m(X)$ *with principal symbol* $a_m(x, \xi)$. *So that* Au *makes* *sense let either $u \in \mathscr{E}'(X)$ or A be properly supported. Finally assume that $a_m(x_0, \xi_0) \neq 0$ and $Au \in C^\infty(X)$. Then $(x_0, \xi_0) \notin WF(u)$.*

Proof. a) By the standard construction of a parametrix in a conic neighbourhood of (x_0, ξ_0) (cf. §5), we obtain a properly supported ΨDO $B \in CL^{-m}(X)$ such that $\sigma_{BA} \equiv 1 \ (mod \, S^{-\infty})$ in this neighbourhood. Obviously $BAu \in C^\infty(X)$ so that replacing A by BA we could obtain $A = I(\mathrm{mod}\, S^{-\infty})$ in the same conic neighbourhood of the point (x_0, ξ_0).

b) Now let $\chi(\xi) = 1$ in a neighbourhood of ξ_0, $\chi(\xi) \in C^\infty(\mathbb{R}^n)$, and $\chi(\xi)$ homogeneous of order 0 in ξ for $|\xi| \geq 1$. Let $\varphi(x) \in C_0^\infty(\mathbb{R}^n)$, $\varphi = 1$ in a neighbourhood of x_0 and the supports of φ and χ be chosen so that

$$\varphi(x)\, \chi(\xi)\, \sigma_A(x, \xi) = \varphi(x)\, \chi(\xi) \qquad (\mathrm{mod}\, S^{-\infty}).$$

From this we obtain

$$\chi(D)\, \varphi(x)\, A - \chi(D)\, \varphi(x) \in L^{-\infty}, \tag{A.1.2}$$

and by $\chi(D)\, \varphi(x)\, Au \in C^\infty(X)$, it follows from (A.1.2) that

$$\chi(D)\, \varphi(x)\, u \in C^\infty(\mathbb{R}^n). \tag{A.1.3}$$

c) Now we show that

$$\chi(D)\, \varphi(x)\, u \in S(\mathbb{R}^n), \tag{A.1.4}$$

from which it follows that $\chi(\xi)\, \widetilde{\varphi u}(\xi) \in S(\mathbb{R}^n)$ and in particular, that $\widetilde{\varphi u}(\xi)$ decreases rapidly in a conic neighbourhood of ξ_0 as required. (A.1.4) follows from the following lemma and (A.1.3)

Lemma A.1.3. *Let $v \in \mathscr{E}'(\mathbb{R}^n)$, $\chi(\xi) \in S^m_{\varrho, 0}$. Then for $\varrho\,(x, \mathrm{supp}\, v) \geq 1$ we have*

$$|D^\alpha \chi(D)\, v(x)| \leq C_{\alpha, N}\, |x|^{-N} \tag{A.1.5}$$

Proof. Since $\xi^\alpha \chi(\xi) \in S^{m+|\alpha|}_{\varrho, 0}$ it all reduces to the case $\alpha = 0$. Further since $v = \sum_{|\alpha| \leq p} D^\alpha v_\alpha$ with continuous v_α, we may reduce to the situation where v is continuous.

We have

$$\chi(D)\, v(x) = \int e^{i(x-y)\cdot \xi}\, \chi(\xi)\, v(y)\, dy\, d\xi. \tag{A.1.6}$$

Integrating by parts and using the formula

$$|x-y|^{-2N}(-\Delta_\xi)^N\, e^{i(x-y)\cdot \xi} = e^{i(x-y)\cdot \xi},$$

from (A.1.6) we obtain

$$\chi(D)\, v(x) = \int e^{i(x-y)\cdot \xi}((-\Delta_\xi)^N\, \chi(\xi))\, |x-y|^{-2N}\, v(y)\, dy\, d\xi, \tag{A.1.7}$$

which makes sense for $\varrho(x, \mathrm{supp}\, v) \geqq 1$. Picking N so large that $(-\Delta_\xi)^N$ $\chi(\xi) \in S_{\varrho,0}^{-n-1}$, we see that the integral (A.1.7) converges absolutely and is estimated by $C\langle x \rangle^{-2N}$ for $\varrho(x, \mathrm{supp}\, v) \geqq 1$. $\quad\square$

Remark A.1.1. The condition $a_m(x_0, \xi_0) \neq 0$ is sometimes called *ellipticity of A at* (x_0, ξ_0). It is easy to formulate and prove the hypoelliptic analogue of Proposition A.1.2. We leave this for the reader as an excercise.

Corollary A.1.2. *For* $A \in CL^m(X)$ *denote*

$$\mathrm{char}\,(A) = \{(x, \xi) \in X \times (\mathbb{R}^n \setminus 0): a_m(x, \xi) = 0\}\,.$$

Then if $Au = f \in C^\infty(X)$ *we have* $WF(u) \subset \mathrm{char}\,(A)$. *In particular, if* $\mathrm{char}\,(A) = \emptyset$ *we have* $u \in C^\infty(X)$.

Corollary A.1.3. *If* $u \in \mathscr{E}'(X)$, *then*

$$WF(u) = \bigcap_{\substack{A \in CL^0(X) \\ Au \in C^\infty(X)}} \mathrm{char}\,(A)\,. \tag{A.1.8}$$

This holds for $u \in \mathscr{D}'(X)$ *if we take the intersection only over properly supported A.*

The importance of Corollary A.1.3 is that (A.1.8) shows how to define $WF(u)$ invariantly as a closed conic subset of T^*X when X is a manifold. We now generalize Proposition A.1.2 even more, by weakening the requirement $Au \in C^\infty(X)$.

Proposition A.1.3. *Again let* $A \in CL^m(X)$, $u \in \mathscr{D}'(X)$ *and either A be properly supported or* $u \in \mathscr{E}'(X)$. *Then, assuming* $a_m(x_0, \xi_0) \neq 0$ *and* $(x_0, \xi_0) \notin WF(Au)$, *we have* $(x_0, \xi_0) \notin WF(u)$. *In other words,*

$$WF(u) \subset \mathrm{char}\,(A) \cup WF(Au)\,. \tag{A.1.9}$$

Proof. By proposition A.1.1 there exists a properly supported $P \in CL^0(X)$, with $\sigma_P \equiv 1 \pmod{S^{-\infty}}$ in a conic neighbourhood of (x_0, ξ_0) and $(PA)(u) \in C^\infty(X)$. But then, from Proposition A.1.2 it obviously follows that $(x_0, \xi_0) \notin WF(u)$. $\quad\square$

Proposition A.1.4 (Pseudolocality of ΨDO). *Let* $u \in \mathscr{D}'(X)$, $A \in L_{\varrho,\delta}^m(X)$ $0 \leq \delta < \varrho \leq 1$ *and assume either A is properly supported or* $u \in \mathscr{E}'(X)$. *Then if* $(x_0, \xi_0) \notin WF(u)$, *we have* $(x_0, \xi_0) \notin WF(Au)$. *In other words,*

$$WF(Au) \subset WF(u)\,. \tag{A.1.10}$$

Proof. The condition $(x_0, \xi_0) \notin WF(u)$ amounts to the existence of a properly supported ΨDO $P \in CL^0(X)$ such that $Pu \in C^\infty(X)$ and $\sigma_P \equiv 1 \pmod{S^{-\infty}}$ in a conic neighbourhood of (x_0, ξ_0). Now let Q be a properly supported ΨDO

$Q \in CL^0(X)$ such that $q_0(x_0, \xi_0) \neq 0$, (q_0 be principal symbol of Q) $\sigma_Q \in S^{-\infty}$ outside of some sufficiently small conic neighbourhood of (x_0, ξ_0) and

$$PQ \equiv Q \, (\mathrm{mod} \, L^{-\infty}) \quad \text{if} \quad QP \equiv Q \, (\mathrm{mod} \, L^{-\infty}).$$

Let us demonstrate that $QAu \in C^\infty(X)$. We have $QA - QAP \in L^{-\infty}$ so it suffices to show $QAPu \in C^\infty(X)$. This however, is obvious since $Pu \in C^\infty(X)$. Now $(x_0, \xi_0) \notin WF(Au)$ follows from Proposition A.1.2. $\square$

Corollary A.1.4. *If $A \in CL^m(X)$, then*

$$WF(Au) \subset WF(u) \subset WF(Au) \cup \mathrm{char}(A). \tag{A.1.11}$$

Corollary A.1.5. *If the operator $A \in CL^m(X)$ is elliptic, then*

$$WF(Au) = WF(u) \tag{A.1.12}$$

Exercise A.1.1. Compute the wave fronts of the following distributions:

a) $\delta(x)$;

b) $\delta(x') \oplus 1(x'')$, $x' \in \mathbb{R}^k$, $x'' \in \mathbb{R}^{n-k}$;

c) δ_S, where S is a smooth submanifold in $\mathbb{R}^n$ ($\langle \delta_S, \varphi \rangle$ is defined as the integral of the function φ restricted to the surface S with respect to the induced measure);

d) $(x + i0)^{-1}$ on $\mathbb{R}^1$;

e) the indicator function of an angle in $\mathbb{R}^2$ (the function which is equal to 1 in the angle and 0 outside it).

A.1.2 Applications: Product of two distributions, trace of a distribution on a submanifold

1) Let $u_j \in \mathscr{D}'(X), j = 1, 2$. What does $u_1 \cdot u_2$ mean? It should be the ordinary product $u_1 \cdot u_2$ if, for example, u_1 and u_2 are continuous or if one of them is smooth, and it should be a natural extension (e.g. by continuity in some sense). It will turn out that we can define $u_1 \cdot u_2$ under the condition

$$WF(u_1) + WF(u_2) \subset X \times (\mathbb{R}^n \setminus 0), \tag{A.1.13}$$

i.e. if there are no $(x, \xi) \in WF(u_1)$ such that $(x, -\xi) \in WF(u_2)$.

Since the product is a bilinear operation, then using a partition of unity we may assume that u_1, u_2 are in $\mathscr{E}'(\mathbb{R}^n)$ and have sufficiently small supports so that $\tilde{u}_1(\xi)$ is rapidly decreasing outside a cone Γ_1 and $\tilde{u}_2(\xi)$ is rapidly decreasing outside a cone Γ_2, whereby $\Gamma_1 + \Gamma_2 \subset \mathbb{R}^n \setminus \{0\}$ (i.e. Γ_1 and Γ_2 do not contain opposite points). In the usual situation one has $\widetilde{u_1 u_2}(\xi) = (\tilde{u}_1 * \tilde{u}_2)(\xi)$, where

$$\tilde{u}_1 * \tilde{u}_2(\xi) = \int \tilde{u}_1(\xi - \eta) \, \tilde{u}_2(\eta) \, d\eta. \tag{A.1.14}$$

In our case this integral converges absolutely also, since either $|\tilde{u}_1(\xi-\eta)|$ or $|\tilde{u}_2(\eta)|$ rapidly tends to zero as $|\eta|\to+\infty$. Put

$$u_1 u_2 = F^{-1}\int\tilde{u}_1(\xi-\eta)\,\tilde{u}_2(\eta)\,d\eta\,. \qquad (A.1.15)$$

Why is this an extension by continuity? We can for instance show that if $\chi\in C_0^\infty(\mathbb{R}^n)$, $\int\chi(x)\,dx=1$, $\chi(x)=\varepsilon^{-n}\chi(\varepsilon^{-1}x)$ and $u^{(\varepsilon)}=u*\chi_\varepsilon(\in C_0^\infty(\mathbb{R}^n))$, then $\lim\limits_{\varepsilon\to 0+}u_1^{(\varepsilon)}u_2^{(\varepsilon)}=u_1 u_2$ in the topology of $\mathscr{D}'(\mathbb{R}^n)$ where $u_1 u_2$ is understood in the sense of (A.1.15).

Example A.1.1. Let $u_k\in\mathscr{D}'(\mathbb{R}^2)$,

$$\langle u_k,\varphi\rangle=\int\chi_k(x_1)\,\varphi(x_1,kx_1)\,dx_1\,,\qquad \chi_k(t)\in C_0^\infty(\mathbb{R}^1)\,,$$

so that $\operatorname{supp}u_k\subset\{(x_1,x_2):x_2=kx_1\}$. Consider the product $u_k\cdot u_0$. We have

$$\tilde{u}_k(\xi)=\int\chi_k(x_1)\,e^{i(\xi_1 x_1+\xi_2 kx_1)}\,dx_1=\tilde{\chi}_k(\xi_1+k\xi_2)\,.$$

From this we see that $WF(u_k)$ is the set of all normals to the line $x_2=kx_1$, lying over $\pi_k^{-1}(\operatorname{supp}\chi_k)$, where $\pi_k:(x_1,kx_1)\to x_1$. Consider now the convolution $\tilde{u}_k*\tilde{u}_0$:

$$(\tilde{u}_0*\tilde{u}_k)(\xi)=\int\tilde{\chi}_0(\xi_1-\eta_1)\,\tilde{\chi}_k(\eta_1+k\eta_2)\,d\eta_1\,d\eta_2$$

$$=\int\tilde{\chi}_0(\eta_1)\,d\eta_1\cdot\int\bar{\chi}_k(k\eta_2)\,d\eta_2=\frac{1}{k}\,\chi_0(0)\,\chi_k(0);$$

from which

$$u_0 u_k=\frac{1}{k}\,\chi_0(0)\cdot\chi_k(0)\cdot\delta(x)\,.$$

The limit as $k\to 0$ does not exist, which is natural, since for $k=0$ condition (A.1.13) fails.

2) Let $u\in\mathscr{D}'(X)$, Y a submanifold of X, NY the family of all normals to Y in T^*X (the normal bundle to Y.) If $WF(u)\cap NY=\emptyset$, then the trace $u|_Y$ is defined naturally in the same sense as for products. Indeed, localizing we may assume that $Y=\{x_1=\ldots=x_k=0\}$. For $u\in C_0^\infty(X)$ we have

$$\widetilde{u|_Y}=\int\tilde{u}(\xi_1,\ldots,\xi_n)d\xi_1\ldots d\xi_k. \qquad (A.1.16)$$

But by hypothesis vectors of the form $(\xi_1,\ldots,\xi_k,0,\ldots,0)$ are not contained in $WF(u)$, so $\tilde{u}$ decrease rapidly in their direction and the integral (A.1.16) is defined. In particular, if $u\in\mathscr{D}'(\mathbb{R}^{n+1})$ satisfies a differential equation of order m of the form

$$a\left(t,x,\frac{\partial}{\partial t},\frac{\partial}{\partial x}\right)u=f\in C^\infty(\mathbb{R}^{n+1}),\qquad t\in\mathbb{R}^1,\ x\in\mathbb{R}^n,$$

where the hyperplane $t = 0$ is non-characteristic, i.e. $a_m(0, x, 1, 0) \neq 0$ then all the restrictions $\left. \dfrac{\partial^k u}{\partial t^k} \right|_{t=0}$ are defined and lie in $\mathscr{D}'(\mathbb{R}^n)$. This means in particular that the Cauchy data make sense. Thereby $u|_{t=t_0}$ is a smooth function of t_0 with values in $\mathscr{D}'(\mathbb{R}^n)$.

A.1.3 The theorem on propagation of singularities

First, we state the simplest version of the theorem on propagation of singularities.

Theorem A.1.1. *Let $P \in CL^m(X)$ have a real principal symbol $p_m(x, \xi)$, $u \in \mathscr{D}'(X)$ and either P is properly supported or $u \in \mathscr{E}'(X)$ so that Pu makes sense. Then if I is any connected interval of a bicharacteristic of the function $p_m(x, \xi)$ not intersecting $WF(Pu)$ then either $I \subset WF(u)$ or $I \cap WF(u) = \emptyset$.*

In other words, in the complement of $WF(Pu)$, the set $WF(u)$ is invariant under the shifts along the trajectories of the Hamiltonian system

$$\begin{cases} \dot{\xi} = -\dfrac{\partial p_m}{\partial x}, \\[2mm] \dot{x} = \dfrac{\partial p_m}{\partial \xi}. \end{cases} \tag{A.1.17}$$

Example A.1.2. Let us prove that the wave equation $\dfrac{\partial^2 u}{\partial x_0^2} - \Delta u = 0$ in $\mathbb{R}^{n+1}$ cannot have solutions with isolated or compactly supported singularities. We have $m = 2$, $p_2(x, \xi) = -\xi_0^2 + |\xi|^2$. The system (A.1.17) has a solution $\xi = \mathrm{const}$, $x_0 = -2\xi_0 t$, $x = 2\xi t$. Let $0 \in \mathrm{sing\,supp}\, u$. Then there exists a point $(0, 0, \xi_0, \xi) \in WF(u)$, so by Proposition A.1.2 it is obvious that $|\xi|^2 = \xi_0^2$. By Theorem A.1.1. $(-2\xi_0 t, 2\xi t, \xi_0, \xi) \in WF(u)$ for any t and, in particular, $(-2\xi_0 t, 2\xi t) \in \mathrm{sing\,supp}\, u$ for any t which also yields the required result.

We now give a proof of Theorem A.1.1. due to V.N. Tulovsky. It is based on the following proposition describing wave fronts in terms of the action of distributions on rapidly oscillating exponentials.

Proposition A.1.5. *Let $u \in \mathscr{D}'(X)$. Then the condition $(x_0, \xi_0) \notin WF(u)$ is equivalent to the following condition;*

A. There exists $\varepsilon > 0$ such that if $\Phi(x, \theta)$ is a smooth real valued function defined for $|x - x_0| < \varepsilon$ and $\theta \in (\mathbb{R}^N \setminus 0)$, is a homogeneous function in θ of degree 1 such that $\Phi_x(x_0, \theta_0) = \xi_0$ for some $\theta_0 \neq 0$, then for an arbitrary symbol $\varphi \in CS^0(\{|x - x_0| < \varepsilon\} \times \mathbb{R}^N)$ vanishing for $|x - x_0| \geq \varepsilon/2$ there exists a conical neighbourhood Γ of the point θ_0 in $\mathbb{R}^N \setminus 0$ such that

$$|\langle u(x), \varphi(x, \theta) e^{-i\Phi(x, \theta)} \rangle| \leq C_N |\theta|^{-N}, \qquad \theta \in \Gamma, \ |\theta| \geq 1. \tag{A.1.18}$$

Proof. 1) Let condition A. be satisfied. Take $\varphi \in C_0^\infty(\mathbb{R}^n)$, such that $\varphi \equiv 1$ in a neighbourhood of x_0, $\varphi = 0$ for $|x - x_0| \geq \varepsilon/2$ and $\Phi(x, \theta) = \langle x, \theta \rangle$ so that $N = n$, $\theta_0 = \xi_0$. We then see that $\langle u(x), \varphi(x)\,e^{-ix\cdot\theta} \rangle = \widehat{\varphi u}(\theta)$ decreases rapidly in a conic neighbourhood of ξ_0. But from this it follows that $(x_0, \xi_0) \notin WF(u)$.

2) Let $(x_0, \xi_0) \notin WF(u)$. We will verify that condition A is satisfied. Let Φ and φ be as described in the condition. Without loss of generality, we can assume $u \in \mathscr{E}'(\mathbb{R}^n)$. Express the left-hand side of (A.1.18) in terms of $\tilde{u}(\xi)$:

$$\langle u(x), \varphi(x, \theta)e^{-i\Phi(x,\theta)} \rangle = \int \tilde{u}(\xi)e^{i[x\cdot\xi - \Phi(x,\theta)]}\varphi(x, \theta)\,dx\,d\xi, \qquad (A.1.19)$$

where the integral is understood as an oscillatory integral. Picking a small conic neighbourhood Γ_1 around ξ_0 in $\mathbb{R}^n$ such that $\tilde{u}(\xi)$ decreases rapidly in Γ_1 we can decompose this integral into the sum $I_1 + I_2$, where in I_1 the integral with respect to ξ is taken over Γ_1, and in I_2 over $\mathbb{R}^n \backslash \Gamma_1$. Let us estimate I_1 and I_2 separately.

a) In I_1 we integrate by parts with respect to x with the help of the exponent $e^{-i\Phi(x,\theta)}$. We put

$$'L = |\Phi_x|^{-2} \sum_{j=1}^{n} i\Phi_{x_j} \frac{\partial}{\partial x_j},$$

to obtain the identity $'Le^{-i\Phi} = e^{-i\Phi}$. The coefficients of $'L$ and L are homogeneous in θ of degree -1. Since $\Phi_x(x_0, \theta_0) = \xi_0 \neq 0$ we have $|\Phi_x(x, \theta)| \neq 0$ if $|x - x_0| < \varepsilon$ and $\theta \in \Gamma$, where Γ is a sufficiently small conic neighbourhood of θ_0. We have

$$I_1 = \int\limits_{\xi \in \Gamma_1} L^N\left(e^{ix\cdot\xi}\,\varphi(x, \theta)\right) \tilde{u}(\xi)\, e^{-i\Phi(x,\theta)}\, dx\, d\xi.$$

Since

$$|L^N[e^{ix\cdot\xi}\,\varphi(x, \theta)]| \leq C_N \langle\xi\rangle^N \langle\theta\rangle^{-N}, \qquad \theta \in \Gamma, \qquad |\theta| \geq 1,$$

then in view of the decrease of $\tilde{u}(\xi)$ in Γ_1 we obtain

$$|I_1| \leq C_N \langle\theta\rangle^{-N}, \qquad \theta \in \Gamma, \qquad |\theta| \geq 1. \qquad (A.1.20)$$

b) To estimate I_2 it is necessary to integrate by parts with respect to x again with the help of the exponent $e^{i[x\cdot\xi - \Phi(x,\theta)]}$. Choosing a sufficiently small conic neighbourhood Γ of θ_0 we have $\xi - \Phi_x(x, \theta) \neq 0, \theta \in \Gamma, \xi \in \mathbb{R}^n \backslash \Gamma_1, |x - x_0| < \varepsilon$, from which

$$\mathrm{grad}_x(x \cdot \xi - \Phi(x, \theta)) \neq 0,$$

for the same x, ξ, θ which allows us to carry out a standard integration by parts. In fact it is obvious that for some $C > 0$

$$|\xi - \Phi_x(x, \theta)| \geq C(|\xi| + |\theta|), \qquad \theta \in \Gamma, \qquad \xi \in \mathbb{R}^n \backslash \Gamma_1, \qquad |x - x_0| < \varepsilon.$$

Putting

$$^tL = -i\,|\xi - \Phi_x(x,\theta)|^{-2} \sum_{j=1}^{n} (\xi_j - \Phi_{x_j}(x,\theta))\, \frac{\partial}{\partial x_j},$$

we obtain, $^tL e^{i[x\cdot\xi - \Phi(x,\theta)]} = e^{i[x\cdot\xi - \Phi(x,\theta)]}$, and hence

$$I_2 = \int_{\xi \in \mathbb{R}^n \setminus \Gamma_1} e^{i[x\cdot\xi - \Phi(x,\theta)]} \, [L^N \varphi(x,\theta)]\, \hat{u}(\xi)\, dx\, d\xi. \tag{A.1.21}$$

Since $\hat{u}(\xi)$ satisfies $|\hat{u}(\xi)| \leq C(1+|\xi|)^{N_1}$ for sufficiently large C, N_1, then for sufficiently large N the integral (A.1.21) becomes absolutely convergent and for $\theta \in \Gamma$, $|x-x_0| < \varepsilon$ can be estimated by $C_p \langle\theta\rangle^{-p}$ for an arbitrary p. This together with (A.1.20) yields the required statement. $\square$

Remark A.1.2. If the point (x_0, ξ_0) and the functions φ and Φ depend on a parameter and all the conditions are satisfied uniformly, then the constant C_N in (A.1.18) can be selected so as to not depend on the parameter.

Proof of theorem A.1.1. 1) For simplicity consider first the operator $P = \frac{1}{i}\frac{\partial}{\partial x_n}$. Let $Pu = f$, $(x_0, \xi_0) \notin WF(f)$. The bicharacteristic of the symbol ξ_n of P, which passes through (x_0, ξ_0), is of the form $(x_0', (x_0)_n + t, \xi_0)$ where $(x_0)_n$ is the n-th coordinate of x_0, x_0' is the collection of its $(n-1)$ first coordinates and t is the parameter along the bicharacteristic. Let I be an interval on this bicharacteristic containing (x_0, ξ_0) and not intersecting $WF(f)$. We will show that either $I \subset WF(u)$ or $I \cap WF(u) = \emptyset$.

Let $\varphi(x,\theta)$ and $\Phi(x,\theta)$ be determined as in Proposition A.1.5. Put

$$\varphi_t(x,\theta) = \varphi(x', x_n - t, \theta), \qquad \Phi_t(x,\theta) = \Phi(x', x_n - t, \theta).$$

Then $\operatorname{supp}\varphi_t$ is close to $(x_0', (x_0)_n + t)$ and near this point the function Φ_t is defined. It is clear that by the choice of φ, Φ one can make φ_t and Φ_t to be any functions for which the conditions of Proposition A.1.5 are fulfilled at $(x_0', (x_0)_n + t, \xi_0)$. We have

$$\frac{d}{dt} \langle u, \varphi_t e^{-i\Phi_t}\rangle = \left\langle u, \frac{\partial}{\partial t}(\varphi_t e^{-i\Phi_t})\right\rangle = \left\langle u, \left(-\frac{\partial}{\partial x_n}\right)(\varphi_t e^{-i\Phi_t})\right\rangle$$

$$= \left\langle \frac{\partial}{\partial x_n} u, \varphi_t e^{-i\Phi_t}\right\rangle = \langle if, \varphi_t e^{-i\Phi_t}\rangle = R(t,\theta), \tag{A.1.22}$$

from which, by the condition $I \cap WF(f) = \emptyset$ and by Proposition A.1.5, it follows that for the values of t of interest to us, the estimate

$$|R(t,\theta)| \leq C_N \langle\theta\rangle^{-N}, \qquad \theta \in \Gamma, \ |\theta| \geq 1, \tag{A.1.23}$$

holds uniformly in t. Here Γ is a sufficiently small conic neighbourhood of θ_0. But from this

$$|\langle u, \varphi_{t_1} e^{-i\Phi_{t_1}}\rangle - \langle u, \varphi_{t_2} e^{-i\Phi_{t_2}}\rangle|$$

$$= \left| \int_{t_1}^{t_2} \frac{d}{dt} \langle u, \varphi_t e^{-i\Phi_t}\rangle \, dt \right| \leq |t_1 - t_2| \, C_N \langle\theta\rangle^{-N}. \tag{A.1.24}$$

Therefore, if for some t the function $\langle u, \varphi_t e^{-i\Phi_t}\rangle$ decreases rapidly for θ in Γ, then this holds for all t, yielding the required result.

2) Now let P be any classical first order ΨDO (i.e. $P \in CL^1(X)$) with real principal symbol $p_1(x, \xi)$. Again we will carry out the computation (A.1.22). For this we only have to find the dependence on the parameter τ for the functions φ and Φ, so that

$$\frac{1}{i} \frac{\partial}{\partial \tau} [\varphi e^{-i\Phi}] - {}^t P [\varphi e^{-i\Phi}] \tag{A.1.25}$$

rapidly decreases for θ in the cone Γ and uniformly in τ, where for $\tau = 0$ the conditions of Proposition A.1.5 are satisfied for the point (x_0, ξ_0).

The condition of decreasing for (A.1.25) leads to the equation for Φ,

$$\frac{\partial \Phi}{\partial \tau} - p_1(x, \Phi_x) = 0, \tag{A.1.26}$$

which we can solve (for small τ) for arbitrary $\Phi|_{\tau=0}$. In doing so, it is important to note that along the bicharacteristic $(x(\tau), \xi(\tau))$ of the function $p_1(x, \xi)$ we will have

$$\Phi_x(x(\tau), \theta_0, \tau) = \xi(\tau) \tag{A.1.27}$$

(cf. §17), provided this is so for $\tau = 0$, which we will assume. For the homogeneous components of φ, we obtain transport equations which are again solved for arbitrary $\varphi|_{\tau=0}$ by analogy with the transport equations for q_{-j} from §20. The solution procedure (cf. §17) shows, that the support of any homogeneous component of φ propagates along the bicharacteristics of the function $p_1(x, \xi)$. Thanks to this the proof in this case for small τ may be ended in a similar way to the case of the operator $i^{-1} \dfrac{\partial}{\partial x_n}$.

Let us now remark, that the necessity of restricting ourselves to small τ is of no importance, since it is enough to prove theorem A.1.1 for arbitrarily small parts of the bicharacteristics, since as an obvious consequence it will then be true also for large connected pieces.

3) Finally consider the case of an operator P of arbitrary degree m. Let Q be a classical elliptic properly supported ΨDO of degree $(1-m)$ with real principal

symbol $q(x, \xi)$. Put $P_1 = PQ$. Then $P_1 \in CL^1(X)$ and the principal symbol of P_1 is of the form

$$p_1(x, \xi) = p_m(x, \xi)\, q(x, \xi).$$

Note that in Theorem A.1.1, in view of Corollary A.1.4, it suffices to consider only null bicharacteristics. But from the relations

$$(p_1)_\xi = (p_m)_\xi\, q + p_m \cdot q_\xi = (p_m)_\xi\, q \qquad \text{for } p_m = 0$$
$$(p_1)_x = (p_m)_x\, q + p_m \cdot q_x = (p_m)_x\, q \qquad \text{for } p_m = 0$$

the null bicharacteristics of p_1 and p_m differ only by a change of parameter. Corollary A.1.5, being taken into account the result of theorem A.1.1 for P follows from the same result about P_1 already proved in 2). $\quad\square$

Problem A.1.1. Let A be a distribution determined by an oscillatory integral

$$\langle A, \varphi \rangle = \int e^{i\Phi(x,\theta)}\, a(x, \theta)\, \varphi(x)\, dx\, d\theta,$$

where Φ is a non-degenerate phase function, $a(x, \xi) \in S^m(X \times \mathbb{R}^N)$ (cf. §1). Show that

$$WF(A) \subset \{(x, \xi) \in X \times (\mathbb{R}^n \setminus 0): \ \exists \theta \in \mathbb{R}^N \setminus 0, \ \Phi'_\theta(x, \theta) = 0, \ \Phi'_x(x, \theta) = \xi \}.$$

Problem A.1.2. Let two distributions u_1, $u_2 \in \mathscr{D}'(X)$ satisfy (A.1.13) allowing them to be multiplied. Show that

$$WF(u_1 u_2) \subset [WF(u_1) + WF(u_2)] \cup WF(u_1) \cup WF(u_2).$$

Problem A.1.3. Prove that for the operator D_n^k, for any positive integer k, the same theorem on propagation of singularities is true as for D_n. What form does Theorem A.1.1 take?

Appendix 2
Quasiclassical Asymptotic Behaviour of Eigenvalues

Observables in quantum mechanics can be represented by operators of the form

$$(A_{(h)}u)(x) = \int e^{i(x-y)\cdot\xi}\, b\left(\frac{x+y}{2}, h\xi\right) u(y)\, dy\, d\xi, \qquad (A.2.1)$$

where the parameter $h > 0$ is the *Planck constant*; the operator $A_{(h)}$ is well defined on $S(\mathbb{R}^n)$ for example, if the function $b(z)$ belongs to $\Gamma_\varrho^m(\mathbb{R}^{2n})$.

Classical mechanics is the limiting case of quantum mechanics, when the Planck constant can be considered to be negligible. This motivates an interest in the asymptotic properties of operators of the form (A.2.1) as $h \to 0$; the corresponding asymptotic analysis is called *quasi-classical* or *semi-classical*.

A.2.1 Basic results

The change of variables $\xi \to h^{-1}\xi$ transforms (A.2.1) into

$$(A_{(h)}u)(x) = \frac{1}{h^n} \int e^{i(x-y)\cdot\xi/h}\, b\left(\frac{x+y}{2},\xi\right) u(y)\, dy\, d\xi; \qquad (A.2.2)$$

where the symbol no longer contains the parameter h, which now is included in the exponent instead.

We will say that $b(z)$ is the Weyl h-*symbol* of $A_{(h)}$ or, briefly, the h-*symbol* (in this appendix, we will not use the τ-symbols of chap. IV, which avoids any confusion). Clearly, the 1-symbol is then the ordinary Weyl symbol.

Between the h- and 1-symbols the following relation exists. Making the change of variables $x \to \sqrt{h}\,x$, $y \to \sqrt{h}\,y$, $\xi \to \sqrt{h}\,\xi$ in (A.2.2), this expression becomes

$$(A_{(h)}u)(x\sqrt{h}) = \int e^{i(x-y)\cdot\xi}\, b\left(\sqrt{h}\,\frac{x+y}{2},\ \sqrt{h}\,\xi\right) u(\sqrt{h}\,y)\, dy\, d\xi. \qquad (A.2.3)$$

In the space of functions on $\mathbb{R}^n$ introduce the dilatation operator

$$T_h\colon f(x) \to h^{n/4} f(\sqrt{h}\,x). \qquad (A.2.4)$$

It is easily seen that T_h is unitary on $L^2(\mathbb{R}^n)$. Using this operator, (A.2.3) can be written as $T_h A_{(h)} u = A_{(1)}^{(h)} T_h u$ or

$$A_{(h)} = T_h^{-1} A_{(1)}^{(h)} T_h, \qquad (A.2.5)$$

where $A_{(1)}^{(h)}$ is the operator with the 1-symbol $b^{(h)}(z) = b(\sqrt{h}\,z)$. Therefore the operator with h-symbol $b(z)$ is unitarily equivalent to the operator with the 1-symbol $b^{(h)}(z)$.

We will be interested in the quasicalssical asymptotic behaviour of the eigenvalues.

Definition A.2.1. Let $A_{(h)}$ be a self-adjoint operator semi-bounded from below. $N_h(\lambda)$ denotes the number of eigenvalues of the operator not exceeding λ (counting multiplicities). If there are points from the continuous spectrum of $A_{(h)}$ in the interval $(-\infty, \lambda]$, then by definition $N_h(\lambda) = +\infty$.

Remark A.2.1. The Glazman variational principle (28.1) remains valid also for $N_h(\lambda)$; the proof (cf. §28) can be taken over with minor changes to the case $N_h(\lambda)$.

To formulate the basic result, we need the following

Proposition A.2.1. *Let $A_{(h)}$ be an operator with the real h-symbol $b(z) \in H\Gamma_\varrho^{m,m_0}$, $m_0 \geq 0$. Then for any fixed $h > 0$ the operator $A_{(h)}$ is essentially self-adjoint.*

Proof. For $m_0 > 0$ the proposition follows from Theorem 26.2. An analysis of the proof of Theorem 26.2 shows that the strict inequality $m_0 > 0$ is only needed in order to ensure $A \pm iI \in HG_\varrho^{m,m_0}$. Under the assumptions of the proposition, for $m_0 = 0$ and $h = 1$ the inclusions $A \pm iI \in HG_\varrho^{m,0}$ follow from the estimates

$$|b(z) \pm i| > b(z),$$

$$|\partial^\gamma (b(z) \pm i)| \leq C_\gamma |b(z)| \, |z|^{-\varrho|\gamma|}$$
$$= C_\gamma |b(z) \pm i| \, |z|^{-\varrho|\gamma|} |b(z)|/|b(z) \pm i| \leq C_\gamma |b(z) \pm i| \, |z|^{-\varrho|\gamma|}.$$

For $h \neq 1$ one has to use (A.2.5) and the fact that $b^{(h)} \in H\Gamma_\varrho^{m,m_0}$ (as for $b(z)$ but with other constants in the estimates of the derivatives).

Let $A_{(h)}$ have a real h-symbol $b(z) \in H\Gamma_\varrho^{m,0}$. As in §30 changing the sign if necessary we may assume that $b(z) \geq C > 0$ for $|z| \geq R_0$. Put

$$V(\lambda) = (2\pi)^{-n} \int\limits_{b(z) < \lambda} dz. \qquad (A.2.6)$$

The main goal of this appendix is the proof of the following theorem.

Theorem A.2.1. *Let $A_{(h)}$ have a real h-symbol $b(z) \in H\Gamma_\varrho^{m,\,0}$, $b(z) \geq C > 0$ for $|z| \geq R_0$. Let λ_0 be such that $V(\lambda_0) < +\infty$. Then, for almost all $\lambda < \lambda_0$ and arbitrary $\varepsilon > 0$ we have the asymptotic formula*

$$N_h(\lambda) = h^{-n}(V(\lambda) + O(h^{1/2-\varepsilon})). \tag{A.2.7}$$

Remark A.2.2. Between the asymptotic formulae in h as $h \to 0$ and the ones in λ as $\lambda \to +\infty$ there is an intimate relation which can be explicitly exhibited when $b(z)$ is homogeneous: $b(tz) = t^s b(z)$, $t > 0$, $s > 0$. In accordance with (A.2.5) the operator with the h-symbol $b(z)$ is conjugate to the operator with symbol $b^{(h)}(z) = h^{s/2} b(z)$, so that $N_h(\lambda) = N(h^{-s/2}\lambda)$.

Remark A.2.3. Theorem A.2.1 is analogous to Theorem 30.1. In the latter we assumed the essential inequality $b(z) \geq C|z|^{m_0}$, $c > 0$, $m_0 > 0$. Theorem A.2.1 states the weaker dependence of the asymptotic behaviour in h on the behaviour of the symbol at infinity.

A.2.2 The idea of proof of Theorem A.2.1.

The proof of the theorem is based on the same considerations as the proof of Theorem 30.1. We will construct an approximate spectral projection $\mathscr{F}_h$ in the following way. Let χ_λ be the indicator function of the interval $(-\infty, \lambda]$; we construct a family of functions $\chi_{h,\lambda}$, converging to χ_λ as $h \to 0$. The operator $\mathscr{F}_h$ has the h-symbol $\chi_{h,\lambda}(b(z))$, where $b(z)$ is the h-symbol of $A_{(h)}$.

We will show that the family $\mathscr{F}_h$ has the following properties:

1°. $\mathscr{F}_h^* = \mathscr{F}_h$;

2°. $\mathscr{F}_h$ is a trace class operator and

$$\| \mathscr{F}_h^2 - \mathscr{F}_h \|_1 = O(h^{-n+\varkappa}) \quad \text{as} \quad h \to 0;$$

3°. $\mathscr{F}_h(A_{(h)} - \lambda I)\, \mathscr{F}_h \leq Ch^\varkappa$;

4°. $(I - \mathscr{F}_h)\,(A_{(h)} - \lambda I)\,(I - \mathscr{F}_h) \geq -Ch^\varkappa$;

5°. $\mathrm{Sp}\, \mathscr{F}_h = h^{-n} V(\lambda)\,(1 + O(h^\varkappa))$;

here $0 < \varkappa < 1/2$ and the function $V(\lambda)$ in 5°, is a positive, non-vanishing function, defined on the interval $[\lambda, \lambda + \varepsilon]$ and differentiable from the right at λ. Note also that $\mathrm{Im}\, \mathscr{F}_h \subset D_{A_{(h)}}$, where $D_{A_{(h)}}$ is the domain of $A_{(h)}$.

In the presence of a family of operators, with properties 1°–5°, theorem 28.1, reformulated in the new terminology is fundamental in obtaining the asymptotic formula (A.2.7). For convenience we formulate the following result.

Proposition A.2.2. *Let $A_{(h)}$ be a family of essentially self-adjoint bounded from below operators; $\mathscr{F}_h$ a family of operators such that $\mathrm{Im}\, \mathscr{F}_h \subset D_{A_{(h)}}$ and having the properties 1°–5°. Then*

$$N_h(\lambda) = h^{-n}(V(\lambda) + O(h^\varkappa)) \; as \quad h \to 0$$

Proof. Similar to the proof of Theorem 28.1.

Exercise A.2.1. Prove Proposition A.2.2.

A.2.3 Symbols and operators with parameters

To study the approximate spectral projection, it is convenient to introduce class of symbols, depending on a parameter (cf. 29.1).

Definition A.2.2. Denote by $\Sigma_{\varrho,\sigma}^{m,\mu}$ the class of functions $a(z,h)$, defined for $z \in \mathrm{IR}^{2n}$, $0 < h \leq h_0$, infinitely differentiable in z and satisfying the estimate

$$|\partial_z^\gamma a(z,h)| \leq C_\gamma \langle z \rangle^{m-\varrho|\gamma|} h^{\mu-\sigma|\gamma|}. \tag{A.2.8}$$

Here $m, \mu, \varrho, \sigma \in \mathrm{IR}$, $\varrho > 0$, $\sigma \leq 1/2$.

Obviously, if $a_j \in \Sigma_{\varrho_j,\sigma_j}^{m_j,\mu_j}$, $j = 1$, 2, then $a_1 a_2 \in \Sigma_{\varrho,\sigma}^{m_1+m_2,\mu_1+\mu_2}$, where $\varrho = \min(\varrho_1, \varrho_2)$, $\sigma = \max(\sigma_1, \sigma_2)$.

Note that for any fixed h, $0 < h \leq h_0$, from $a \in \Sigma_{\varrho,\sigma}^{m,\mu}$ it follows that $a(z,h) \in \Gamma_\varrho^m(\mathrm{IR}^{2n})$ and (A.2.2) properly defines a class of operators $A(h)$ depending on a parameter and acting on $S(\mathrm{IR}^n)$ (the parameter dependence appears either in the symbol or in the exponent). The corresponding class of operators will be denoted by $S_{\varrho,\sigma}^{m,\mu}$.

From definition A.2.2 it follows, that for $\sigma > 0$ the derivatives of the symbol can be estimated by powers of h which are increasing as $h \to 0$. However the influence of the increasing powers of h disappears under the action of the corresponding h-symbol of the operator. In particular, one has

Proposition A.2.3. *Let $A(h) \in S_{\varrho,\sigma}^{0,\mu}$, $\mu > 0$, $\sigma < 1/2$. Then $A(h)$ is bounded in $L^2(\mathrm{IR}^n)$ uniformly in h for $0 < h \leq h_0$.*

Proof. Let $a(z,h)$ be the h-symbol of the operator $A(h)$ and $A_{(1)}^{(h)}(h)$ the operator with 1-symbol $a^{(h)}(z,h) = a(\sqrt{h}\,z,h)$. By (A.2.5) the operators $A(h)$ and $A_{(1)}^{(h)}(h)$ are unitarily equivalent, i.e.

$$\|A(h)\| = \|A_{(1)}^{(h)}(h)\|. \tag{A.2.9}$$

Let us now show the boundedness of the operator $A_{(1)}^{(h)}(h)$. Selecting $0 < \varrho' < \min(\varrho, 1 - 2\sigma)$ we have

$$|\partial_z^\gamma a^{(h)}(z,h)| = h^{|\gamma|/2} |\partial_y^\gamma a(y,h)|_{y=\sqrt{h}\,z}|$$

$$\leq C_\gamma h^{|\gamma|/2} \langle \sqrt{h}\,z \rangle^{-\varrho|\gamma|} h^{\mu-\sigma|\gamma|} \leq C_\gamma h^{|\gamma|(1/2-\sigma)+\mu} \langle \sqrt{h}\,z \rangle^{-\varrho'|\gamma|} \tag{A.2.10}$$

$$\leq C_\gamma' h^{|\gamma|(1/2-\sigma-\varrho'/2)+\mu} \langle z \rangle^{-\varrho'|\gamma|}.$$

In (A.2.10) we used the obvious inequality

$$(1 + \sqrt{h}\,|z|)^{-\varkappa} \leq C(\sqrt{h} + \sqrt{h}\,|z|)^{-\varkappa} = C h^{-\varkappa/2} \langle z \rangle^{-\varkappa} \qquad \text{for} \quad \varkappa > 0.$$

Thus, $a^{(h)}(z,h) \in \Gamma_\varrho^0(\mathbb{R}^{2n})$ uniformly in h and by Theorem 24.3 the operator $A_{(1)}^{(h)}(h)$ is bounded in $L^2(\mathbb{R}^n)$ uniformly in h. $\square$

Let us now introduce expressions for the trace of an operator in terms of its h-symbol. It follows from (A.2.5) that

$$\mathrm{Sp}\, A_{(h)} = \mathrm{Sp}\,(T_h^{-1} A_{(1)}^{(h)} T_h) = \mathrm{Sp}\, A_{(1)}^{(h)}. \tag{A.2.11}$$

Using (27.2) we obtain

$$\mathrm{Sp}\, A_{(h)} = \int b^{(h)}(x,\xi)\,dx\,d\xi = h^{-n} \int b(x,\xi)\,dx\,d\xi. \tag{A.2.12}$$

With the help of similar arguments the estimates of the trace class norm (27.12) can be transferred to the case of h-symbols:

Proposition A.2.4. *There exist constants C and N such that for the operator $A(h)$ with the h-symbol $b(z,h)$ the following estimate of the trace class norm holds*

$$\|A(h)\|_1 \leqq Ch^{-n} \sum_{|\gamma| \leqq N} h^{|\gamma|/2} \int |\partial_z^\gamma b(z,h)|\,dz. \tag{A.2.13}$$

A.2.4 The h-anti-Wick symbols

For the operators (A.2.2), where the action depends on the parameter h, we have the following analogue of the anti-Wick symbols introduced in §24.

The whole construction in section 24.1 can be carried out if, instead of $\Phi_0(x)$ we take the function

$$\Psi_0(x) = (\pi h)^{-n/4} e^{-x^2/(2h)} = (T_h^{-1}\Phi_0)(x)$$

as the starting point. An operator $A^{(h)}$ with an h-anti-Wick symbol $a(x,\xi)$ is defined by analogy with (24.9):

$$A_{(h)} = \int a(x,\xi)\,Q_{x,\xi}\,dx\,d\xi, \tag{A.2.14}$$

where $Q_{x,\xi} = T_h^{-1} P_{x,\xi} T_h$ are the projection operators, playing the role of $P_{x,\xi}$ in definition (24.9).

All the results from §24 can with no effort be extended to the case of h-anti-Wick symbols. In particular, it follows from (A.2.14) that an operator is non-negative whenever its h-anti-Wick symbol is non-negative.

It is easy to compute that the kernel of the projection operator Q_0 on the vector Ψ_0 is equal to $(\pi h)^{-n/2} e^{-(x^2+y^2)/(2h)}$, whereas the h-Weyl symbol $\sigma_0(x,\xi)$ of Q_0 is equal to

$$\sigma_0(x,\xi) = 2^n e^{-(x^2+\xi^2)/h} \tag{A.2.15}$$

Using (A.2.14) and (A.2.15) we obtain a formula connecting the h-Weyl symbol $b(z)$ and the h-anti-Wick symbol $a(z)$:

$$b(z) = (\pi h)^{-n} \int a(z') \, e^{-\frac{|z-z'|^2}{h}} \, dz' . \tag{A.2.16}$$

From (A.2.16), by the same reasoning as in the proof of Theorem 24.1 we can obtain the following analogue of Lemma 29.1.

Lemma A.2.1. *Let $B(h)$ be an operator with the h-anti-Wick symbol $a(z, h) \in \Sigma_{\varrho,\sigma}^{m,\mu}$ and $b(z, h)$ its symbol. Then*

$$a - b = \sum_{0 < |\gamma| < N} h^{|\gamma|/2} \, c_\gamma \, (\partial_z^\gamma a) + r_N ,$$

where $c_\gamma = 0$ for odd $|\gamma|$ and $r_N \in \Sigma_{\varrho,\sigma}^{m-\varrho N, \mu+(1/2-\sigma)N}$.

A.2.5 The composition formula

For operators of the classes $S_{\varrho,\sigma}^{m,\mu}$ the composition formula holds in the following form.

Theorem A.2.2. *Let $a_j \in \Sigma_{\varrho_j,\sigma_j}^{m_j,\mu_j}$, $j = 1, 2$; let $A_j(h)$ be the corresponding operators. Then*

$$A_1(h) \circ A_2(h) \in S_{\varrho,\sigma}^{m_1+m_2,\mu_1+\mu_2},$$

where $\varrho = \min(\varrho_1, \varrho_2)$, $\sigma = \max(\sigma_1, \sigma_2)$, where for the h-symbol of the composition $b(z, h)$ we have

$$b = \sum_{|\alpha+\beta| < N} \frac{(-1)^{|\beta|}}{\alpha! \beta!} \left(\frac{h}{2}\right)^{|\alpha+\beta|} (\partial_\xi^\alpha D_x^\beta a_1)(\partial_\xi^\beta D_x^\alpha a_2) + h^N r_N , \tag{A.2.17}$$

where

$$r_N \in \Sigma_{\varrho,\sigma}^{m_1+m_2-N(\varrho_1+\varrho_2),\,\mu_1+\mu_2-N(\sigma_1+\sigma_2)} . \tag{A.2.18}$$

Proof. We may obtain the proof just by copying the proof of Theorem 29.1. So that we do not repeat the calculations of Section 29.1 here, we will, wherever possible, refer to the proof of Theorem 29.1.

Using (A.2.5), we obtain:

$$A_1(h) \circ A_2(h) = T_h^{-1} (A_1(h))_{(1)}^{(h)} T_h \circ T_h^{-1} (A_2(h))_{(1)}^{(h)} T_h$$
$$= T_h^{-1} [(A_1(h))_{(1)}^{(h)} \circ (A_2(h))_{(1)}^{(h)}] T_h .$$

For the symbol $b'(x, \xi, h)$ of the composition of the operators $(A_j(h))_{(1)}^{(h)}$ with symbols $a_j^{(h)}(z, h) = a_j(\sqrt{h}\, z, h)$ we have according to (29.4)–(29.18) (in which the nature of the dependence on the parameter is not used), the following representation

$$b'(x, \xi, h) = \sum_{|\alpha+\beta|<N} \frac{(-1)^{|\beta|}}{\alpha!\,\beta!} \, 2^{-|\alpha+\beta|} (\partial_\xi^\alpha D_x^\beta a_1^{(h)})\,(\partial_\xi^\beta D_x^\alpha a_2^{(h)}) + \tilde{r}_N\,. \qquad (A.2.19)$$

Note that

$$(\partial_\xi^\alpha D_x^\beta a_1^{(h)})\,(\partial_\xi^\beta D_x^\alpha a_2^{(h)}) = h^{|\alpha+\beta|}\,[(\partial_\xi^\alpha D_x^\beta a_1)\,(\partial_\xi^\beta D_x^\alpha a_2)]^{(h)}\,.$$

Therefore, passing from the Weyl 1-symbol in the equality (A.2.19) to the h-Weyl symbol according to (A.2.5), we obtain the terms in the sum (A.2.17).

The estimate of the remainder (we are talking about the Weyl 1-symbol) is reduced, as in section 29.1, to the uniform estimate in $\tau_1, \tau_2 \in [0, 1]$ of integrals of the form

$$I(x, \xi) = \int e^{2i(y\cdot\zeta - z\cdot\eta)}\,[\partial_\xi^\alpha D_x^\beta a_1^{(h)}(x+\tau_1 y, \xi+\eta, h)]$$

$$\times [\partial_\xi^\beta D_x^\alpha a_2^{(h)}(x+\tau_2 z, \xi+\zeta, h)]\,d\eta\,d\zeta\,dy\,dz\,, \qquad (A.2.20)$$

$$|\alpha+\beta| \geqq N\,.$$

By analogy with the above, we obtain that the Weyl 1-symbol $I(x, \xi)$ corresponds to the Weyl h-symbol

$$J(x, \xi) = h^{|\alpha+\beta|}\int e^{2i(y\cdot\zeta - z\cdot\eta)}\,[\partial_\xi^\alpha D_x^\beta a_1(x+\tau_1 \sqrt{h}\,y, \xi+\sqrt{h}\,\eta, h)]$$

$$\times [\partial_\xi^\beta D_x^\alpha a_2(x+\tau_2 \sqrt{h}\,z, \xi+\sqrt{h}\,\zeta, h)]\,d\eta\,d\zeta\,dy\,dz\,, \qquad (A.2.21)$$

$$|\alpha+\beta| \geqq N\,.$$

It is necessary to prove that

$$R_N = h^{-N}J \in \sum_{\varrho,\sigma}{}^{m_1+m_2-N(\varrho_1+\varrho_2),\,\mu_1+\mu_2-N(\sigma_1+\sigma_2)}$$

uniformly in $\tau_1, \tau_2 \in [0, 1]$.

Differentiating (A.2.21) with respect to x and ξ, we see that the derivative $\partial_x^\gamma \partial_\xi^\delta R_N$ is a linear combination of expressions of the form

$$h^{|\alpha+\beta|-N}\int e^{2i(y\cdot\zeta - z\cdot\eta)}\,[\partial_x^{\alpha+\gamma'}\partial_\xi^{\beta+\delta'} a_1(x+\tau_1 \sqrt{h}\,y, \xi+\sqrt{h}\,\eta, h)]$$

$$\times [\partial_x^{\beta+\gamma''}\partial_\xi^{\alpha+\delta''} a_2(x+\tau_2 \sqrt{h}\,z, \xi+\sqrt{h}\,\zeta, h)]\,d\eta\,d\zeta\,dy\,dz\,, \qquad (A.2.22)$$

$$\gamma' + \gamma'' = \gamma\,, \qquad \delta' + \delta'' = \delta\,.$$

In the same way as in Section 29.1, if we integrate by parts we obtain decreasing factors of the type $\langle y, \eta\rangle^{-2M}$, $\langle z, \zeta\rangle^{-2M}$. It is only necessary to take into account that in differentiating $a_1(a_2)$ in y and η (z and ζ) there appears a factor $h^{1/2}$. Therefore (A.2.22) lead to a linear combination of integrals of the type

$$h^{|\alpha+\beta|-N+|\varkappa_1+\varkappa_2|/2}\int e^{2i(y\cdot\zeta-z\cdot\eta)}\langle y,\eta\rangle^{-2M}\langle z,\zeta\rangle^{-2M}$$

$$\times[\partial_{x,\xi}^{v_1+\varkappa_1}a_1(x+\tau_1\sqrt{h}\,y,\xi+\sqrt{h}\eta,h)]$$

$$\times[\partial_{x,\xi}^{v_2+\varkappa_2}a_2(x+\tau_2\sqrt{h}\,z,\xi+\sqrt{h}\zeta,h)]\,d\eta\,d\zeta\,dy\,dz\,,\qquad\text{(A.2.23)}$$

where

$$v_1=\alpha+\beta+\gamma'+\delta'\,,\qquad\qquad |v_1|\geq N+|\gamma'+\delta'|\,;$$
$$v_2=\alpha+\beta+\gamma''+\delta''\,,\qquad\qquad |v_2|\geq N+|\gamma''+\delta''|\,.\qquad\text{(A.2.24)}$$

The absolute value of (A.2.23) can be estimated by

$$h^s\int\langle x+\tau_1\sqrt{h}\,y,\xi+\sqrt{h}\eta,h\rangle^{m_1-\varrho_1 N-\varrho_1|\gamma'+\delta'|}$$

$$\times\langle x+\tau_2\sqrt{h}\,z,\xi+\sqrt{h}\zeta,h\rangle^{m_2-\varrho_2 N-\varrho_2|\gamma''+\delta''|}$$

$$\times\langle y,\eta\rangle^{-2M}\langle z,\zeta\rangle^{-2M}\,d\eta\,d\zeta\,dy\,dz\,,\qquad\text{(A.2.25)}$$

where

$$s=|\alpha+\beta|-N+|\varkappa_1+\varkappa_2|/2+\mu_1-\sigma_1|v_1+\varkappa_1|+\mu_2-\sigma_2|v_2+\varkappa_2|\,.$$

If one takes into account the relations

$$\sigma_1,\,\sigma_2\leq\sigma<1/2\,,\quad|\alpha+\beta|\geq N\,,\quad\gamma'+\gamma''=\gamma\,,\quad\delta'+\delta''=\delta$$

and (A.2.24), then the exponent s can be estimated as follows:

$$s=\mu_1+\mu_2-\sigma_1|\gamma'+\delta'|-\sigma_2|\gamma''+\delta''|-N(\sigma_1+\sigma_2)$$

$$+(N-|\alpha+\beta|)\,(\sigma_1+\sigma_2-1)+|\varkappa_1|\,(1/2-\sigma_1)+|\varkappa_2|\,(1/2-\sigma_2)$$

$$\geq\mu_1+\mu_2-\sigma|\gamma+\delta|-N(\sigma_1+\sigma_2)\,.$$

We see that the power of h in (A.2.25) corresponds to the statement of the theorem. Next, estimate the integral in (A.2.25) in terms of the necessary power of $\langle x,\xi\rangle$. Note first that it splits into the product of integrals

$$I=\int\langle x+\tau_1\sqrt{h}\,y,\xi+\sqrt{h}\eta\rangle^{m_1-\varrho_1(N+|\gamma'+\delta'|)}\langle y,\eta\rangle^{-2M}\,d\eta\,dy\,,\qquad\text{(A.2.26)}$$

$$I'=\int\langle x+\tau_2\sqrt{h}\,z,\xi+\sqrt{h}\zeta\rangle^{m_2-\varrho_2(N+|\gamma''+\delta''|)}\langle z,\zeta\rangle^{-2M}\,d\zeta\,dz\,.\qquad\text{(A.2.26}')$$

We will estimate the integral (A.2.26). Assume that $m_1-\varrho_1 N<0$; if this is not the case, then in the expansion (A.2.17) we can take N' terms so that $m_1-\varrho_1 N'<0$ and examine the remainder $r_{N'}$, representing the remainder $r_{N'}$

as r_N plus the finite sum within the limits $N \leqq |\alpha+\beta| \leqq N'$. Using the obvious inequality

$$\langle x+\tau_1 \sqrt{h}y, \xi+\sqrt{h}\eta\rangle^{-\varkappa} \leqq \langle x,\xi\rangle^{-\varkappa}\langle\tau_1\sqrt{h}y, \sqrt{h}\eta\rangle^{\varkappa}, \qquad \varkappa>0. \qquad (A.2.27)$$

a special case of the inequality

$$(1+|x+y|+|\xi+\eta|)^m \leqq (1+|x|+|\xi|)^m (1+|y|+|\eta|)^{|m|},$$

and applying (A.2.27) with $\varkappa=|m_1-\varrho_1 N-\varrho_1|\gamma'+\delta'||$, we obtain for the integral (A.2.26) the estimate

$$I \leqq \langle x,\xi\rangle^{m_1-\varrho_1 N-\varrho_1|\gamma'+\delta'|} \int \langle\tau_1\sqrt{h}y, \sqrt{h}\eta\rangle^{\varkappa}\langle y,\eta\rangle^{-2M} d\eta\, dy$$

$$\leqq \langle x,\xi\rangle^{m_1-\varrho_1 N-\varrho_1|\gamma'+\delta'|} \int \langle y,\eta\rangle^{\varkappa-2M} d\eta\, dy$$

$$= C_1 \langle x,\xi\rangle^{m_1-\varrho_1 N-\varrho_1|\gamma'+\delta'|}. \qquad (A.2.28)$$

Combining (A.2.28) with the corresponding estimate for (A.2.26′), we obtain the necessary power of $\langle x,\xi\rangle$ in the estimate of (A.2.25) so that the inclusion (A.2.18) is proved. $\quad\square$

A.2.6 Proof of Theorem A.2.1

The plan of the proof is as follows: first we construct an approximate spectral projection and successively verify the properties needed to apply Proposition A.2.2.

1. Let $\chi(t,\lambda,\delta)$ be the function introduced in Section 28.6:

$$\chi(t,\lambda,\delta) = \begin{cases} 1 & \text{for } t\leqq\lambda, \\ 0 & \text{for } t\geqq\lambda+2\delta, \end{cases} \qquad (A.2.29)$$

$$|(\partial/\partial t)^k \chi(t,\lambda,\delta)| \leqq C_k \delta^{-k}. \qquad (A.2.30)$$

Set $\varkappa = 1/2-\varepsilon$, $\varepsilon>0$ and

$$e(z,h) = \chi(b(z),\lambda,h^{\varkappa}). \qquad (A.2.31)$$

(We omit the unimportant argument λ of $e(z,h)$). The operator $\mathscr{F}_h$ is defined as the operator with h-symbol $e(z,h)$.

The function $e(z,h)$ is infinitely differentiable with respect to z and

$$e(z,h) = \begin{cases} 1 & \text{for } b(z)\leqq\lambda, \\ 0 & \text{for } b(z)\geqq\lambda+2h^{\varkappa}. \end{cases}$$

Let us estimate the z-derivatives of $e(z,h)$. Differentiating (A.2.31) with respect to z gives:

$$\partial_z^\gamma e(z,h) = \sum_{\gamma_1 + \ldots + \gamma_k = \gamma} c_{\gamma_1, \ldots, \gamma_k} (\partial^{\gamma_1} b(z)) \ldots (\partial^{\gamma_k} b(z)) \frac{\partial^k \chi}{\partial t^k}(t, \lambda, h)\big|_{t=b(z)}.$$

$$\text{(A.2.32)}$$

The summation in (A.2.32) runs over all possible decompositions of γ into a sum $\gamma_1 + \ldots + \gamma_k$, where $k \le |\gamma|$. Taking (A.2.30) and $b \in H\Gamma_\varrho^{m,0}$ into account, we obtain for an individual term in (A.2.32) the estimate

$$\left| (\partial^{\gamma_1} b) \ldots (\partial^{\gamma_k} b) \left(\frac{\partial^k \chi(t, \lambda, h)}{\partial t^k}\bigg|_{t=b(z)} \right) \right| \le Ch^{-k\varkappa} |b|^k$$

$$\times \prod_{i=1}^k |\partial^{\gamma_i} b / b| \le Ch^{-k\varkappa} |b|^k (1 + |z|)^{-\varrho|\gamma|}. \qquad \text{(A.2.33)}$$

Note also that on the support of $e(z,h)$

$$|b| < \lambda + h^\varkappa. \qquad \text{(A.2.34)}$$

Therefore, from (A.2.32)–(A.2.34) we obtain the estimate $|\partial_z^\gamma e(z,h)| \le Ch^{-k\varkappa}\langle z \rangle^{-\varrho|\gamma|}$, i.e.

$$e(z,h) \in \Sigma_{\varrho, \varkappa}^{0,0}. \qquad \text{(A.2.35)}$$

2. We need now to verify, that all the conditions of Proposition A.2.2 are fulfilled for $\mathscr{F}_h$. $\mathscr{F}_h$ is symmetric due to the real-valuedness of the symbol, and bounded by (A.2.35) and proposition A.2.3, hence $\mathscr{F}_h^* = \mathscr{F}_h$. The fact that $\mathscr{F}_h$ is of trace class follows from Proposition A.2.4.

We denote the h-symbol of an arbitrary operator A by $\sigma(A)$. In order to estimate the trace class norm $\| \mathscr{F}_h^2 - \mathscr{F}_h \|_1$, we need to compute the h-symbol of the operator $\mathscr{F}_h^2 - \mathscr{F}_h$.

By the composition formula (A.2.17)

$$\sigma(\mathscr{F}_h^2 - \mathscr{F}_h) = -e(z,h) + e(z,h)^2 + \sum_{|\alpha| + |\beta| < N} c_{\alpha\beta} h^{|\alpha + \beta|} (D_x^\alpha \partial_\xi^\beta e)$$

$$\times (D_x^\beta \partial_\xi^\alpha e) + h^N r_N, \quad h^N r_N \in \Sigma_{\varrho, \varkappa}^{-2\varrho N, (1 - 2\varkappa)N}. \qquad \text{(A.2.36)}$$

All terms on the right-hand side of (A.2.36) except r_N are supported on $\{z: \lambda < b(z) < \lambda + 2h^\varkappa\}$, so the trace class norms of the corresponding operators by Proposition A.2.4, can be estimated by

$$C[V(\lambda + 2h^\varkappa) - V(\lambda)] h^{-n}. \qquad \text{(A.2.37)}$$

The trace class norm of the remainder can be estimated by $o(h^{-n+\varkappa})$ for sufficiently large N.

Now, note that $V(\lambda)$ is a non-decreasing function; by the well-known Lebesgue theorem it is almost everywhere differentiable. In what follows, we assume that the λ under consideration belongs to the set of full measure where V is differentiable. Then (A.2.37) is transformed into the desired estimate

$$\|\mathscr{F}_h^2 - \mathscr{F}_h\|_1 \leq C\,[V'(\lambda)\,h^{\varkappa} + o\,(h^{\varkappa})]\,h^{-n} \leq Ch^{-n+\varkappa}. \qquad (A.2.38)$$

In addition, by (A.2.12) we have

$$\mathrm{Sp}\,\mathscr{F}_h = (2\pi h)^{-n}\int e\,(z,h)\,dz = h^{-n}[V(\lambda) + O(V(\lambda) - V(\lambda + 2h^{\varkappa}))]$$

$$= h^{-n}V(\lambda)\,(1 + O(h^{\varkappa})). \qquad (A.2.39)$$

Consequently, we have satisfied $1°$, $2°$ and $5°$ of Proposition A.2.2 ($\mathscr{F}_h^* = \mathscr{F}_h$, (A.2.38) and (A.2.39)).

3. Now let us verify $3°$ of Proposition A.2.2. For this we need the h-symbol of $\mathscr{F}_h(A_{(h)} - \lambda I)\,\mathscr{F}_h$. So we begin by computing it. We have

$$\sigma\,(\mathscr{F}_h(A_{(h)} - \lambda I)) = e\,(z,h)\,(b\,(z) - \lambda)$$

$$+ \sum_{1 \leq |\alpha + \beta| < N} h^{|\alpha + \beta|}c_{\alpha\beta}\,(D_x^{\alpha}\partial_{\xi}^{\beta}e)\,(D_x^{\beta}\partial_{\xi}^{\alpha}b) + r_N\,, \qquad (A.2.40)$$

$$r_N \in \Sigma_{\varrho,\varkappa}^{m-2N\varrho,N(1-\varkappa)}.$$

The operator R_N, corresponding to the symbol r_N, is bounded for $m - 2N\varrho < 0$ and $\|R_N\| = 0\,(h^{N(1-\varkappa)}) = o\,(h^{\varkappa})$. Let us show that

$$\varphi_{\alpha\beta} \in \Sigma_{\varrho,\varkappa}^{-2|\alpha+\beta|\varrho,\,\varkappa + (1-2\varkappa)|\alpha+\beta|}, \qquad |\alpha + \beta| > 0. \qquad (A.2.41)$$

for $\varphi_{\alpha\beta} = h^{|\alpha+\beta|}(\partial_x^{\alpha}\partial_{\xi}^{\beta}e)\,(\partial_x^{\beta}\partial_{\xi}^{\alpha}b)$.

For this we construct in a standard way a function $\varphi\,(z,h) \in C^{\infty}(\mathbb{R}^{2n})$, for which $0 \leq \varphi\,(z,h) \leq 1$,

$$\varphi\,(z,h) = \begin{cases} 1, & \text{for } z \in \mathrm{supp}\,\partial_z e\,, \\ 0, & \text{for } b\,(z) \geq \lambda + 3h^{\varkappa},\ b\,(z) \leq \lambda - h^{\varkappa} \end{cases}$$

and which, like $e\,(z,h)$ in the first part of this proof, is determined with the help of the smoothed characteristic function $\chi\,(t, \lambda, \lambda + 2h^{\varkappa})$ of the interval $(\lambda, \lambda + 2h^{\varkappa})$ by the formula $\varphi\,(z,h) = \chi\,(b\,(z), \lambda, \lambda + 2h^{\varkappa})$.

By analogy with (A.2.33) and (A.2.34) we verify that

$$\varphi \in \Sigma_{\varrho,\varkappa}^{0,0}. \qquad (A.2.42)$$

Now note that for $|\alpha + \beta| > 0$

$$\varphi_{\alpha\beta} = h^{|\alpha+\beta|}(\partial_x^{\alpha}\partial_{\xi}^{\beta}e)\,[\partial_x^{\beta}\partial_{\xi}^{\alpha}(\varphi \cdot (b - \lambda))]. \qquad (A.2.43)$$

Let us show that $\varphi \cdot (b - \lambda) \in \Sigma_{\varrho, \varkappa}^{0, \varkappa}$. Obviously $|\varphi \cdot (b - \lambda)| \leq Ch^{\varkappa}$ and computing the derivative with respect to z we obtain

$$\partial_z^{\gamma}[\varphi \cdot (b - \lambda)] = (\partial^{\gamma}\varphi)\,(b - \lambda) + \sum_{|\alpha|>0} c_{\alpha}(\partial^{\gamma - \alpha}\varphi)\,(\partial^{\alpha}(b - \lambda))\,. \tag{A.2.44}$$

Owing to (A.2.42) for the first term we have

$$|(\partial^{\gamma}\varphi(z, h))\,(b(z) - \lambda)| \leq Ch^{-\varkappa|\gamma|}\,\langle z \rangle^{-\varrho|\gamma|}\,h^{\varkappa}\,; \tag{A.2.45}$$

and for the other summands of (A.2.44) we get

$$\begin{aligned}
|(\partial^{\gamma - \alpha}\varphi)(\partial^{\alpha}(b - \lambda))| &= |\partial^{\gamma - \alpha}\varphi| \cdot |b| \cdot |(\partial^{\alpha}b)/b| \\
&\leq Ch^{-\varkappa|\gamma - \alpha|}\langle z \rangle^{-\varrho|\gamma - \alpha| - \varrho|\alpha|} \leq Ch^{\varkappa - \varkappa|\gamma|}\langle z \rangle^{-\varrho|\gamma|}\,.
\end{aligned} \tag{A.2.46}$$

The estimates (A.2.45) and (A.2.46) show that $\varphi(b - \lambda) \in \Sigma_{\varrho, \varkappa}^{0, \varkappa}$, from which, taking (A.2.43) into account, (A.2.41) follows. Thus, the finite sum in (A.2.40) belongs to $\Sigma_{\varrho, \varkappa}^{-2\varrho, 1 - \varkappa}$ and the operator $\mathscr{F}_h(A_{(h)} - \lambda I)$ can be written in the form

$$\mathscr{F}_h(A_{(h)} - \lambda I) = Q_1 + R\,, \tag{A.2.47}$$

where $\|R\| = 0\,(h^{1 - \varkappa}) = o\,(h^{\varkappa})$ and $\sigma(Q_1) = e \cdot (b - \lambda)$.

Using (A.2.47) we have

$$\mathscr{F}_h(A_{(h)} - \lambda I)\,\mathscr{F}_h = Q_1\,\mathscr{F}_h + R_1\,, \qquad \|R_1\| = o\,(h^{\varkappa})\,. \tag{A.2.48}$$

We will compute the symbol of $Q_1\,\mathscr{F}_h$:

$$\begin{aligned}
\sigma(Q_1\,\mathscr{F}_h) = e^2 \cdot (b - \lambda) &+ \sum_{0 < |\alpha + \beta| < N} c_{\alpha\beta}\,h^{|\alpha + \beta|}(\partial_x^{\alpha}\partial_{\xi}^{\beta}[e \cdot (b - \lambda)]) \\
&\times (\partial_x^{\beta}\partial_{\xi}^{\alpha}e) + r_N\,.
\end{aligned} \tag{A.2.49}$$

Introducing as before the function φ and using $e \cdot \varphi(b - \lambda) \in \Sigma_{\varrho, \varkappa}^{0, \varkappa}$, we see that the norm of the operator corresponding to the finite sum and the remainder in (A.2.49), can also be estimated via $O\,(h^{1 - \varkappa})$.

Consider the principal part of the symbol $-\sigma(\mathscr{F}_h(A_{(h)} - \lambda I)\,\mathscr{F}_h)$, viz. the function $q(z, h) = -e(z, h)^2\,(b(z) - \lambda)$. For q and its derivatives with respect to z, the following estimates hold

$$q(z, h) \geq -Ch^{\varkappa}\,; \tag{A.2.50}$$

$$\begin{aligned}
|\partial^{\gamma}q| &= \left| \sum_{|\alpha|>0} c_{\alpha}(\partial^{\alpha}b)(\partial^{\gamma - \alpha}e^2) + 2(b - \lambda)\sum_{\alpha} c_{\alpha}(\partial^{\gamma - \alpha}e)(\partial^{\alpha}e) \right| \\
&\leq \sum_{|\alpha|>0} c_{\alpha}h^{-\varkappa|\gamma - \alpha|}\langle z \rangle^{-\varrho|\gamma|}\,|\lambda| + Ch^{\varkappa}h^{-|\gamma|\varkappa}\langle z \rangle^{-\varrho|\gamma|} \\
&\leq Ch^{\varkappa(1 - |\gamma|)}\langle z \rangle^{-\varrho|\gamma|}\,.
\end{aligned} \tag{A.2.51}$$

We shall denote by P the operator having Weyl h-symbol $q(z, h)$ and by Q the one having h-anti-Wick symbol $q(z, h)$. From (A.2.51) it follows that

$$h^{|\gamma|/2} \, \partial^\gamma q \in \Sigma^{-\varrho|\gamma|, \, \varkappa + (1/2 - \varkappa)|\gamma|}_{\varrho, \, \varkappa},$$

and therefore, by Lemma A.2.1, $\sigma(P - Q) \in \Sigma^{-2\varrho, \, 1 - \varkappa}_{\varrho, \, \varkappa}$ and thus $\|P - Q\| = O(h^{1 - \varkappa})$. It is furthermore obvious from (A.2.50) that $Q \geq - Ch^\varkappa$ and since $P = Q + (P - Q)$ it follows that $P \geq - Ch^\varkappa$. Thus, we have for the principal part and consequently for the whole operator $\mathscr{F}_h (A_{(h)} - \lambda I) \, \mathscr{F}_h$ the estimate

$$\mathscr{F}_h (A_{(h)} - \lambda I) \, \mathscr{F}_h \leq Ch^\varkappa. \tag{A.2.52}$$

4. Now we will verify that

$$(I - \mathscr{F}_h)(A_{(h)} - \lambda I) \, (I - \mathscr{F}_h) \geq - Ch^\varkappa. \tag{A.2.53}$$

The symbol of the left hand side of (A.2.53) (after getting rid of parentheses) is

$$\sigma \, (\mathscr{F}_h (A_{(h)} - \lambda I) \, \mathscr{F}_h) - \sigma \, (\mathscr{F}_h (A_{(h)} - \lambda I))$$
$$- \sigma \, ((A_{(h)} - \lambda I) \, \mathscr{F}_h) + (b(z) - \lambda). \tag{A.2.54}$$

In step 3 of this proof, it was shown that in the first two summands of (A.2.54) the principal terms are distinguished $e^2 (b - \lambda)$ and $e(\lambda - b)$, and the operators corresponding to the remainders are estimated in norm by $O(h^{1 - \varkappa})$. The third summand in (A.2.54) is analogous to the second. Thus we obtain

$$\sigma \, ((I - \mathscr{F}_h)(A_{(h)} - \lambda I) \, (I - \mathscr{F}_h)) = (1 - e)^2 (b - \lambda) + r, \tag{A.2.55}$$

and the operator R with the h-symbol r admits the estimate

$$\|R\| = O(h^{1 - \varkappa}).$$

Now consider the operator P with h-symbol

$$q(z, h) = (1 - e(z, h))^2 (b(z) - \lambda).$$

Arguments similar to the ones used in section 29.3 in proving the positivity of an operator with positive symbol show that

$$P = Q_k + A_k. \tag{A.2.56}$$

Here the operator Q_k has the h-anti-Wick symbol

$$q_k(z, h) = q(z, h) + \sum_{2 \leq |\gamma| < N} c_\gamma h^{|\gamma|/2} \partial^\gamma_z q(z, h) + r_N(z, h), \tag{A.2.57}$$

where $r_N \in \Sigma_{\varrho,\varkappa}^{m-2\varrho N,(1/2-\varkappa)N}$; and therefore the operator with the h-symbol r_N can be estimated in norm by $O(h^{1-\varkappa})$.

For the operator A_k in (A.2.56) we have $A_k \in S_{\varrho,\varkappa}^{m-2\varrho k,(1-2\varkappa)k}$ and therefore $\|A_k\| = O(h^{1-\varkappa})$ with an apropriate choice of k.

We keep the notation q_k for the right-hand side of (A.2.57) but without the remainder r_N and Q_k denotes the corresponding operator with the h-anti-Wick symbol q_k.

Let $\varphi(z,h)$ be the function introduced in the preceeding part of this proof. Decompose q_k into two parts

$$q_k = \varphi q_k + (1-\varphi) q_k. \tag{A.2.58}$$

In step 3 it was shown that $\varphi(b-\lambda) \in \Sigma_{\varrho,\varkappa}^{0,\varkappa}$ from which it is obvious that $\varphi(1-e)^2(b-\lambda) \in \Sigma_{\varrho,\varkappa}^{0,\varkappa}$. With calculations, analogous to (A.2.44)–(A.2.46) in step 3 it can be shown that

$$\varphi \cdot h^{|\gamma|/2} \partial_z^\gamma q(z,h) \in \Sigma_{\varrho,\varkappa}^{-\varrho|\gamma|,\varkappa+(1/2-\varkappa)|\gamma|}.$$

Consequently, the operator having the h-anti-Wick symbol φq_k can be estimated in norm by $O(h^\varkappa)$.

Finally we will show that for small h

$$(1-\varphi) q_k \geqq 0 \tag{A.2.59}$$

and that therefore $Q_k \geqq -Ch^\varkappa$. For this note that

$$(1-\varphi) q_k = \begin{cases} (1-\varphi)\left(b-\lambda+\displaystyle\sum_{2\leqq|\gamma|<N} c_\gamma h^{|\gamma|/2}\partial^\gamma b\right), & z \in D_h, \\ 0 & , & z \notin D_h, \end{cases} \tag{A.2.60}$$

where $D_h = \{z: b(z,h) \geqq \lambda + 2h^\varkappa\}$. Relation (A.2.60) is obvious, since $(1-e(z,h)) \times (1-\varphi(z,h)) = 0$ for $z \notin D_h$ and $1-e(z,h) = 1$ for $z \in D_h$. Therefore in D_h

$$(1-\varphi) q_k = (1-\varphi)(b-\lambda)\left(1+\sum_\gamma c_\gamma h^{|\gamma|/2}(\partial^\gamma b)/(b-\lambda)\right). \tag{A.2.61}$$

Now note, that in D_h

$$b(z)/(b(z)-\lambda) = (1-\lambda/b(z))^{-1} \leqq (1-\lambda/(\lambda+2h^\varkappa))^{-1} \leqq Ch^{-\varkappa};$$

from which it follows that

$$|h^{|\gamma|/2}(\partial^\gamma b)/(b-\lambda)| \leqq Ch^{|\gamma|/2-\varkappa}\langle z\rangle^{-\varrho|\gamma|}.$$

Consequently, the sum on the right hand side of (A.2.61) is estimated by $Ch^{1-\varkappa} = o(1)$, i.e.

$$(1-\varphi)\,q_k = (1-\varphi)\,(b-\lambda)\,(1+o(1)) \geqq 0\,.$$

Thus we have verified the requirements $1°-5°$ of Proposition A.2.2 (the relations (A.2.38), (A.2.39), (A.2.52) and (A.2.53)) and we may apply the proposition. Hence Theorem A.2.1 is proved. $\square$

A.2.7 The behaviour of $N_h(\lambda)$ for $V(\lambda) = +\infty$

Theorem A.2.1 discusses the quasi-classical asymptotic behaviour for the eigenvalues $\lambda < \lambda_0$, where $V(\lambda_0) < +\infty$. The membership $b(z) \in H\Gamma_\varrho^{m,0}$ does not rule out the existence of λ such that $V(\lambda) = +\infty$ (for symbols of $H\Gamma_\varrho^{m,m_0}$, $m_0 > 0$, this situation cannot occur). In this case, the following theorem serves as a supplement to Theorem A.2.1;

Theorem A.2.3. *Let $b(z)$ satisfy the conditions of Theorem A.2.1 and $V(\lambda) = +\infty$. Then, for any $\varepsilon_0 > 0$*

$$\lim_{h \to 0} h^n N_h(\lambda + \varepsilon_0) = +\infty\,.$$

Proof. For each N we will construct a space $H_{\tilde N}$ such that the inequality

$$((A_{(h)} - \lambda I)\,\xi, \xi) \leqq Ch^\varkappa(\xi, \xi)\,, \qquad \xi \in H_{\tilde N}\,, \tag{A.2.62}$$

holds and, for sufficiently small h

$$\dim H_{\tilde N} \geqq h^{-n} N\,. \tag{A.2.63}$$

By the Glazman lemma, we obtain then from (A.2.62) and (A.2.63) the inequality $N_h(\lambda + ch^\varkappa) \geq h^{-n} N$ which implies the result of the theorem.

Introduce the set $\Omega^\lambda = \{z: b(z) \leqq \lambda\}$. It is obvious from the definition of $V(\lambda)$, that $V(\lambda) = (2\pi)^{-n}\,\mathrm{mes}\,\Omega^\lambda$, so that under the conditions of the theorem we have $\mathrm{mes}\,\Omega^\lambda = +\infty$.

Now let Ω_ε be a family of open sets with smooth boundaries, satisfying the following conditions

(1) Ω_ε are bounded, $\Omega_\varepsilon \subset \Omega^\lambda$;

(2) $W_\varepsilon = (2\pi)^{-n}\,\mathrm{mes}\,\Omega_\varepsilon \to +\infty$ as $\varepsilon \to +0$;

(3) $x \in \Omega_\varepsilon$, implies $\inf\limits_{y \in \mathbb{R}^{2n} \setminus \Omega^\lambda} |x - y| \geqq \varepsilon$ i.e. the distance between Ω_ε and $\partial\Omega^\lambda$ is not less than ε.

Now construct a smoothed characteristic function of Ω_ε (this is possible along the lines of the construction in 28.6).

Let $2h^{\varkappa} < \varepsilon$ (as always $0 < \varkappa < 1/2$) and $\psi_{\varepsilon}(z, h)$ the characteristic function of the $(h^{\varkappa})$-thickening of the set Ω_{ε}. Put

$$\chi_{\varepsilon}(z, h) = h^{-2n\varkappa} \int \psi_{\varepsilon}(y, h) \, \chi_0((y - z) h^{-\varkappa}) \, dy \, ,$$

where $\chi_0(v) \in C_0^{\infty}(\mathbb{R}^{2n})$, $\chi_0 \geq 0$, $\chi_0(v) = 0$ for $|v| \geq 1$ and $\int \chi_0(v) \, dv = 1$. It is obvious that

$$\operatorname{supp} \chi_{\varepsilon} \subset \Omega^{\lambda} \, . \tag{A.2.64}$$

In addition, it is easily verified that $|\partial_z^{\gamma} \chi_{\varepsilon}(z, h)| < Ch^{-\varkappa|\gamma|}$, and from this estimate and the compactness of the support of χ_{ε} we have

$$\chi_{\varepsilon}(z, h) \in \Sigma_{\varrho, \varkappa}^{-\infty, 0} \, . \tag{A.2.65}$$

Let $\mathscr{F}_h$ be an approximate spectral projection as constructed in A.2.6. Denote by F_h the operator with the h-anti-Wick symbol $e(z, h)$, by $\mathscr{E}_{\varepsilon, h}$ the operator with the h-symbol $\chi_{\varepsilon}(z, h)$ and by $E_{\varepsilon, h}$ the operator with the h-anti-Wick symbol $\chi_{\varepsilon}(z, h)$. The following relations hold between the operators $\mathscr{F}_h$, F_h, $\mathscr{E}_{\varepsilon, h}$ and $E_{\varepsilon, h}$:

$$\mathscr{F}_h - F_h = R \in S_{\varrho, \varkappa}^{-2\varrho, \, 1 - 2\varkappa}, \qquad \|R\| \leq Ch^{1 - 2\varkappa}; \tag{A.2.66}$$

$$F_h \geq E_{\varepsilon, h}; \tag{A.2.67}$$

$$\mathscr{E}_{\varepsilon, h} - E_{\varepsilon, h} = R' \in S_{\varrho, \varkappa}^{-\infty, \, 1 - 2\varkappa}, \qquad \|R'\| \leq Ch^{1 - 2\varkappa}. \tag{A.2.68}$$

Here (A.2.66) and (A.2.68) follow from Lemma A.2.1, whereas (A.2.67) is obvious since $\chi_{\varepsilon} \leq e$ due to (A.2.64).

Now consider the operator $\mathscr{E}_{\varepsilon, h}$. In the same way as in part 2 of the proof of Theorem A.2.1 looking at the h-symbol of the operator $\mathscr{E}_{\varepsilon, h}^2 - \mathscr{E}_{\varepsilon, h}$ we obtain the

$$\| \mathscr{E}_{\varepsilon, h}^2 - \mathscr{E}_{\varepsilon, h} \|_1 \leq C (W_{\varepsilon, h} - W_{\varepsilon}) h^{-n}, \tag{A.2.69}$$

where $W_{\varepsilon, h}$ is the volume of the $(2h^{\varkappa})$-thickening of the set Ω_{ε}. But $W_{\varepsilon, h} - W_{\varepsilon} = O(h^{\varkappa})$ since the open set Ω_{ε} is bounded and has smooth boundary. Therefore

$$\| \mathscr{E}_{\varepsilon, h}^2 - \mathscr{E}_{\varepsilon, h} \|_1 \leq Ch^{-n+\varkappa}. \tag{A.2.70}$$

Similarly we obtain

$$\operatorname{Sp} \mathscr{E}_{\varepsilon, h} = h^{-n} W_{\varepsilon}(1 + O(h^{\varkappa})). \tag{A.2.71}$$

From (A.2.70) and (A.2.71), as in Lemmas 28.2 and 28.3, we may obtain an asymptotic expression for the number $\tilde{N}$ of eigenvalues of $\mathscr{E}_{\varepsilon, h}$ belonging to $[1/2, 3/2]$:

$$\tilde{N} = h^{-n} W_{\varepsilon}(1 + O(h^{\varkappa})). \tag{A.2.72}$$

The space spanned by the corresponding eigenvectors is denoted by $\mathscr{H}_{\tilde{N}}$. Now we will prove that $H_{\tilde{N}} = \mathscr{F}_h \mathscr{H}_{\tilde{N}}$ satisfies (A.2.62) and (A.2.63).

Let $\eta \in \mathscr{H}_{\tilde{N}}$, then for $\mathscr{F}_h$ we have the estimate

$$(\mathscr{F}_h \eta, \eta) \geq (1/2 + O(h^{1-\varkappa}))\,(\eta, \eta). \tag{A.2.73}$$

Indeed, using (A.2.66)–(A.2.68), we get

$$(\mathscr{F}_h \eta, \eta) = (F_h \eta, \eta) + (R\eta, \eta) \geq (E_{\varepsilon,h} \eta, \eta) + (R\eta, \eta)$$
$$= (\mathscr{E}_{\varepsilon,h} \eta, \eta) + ((R + R')\eta, \eta) \geq (1/2 + O(h^{1-\varkappa}))\,(\eta, \eta).$$

By the Cauchy-Schwarz inequality we have

$$(\mathscr{F}_h \eta, \eta) \leq \sqrt{(\mathscr{F}_h \eta, \mathscr{F}_h \eta)(\eta, \eta)}\,;$$

from which, by (A.2.73), it follows that

$$\sqrt{(\mathscr{F}_h \eta, \mathscr{F}_h \eta)} \geq (\mathscr{F}_h \eta, \eta)/\sqrt{(\eta, \eta)} \geq \sqrt{(\eta, \eta)}\,(1/2 + O(h^{1-\varkappa})),$$

or

$$(\mathscr{F}_h \eta, \mathscr{F}_h \eta) \geq [1/4 + O(h^{1-\varkappa})]\,(\eta, \eta). \tag{A.2.74}$$

Now let $\xi \in H_{\tilde{N}}$; then $\xi = \mathscr{F}_h \eta$, where $\eta \in \mathscr{H}_{\tilde{N}}$. We recall the inequality, obtained in proving Theorem A.2.1,

$$\mathscr{F}_h (A_{(h)} - \lambda I)\,\mathscr{F}_h \leq Ch^{\varkappa} \tag{A.2.75}$$

(which is independent of the behaviour of $V(\lambda)$). From (A.2.74) and (A.2.75) it follows that

$$((A_{(h)} - \lambda I)\,\xi, \xi) = ((A_{(h)} - \lambda I)\,\mathscr{F}_h \eta, \mathscr{F}_h \eta) \leq Ch^{\varkappa}(\eta, \eta)$$
$$\leq (4 + O(h^{1-\varkappa}))\,(\mathscr{F}_h \eta, \mathscr{F}_h \eta)\,Ch^{\varkappa} = O(h^{\varkappa})(\xi, \xi).$$

Thus, on $H_{\tilde{N}}$, we have

$$A_{(h)} - (\lambda + O(h^{\varkappa}))\,I \leq 0.$$

In addition, from (A.2.74), it follows that $\mathscr{F}_h$ is injective on $\mathscr{H}_{\tilde{N}}$, hence

$$\dim H_{\tilde{N}} = \dim \mathscr{H}_{\tilde{N}} = h^{-n} W_{\varepsilon}(1 + O(h^{\varkappa})).$$

The proof of the theorem is now completed since the volume W_{ε} may be chosen as large as we like. $\square$

Appendix 3
Hilbert-Schmidt and Trace Class Operators

A.3.1 Hilbert-Schmidt operators and the Hilbert-Schmidt norm

Definition A.3.1. Let H_1 and H_2 be two Hilbert spaces. A bounded linear operator $K\colon H_1 \to H_2$ is called a *Hilbert-Schmidt operator* if for some orthonormal basis $\{e_\alpha\}$ in H_1 we have

$$\sum_\alpha \|Ke_\alpha\|^2 < +\infty. \tag{A.3.1}$$

The set of all Hilbert-Schmidt operators $K\colon H_1 \to H_2$ is denoted by $S_2(H_1, H_2)$, or $S_2(H)$ in case $H_1 = H_2 = H$. The following proposition describes the basic features of these operators.

Proposition A.3.1. 1) *The left hand side of (A.3.1) is independent of the choice of orthonormal basis $\{e_\alpha\}$ (the square root of the left-hand side is called the Hilbert-Schmidt norm of the operator K and denoted by $\|K\|_2$).*

2) $\|K^*\|_2 = \|K\|_2$.

3) $\|K\| \leq \|K\|_2$, *where $\|K\|$ is the usual operator norm.*

4) *every operator $K \in S_2(H_1, H_2)$ is compact.*

5) *if K is a compact self-adjoint operator in the Hilbert space H, then*

$$\|K\|_2^2 = \sum_{j=1}^\infty \lambda_j^2, \tag{A.3.2}$$

where $\lambda_1, \lambda_2, \ldots$ are all the non-zero eigenvalues of K counting multiplicities.

6) *If $\{e_\alpha\}$ is an orthonormal basis in H_1, then the scalar product*

$$(K, L)_2 = \sum_\alpha (Ke_\alpha, Le_\alpha) \tag{A.3.3}$$

with $K, L \in S_2(H_1, H_2)$ is independent of the choice of basis $\{e_\alpha\}$ and defines on $S_2(H_1, H_2)$ a Hilbert space structure with the corresponding norm $\|\cdot\|_2$.

7) *If $\mathscr{L}(H_j)$, $j = 1, 2$, is the algebra of all bounded operators in H_j, then $S_2(H_1, H_2)$ is a left $\mathscr{L}(H_2)$-module and a right $\mathscr{L}(H_1)$-module, moreover*

$$\|AK\|_2 \leq \|A\|\,\|K\|_2, \quad A \in \mathscr{L}(H_2), \quad K \in S_2(H_1, H_2), \tag{A.3.4}$$

$$\|KB\|_2 \leq \|K\|_2\,\|B\|, \quad B \in \mathscr{L}(H_1), \quad K \in S_2(H_1, H_2). \tag{A.3.5}$$

$S_2(H)$ is in particular, a two-sided ideal in $\mathscr{L}(H)$.

Proof. Let $\{f_\beta\}$ be an orthonormal basis in H_2. Then

$$\sum_\alpha \|Ke_\alpha\|^2 = \sum_{\alpha,\beta} |(Ke_\alpha, f_\beta)|^2 = \sum_{\alpha,\beta} |(e_\alpha, K^*f_\beta)| = \sum_\beta \|K^*f_\beta\|^2, \quad (A.3.6)$$

from which 1) and 2) follow. If now $x = \Sigma\, x_\alpha e_\alpha \in H_1$, then

$$\|Kx\|^2 = \left\| \sum_\alpha x_\alpha Ke_\alpha \right\|^2 \leq \left(\sum_\alpha |x_\alpha|\, \|Ke_\alpha\| \right)^2$$

$$\leq \left(\sum_\alpha |x_\alpha|^2 \right) \left(\sum_\alpha \|Ke_\alpha\|^2 \right) = \|K\|_2^2\, \|x\|^2,$$

from which 3) follows. To prove 4) we approximate $K \in S_2(H_1, H_2)$ by finite-dimensional operators (operators of finite rank). Namely, let $e_1, e_2, \ldots$ be all the vectors of an orthonormal basis $\{e_\alpha\}$, such that $Ke_\alpha \neq 0$ (it follows from (A.3.1) that this set is at most countable). Then clearly

$$Kx = \sum_{j=1}^{\infty} (x, e_j)\, Ke_j, \quad (A.3.7)$$

and putting $K_N x = \sum_{j=1}^{N} (x, e_j)\, Ke_j$ we obtain $\|K - K_N\|^2 \leq \|K - K_N\|_2^2$

$$= \sum_{j=N+1}^{\infty} \|Ke_j\|^2 \to 0 \text{ as } N \to \infty \text{ as required.}$$

Statement 5) follows from 1) if we choose a basis of eigenvectors of K. Now we will prove 6). Note first that

$$|(Ke_\alpha, Le_\alpha)| \leq \|Ke_\alpha\|\, \|Le_\alpha\| \leq \tfrac{1}{2}\, (\|Ke_\alpha\|^2 + \|Le_\alpha\|^2),$$

from which the convergence of the series (A.3.3) follows for $K, L \in S_2(H_1, H_2)$. It is clear that $(K, K)_2 = \|K\|_2^2$ and hence $(K, L)_2$ is independent of the choice of basis.

Note that the algebraic properties of the scalar product are clearly satisfied by (A.3.3) and to proof 6) it only remains to demonstrate the completeness of $S_2(H_1, H_2)$ with respect to $\|\cdot\|_2$. For this it is most convenient to establish an isomorphism between $S_2(H_1, H_2)$ and $l^2(M_1 \times M_2)$, where M_j is a set of cardinality $\dim H_j$. The space $l^2(M)$ for an arbitrary set M consists of the functions on M different from zero in at most countably many points and such that $\sum_{\alpha \in M} |f(\alpha)|^2 < +\infty$. We may view this space as $L^2(M)$ if M has the σ-algebra generated by one-point sets with measure 1 for each one-point set.

The required isomorphism between $S_2(H_1, H_2)$ and $l^2(M_1 \times M_2)$ is established by associating with each $K \in S_2(H_1, H_2)$ its matrix $K_{\alpha\beta} = (Ke_\alpha, f_\beta)$, where $\{e_\alpha\}$, $\{f_\beta\}$ are orthonormal bases in the spaces H_1 and H_2. The fact that this association is an isometric isomorphism is clear from the calculation (A.3.6).

Now we will prove 7). Since $\| AKe_\alpha \| \leq \| A \| \, \| Ke_\alpha \|$, the estimate (A.3.4) and therefore the fact that $S_2(H_1, H_2)$ is a left $\mathscr{L}(H_2)$-module, are clear from the definitions. To prove (A.3.5), guaranteeing the possibility of introducing on $S_2(H_1, H_2)$ a right $\mathscr{L}(H_1)$-module structure pass to the adjoint operator:

$$\| KB \|_2 = \| (KB)^* \|_2 = \| B^* K^* \|_2 \leq \| B^* \| \, \| K^* \|_2 = \| B \| \, \| K \|_2 \,,$$

as required. $\square$

Now let X and Y be two spaces with positive measures and $H_1 = L^2(Y)$, $H_2 = L^2(X)$. In this situation, the operators $K \in S_2(H_1, H_2)$ are described as follows

Proposition A.3.2. *The operators $K \in S_2(H_1, H_2)$ are exactly those which can be represented as*

$$(Kf)(x) = \int_Y K(x, y)\, f(y)\, dy \tag{A.3.8}$$

with a kernel $K(x, y) \in L^2(X \times Y)$. We then also have

$$\| K \|_2^2 = \iint_{X \times Y} | K(x, y) |^2 \, dx\, dy \tag{A.3.9}$$

(in these formulas dx and dy denote the measures on X and Y respectively).

Proof. Let $\{e_\alpha(y)\}$ and $\{f_\beta(x)\}$ be orthonormal bases in $L^2(Y)$ and $L^2(X)$, and $K \in S_2(H_1, H_2)$. Note that $\{f_\beta(x)\,\overline{e_\alpha(y)}\}$ constitutes a complete orthonormal basis in $L^2(X \times Y)$ and if we put

$$K(x, y) - \sum_{\alpha,\,\beta} (Ke_\alpha, f_\beta)\, f_\beta(x)\,\overline{e_\alpha(y)}, \tag{A.3.10}$$

then $K(x, y) \in L^2(X \times Y)$ and the operator of the form (A.3.8) coincides with K, since these operators both have the same matrices in the bases $\{e_\alpha\}$ in H_1 and $\{f_\beta\}$ in H_2. The Parseval identity guarantees (A.3.9).

Conversely, if $K(x, y) \in L^2(X \times Y)$, then decomposing $K(x, y)$ in the basis $\{f_\beta(x)\,\overline{e_\alpha(y)}\}$, we obtain

$$K(x, y) = \sum_{\alpha,\,\beta} c_{\alpha\beta} f_\beta(x)\,\overline{e_\alpha(y)}, \qquad \sum_{\alpha,\,\beta} |c_{\alpha\beta}|^2 < +\infty \,.$$

But from this it is obvious that in the base $\{e_\alpha\}$ and $\{f_\beta\}$, the matrix of K is $c_{\alpha\beta} = (Ke_\alpha, f_\beta)$ which implies that $K \in S_2(H_1, H_2)$. $\square$

A.3.2 Trace class operators and the trace

Definition A.3.2. Let H be a Hilbert space and $S_2(H)$ the ideal of Hilbert-Schmidt operators on H. Let $S_1(H) = (S_2(H))^2$ be the two-sided ideal in $\mathscr{L}(H)$, the square of $S_2(H)$, consisting of operators which can be written as finite sums

$$A = \sum_j B_j C_j, \quad B_j, \ C_j \in S_2(H) \tag{A.3.11}$$

The ideal $S_1(H)$ is called the *trace class* and the elements of $S_1(H)$ are called *trace class operators* on H.

Proposition A.3.3. 1) *Let $A \in S_1(H)$ and $\{e_\alpha\}$ an orthonormal basis in H. Then*

$$\sum_\alpha |(Ae_\alpha, e_\alpha)| < +\infty \tag{A.3.12}$$

and

$$\operatorname{Sp} A = \sum_\alpha (Ae_\alpha, e_\alpha) \tag{A.3.13}$$

is independent of the choice of orthonormal basis $\{e_\alpha\}$. This expression is called the trace (Spur in German) *of the operator A. The trace is a linear functional on $S_1(H)$ with $\operatorname{Sp} A \geq 0$ for $A \geq 0$. We may rewrite the scalar product $(K, L)_2$ using the trace, as*

$$(K, L)_2 = \operatorname{Sp}(L^* K), \quad K, \ L \in S_2(H) \tag{A.3.14}$$

2) *If A is a compact self-adjoint operator with non-zero eigenvalues $\lambda_1, \lambda_2, \dots$ (counting multiplicity), then $A \in S_1(H)$ if and only if*

$$\sum_{j=1}^\infty |\lambda_j| < +\infty. \tag{A.3.15}$$

and

$$\operatorname{Sp} A = \sum_{j=1}^\infty \lambda_j. \tag{A.3.16}$$

3) *If $A \in S_1(H)$, then*

$$\operatorname{Sp} A^* = \overline{\operatorname{Sp} A} \tag{A.3.17}$$

4) *If $A \in S_1(H)$ and $B \in \mathscr{L}(H)$, then*

$$\operatorname{Sp}(AB) = \operatorname{Sp}(BA) \tag{A.3.18}$$

Proof. 1) if A is expressed in the form (A.3.11) then

$$(Ae_\alpha, e_\alpha) = \sum_j (B_j C_j e_\alpha, e_\alpha) = \sum_j (C_j e_\alpha, B_j^* e_\alpha),$$

which implies (A.3.12) and

$$\sum_\alpha (Ae_\alpha, e_\alpha) = \sum_j (C_j, B_j^*)_2 , \qquad (A.3.19)$$

from which it is obvious, in particular, that $\mathrm{Sp}\, A$ is independent of the choice of basis.

(A.3.14) is obvious.

2) Now let $A^* = A$. If $A \in S_1(H)$, condition (A.3.15) and (A.3.16) holds, because we may take for $\{e_\alpha\}$ a basis of eigenvectors. Conversely, if (A.3.15) holds and if $\{e_\alpha\}$ is a basis of eigenvectors, $Ae_\alpha = \lambda_\alpha e_\alpha$, then defining B and C by the formulas

$$Be_\alpha = \sqrt{|\lambda_\alpha|}\, e_\alpha , \qquad Ce_\alpha = \lambda_\alpha / \sqrt{|\lambda_\alpha|}\, e_\alpha ,$$

we see that B, $C \in S_2(H)$ and $A = BC$ so that $A \in S_1(H)$.

3) Let us verify (A.3.17). Writing A in the form (A.3.11), we have $A^* = \sum_j C_j^* B_j^*$ and from (A.3.19)

$$\mathrm{Sp}\, A^* = \sum_j (B_j^*, C_j)_2 = \sum_j \overline{(C_j, B_j^*)_2} = \overline{\mathrm{Sp}\, A} ,$$

which proves (A.3.17).

4) Finally we will verify (A.3.18). First let B be unitary. Then AB and BA are unitarily equivalent since $AB = B^{-1}(BA)B$. Hence (A.3.18) for B unitary is a consequence of the independence of the trace on the choice of basis. To prove (A.3.18) in general, note that both parts of (A.3.18) are linear in B and the following statement holds

Lemma A.3.1. *An arbitrary operator $B \in \mathscr{L}(H)$ can be expressed as a linear combination of four unitary operators.*

Proof. Since we may write

$$B = B_1 + iB_2 , \qquad B_1^* = B_1 = \frac{B + B^*}{2} , \qquad B_2^* = B_2 = \frac{B - B^*}{2i} ,$$

it suffices to verify that a self-adjoint operator may be expressed as a linear combination of two unitary ones. We may assume that $\|B\| \leq 1$. But then the desired expression takes the form

$$B = \tfrac{1}{2}\,[B + i\sqrt{I - B^2}\,] + \tfrac{1}{2}\,[B - i\sqrt{I - B^2}\,] .$$

Therefore Lemma A.3.1 and hence Proposition A.3.3 are proved. $\square$

A.3.3 The polar decomposition of an operator

Let H_1 and H_2 be Hilbert spaces. Recall that a bounded operator $U: H_1 \to H_2$ is called a *partial isometry* if it maps isometrically $(\text{Ker } U)^\perp$ onto Im U. It follows that

$$U^*U = E, \quad UU^* = F \tag{A.3.20}$$

where E is the orthogonal projection onto $(\text{Ker } U)^\perp$ in H_1 and F is the orthogonal projection onto Im U in H_2 (in this case Im U is a closed subspace of H_2). If Ker $U = 0$ and Im $U = H_2$, then U is a unitary operator.

Definition A.3.3. The *polar decomposition* of a bounded operator A: $H_1 \to H_2$ is the representation of A in the form

$$A = US \tag{A.3.21}$$

where S is bounded self-adjoint and non-negative on H_1 and $U: H_1 \to H_2$ is a partial isometry such that

$$\text{Ker } U = \text{Ker } S = (\text{Im } S)^\perp \tag{A.3.22}$$

Proposition A.3.4. *The polar decomposition of a bounded operator A: $H_1 \to H_2$ exists and is unique.*

Sketch of the Proof. From (A.3.21) we have $A^* = SU^*$, from which $A^*A = SU^*US = SES$. But $ES = S$ by (A.3.22) so that

$$A^*A = S^2 \tag{A.3.23}$$

and hence

$$S = \sqrt{A^*A} \tag{A.3.24}$$

($\sqrt{A^*A}$ is defined by means of the spectral decomposition theorem).

In fact let $C = A^*A$ and let B be any bounded selfadjoint operator in H_1 such that $B^2 = C$ and $B \geq 0$. We will prove that $B = S$, where S is given by (A.3.24). Being a function of C, S commutes with every operator commuting with C. In particular, $BS = SB$ because $BC = CB = B^3$. Hence

$$(S-B)\, S\, (S-B) + (S-B)\, B\, (S-B) = (S^2 - B^2)\, (S-B) = 0 .$$

Both terms on the left-hand side are non-negative operators so both vanish. Hence so does their difference $(S-B)^3$ and therefore $(S-B)^4 = 0$. This obviously implies $S - B = 0$ because $S - B$ is selfadjoint.

For the details concerning spectral and polar decompositions the reader may consult e.g. F. Riesz, B.Sz.-Nagy [1]. Further formula (A.3.21) defines U

uniquely in view of (A.3.22) thus proving the uniqueness of the polar decomposition.

To show the existence construct S by the formula (A.3.24) and define U by

$$Ux = 0 \qquad \text{for } x \perp \operatorname{Im} S \tag{A.3.25}$$

$$U(Sx) = Ax \qquad \text{for } x \in H_1 \tag{A.3.26}$$

To verify the correctness of this definition, it suffices to show that if $Sx = 0$ then $Ax = 0$. But this follows immediately from (A.3.23) together with

$$\|Sx\|^2 = (Sx, Sx) = (S^2 x, x) = (A^* A x, x) = (Ax, Ax) = \|Ax\|^2 ,$$

which shows that U is a partial isometry. $\square$

Definition A.3.4. If $A = US$ is the polar decomposition of A, we will write $S = |A|$.

Proposition A.3.5. *Let J be an arbitrary left ideal in the algebra $\mathscr{L}(H)$. Then $A \in J$ if and only if $|A| \in J$.*

Proof. This is clear since $A = U|A|$, $U^* A = |A|$. $\square$

Corollary A.3.1. *We have*

$$A \in S_2(H) \Leftrightarrow |A| \in S_2(H) ,$$
$$A \in S_1(H) \Leftrightarrow |A| \in S_1(H) .$$

By using the polar decomposition, we may, as a complement to 4) of Proposition A.3.3 prove

Proposition A.3.6. *If A, $B \in S_2(H)$, then*

$$\operatorname{Sp}(AB) = \operatorname{Sp}(BA) .$$

Proof. Using the identities

$$4AB^* = (A+B)(A+B)^* - (A-B)(A-B)^*$$
$$\qquad + i(A+iB)(A+iB)^* - i(A-iB)(A-iB)^* ,$$
$$4B^*A = (A+B)^*(A+B) - (A-B)^*(A-B)$$
$$\qquad + i(A+iB)^*(A+iB) - i(A-iB)^*(A-iB) ,$$

we see that it suffices to verify that

$$\operatorname{Sp}(AA^*) = \operatorname{Sp}(A^*A) , \qquad A \in S_2(H) \tag{A.3.27}$$

However, using the polar decomposition $A = US$, we see that $S \in S_2(H)$ and hence $S^2 \in S_1(H)$ and since

$$A^*A = S^2 , \qquad AA^* = US^2 U^* ,$$

then in view of part 4) of Proposition A.3.3

$$\mathrm{Sp}\,(AA^*) = \mathrm{Sp}\,(US^2U^*) = \mathrm{Sp}\,(U^*US^2) = \mathrm{Sp}\,S^2 = \mathrm{Sp}\,(A^*A),$$

as required. $\square$

A.3.4 The trace class norm

Definition A.3.5. The *trace class norm* of the operator $A \in S_1(H)$ is the expression

$$\|A\|_1 = \mathrm{Sp}\,|A|. \tag{A.3.28}$$

Proposition A.3.7. 1) *The following inequalities hold*

$$\|A\|_2 \leq \|A\|_1, \qquad A \in S_1(H); \tag{A.3.29}$$

$$\|BA\|_1 \leq \|B\|\,\|A\|_1, \qquad A \in S_1(H), \quad B \in \mathscr{L}(H); \tag{A.3.30}$$

$$\|AB\|_1 \leq \|A\|_1\,\|B\|, \qquad A \in S_1(H), \quad B \in \mathscr{L}(H); \tag{A.3.30'}$$

$$|\mathrm{Sp}\,A| \leq \|A\|_1, \qquad A \in S_1(H), \tag{A.3.31}$$

as well as the relations

$$\|A^*\|_1 = \|A\|_1, \qquad A \in S_1(H); \tag{A.3.32}$$

$$\|A\|_1 = \sup_{\substack{B \in \mathscr{L}(H) \\ \|B\| \leq 1}} |\mathrm{Sp}\,(BA)|, \qquad A \in S_1(H). \tag{A.3.33}$$

2) *The trace class norm defines a Banach space structure on* $S_1(H)$.

Proof. 1) a) Let us prove (A.3.29). Suppose $A = US$ is the polar decomposition of A. Then

$$\|A\|_1 = \|S\|_1 = \mathrm{Sp}\,S, \tag{A.3.34}$$

$$\|A\|_2^2 = \mathrm{Sp}\,(A^*A) = \mathrm{Sp}\,S^2, \tag{A.3.35}$$

so that (A.3.29) is equivalent to the inequality

$$\mathrm{Sp}\,S^2 \leq (\mathrm{Sp}\,S)^2 \tag{A.3.36}$$

which becames evident if we express $\mathrm{Sp}\,S^2$ and $\mathrm{Sp}\,S$ in terms of an eigenbasis of S.

b) To prove (A.3.32) note that $A^* = SU^*$, $AA^* = US^2U^*$ and $|A^*| = USU^*$, hence

$$\mathrm{Sp}\,|A^*| = \mathrm{Sp}\,(USU^*) = \mathrm{Sp}\,(U^*US) = \mathrm{Sp}\,S = \mathrm{Sp}\,|A|.$$

c) Now we will prove (A.3.31). Suppose $\{e_\alpha\}$ is an orthonormal basis of eigenvectors of S with eigenvalues s_α,

$$Se_\alpha = s_\alpha e_\alpha . \tag{A.3.37}$$

We have

$$\operatorname{Sp} A = \sum (USe_\alpha, e_\alpha) = \sum s_\alpha (Ue_\alpha, e_\alpha), \tag{A.3.38}$$

and since $|(Ue_\alpha, e_\alpha)| \leqq 1$, then clearly $|\operatorname{Sp} A| \leqq \sum_\alpha s_\alpha = \operatorname{Sp} S = \|A\|_1$.

d) To prove (A.3.30) let $BA = VT$ be the polar decomposition of BA. Then

$$\|BA\|_1 = \operatorname{Sp} T = \operatorname{Sp}(V^*BA) = \operatorname{Sp}(V^*BUS).$$

Now note that $\|V^*BU\| \leqq \|B\|$. The remaining argument is as in c).

e) The estimate (A.3.30') follows from the estimate (A.3.30), since $\|AB\|_1 = \|(AB)^*\|_1 = \|B^*A^*\|_1$.

f) The relation (A.3.33) now readily follows from the estimate

$$|\operatorname{Sp}(BA)| \leqq \|BA\|_1 \leqq \|B\| \, \|A\|_1 ,$$

where we have equality for $B = U^*$.

2) a) We will prove that the trace class norm $\|\cdot\|_1$ has the usual properties of a norm:

$$\|A' + A''\|_1 \leqq \|A'\|_1 + \|A''\|_1 , \qquad A', \ A'' \in S_1(H);$$
$$\|\lambda A\|_1 = |\lambda| \, \|A\|_1 , \qquad A \in S_1(H), \qquad \lambda \in \mathbb{C};$$
$$\|A\|_1 = 0 \Leftrightarrow A = 0 .$$

Here only the first relation is non-trivial. To prove it we use (A.3.33)

$$\begin{aligned}
\|A' + A''\|_1 &= \sup_{\|B\| \leqq 1} |\operatorname{Sp} B(A' + A'')| = \sup_{\|B\| \leqq 1} |\operatorname{Sp}(BA') + \operatorname{Sp}(BA'')| \\
&\leqq \sup_{\|B\| \leqq 1} (|\operatorname{Sp}(BA')| + |\operatorname{Sp}(BA'')|) \\
&\leqq \sup_{\substack{\|B'\| \leqq 1 \\ \|B''\| \leqq 1}} (|\operatorname{Sp}(B'A')| + |\operatorname{Sp}(B''A'')|) = \sup_{\|B'\| \leqq 1} |\operatorname{Sp}(B'A')| \\
&\quad + \sup_{\|B''\| \leqq 1} |\operatorname{Sp}(B''A'')| = \|A'\|_1 + \|A''\|_1 .
\end{aligned}$$

b) We want to verify the completeness of $S_1(H)$ in the norm $\|\cdot\|_1$. Let $n = 1, 2, \ldots, A_n \in S_1(H)$ and $\|A_n - A_m\|_1 \to 0$ as $n, m \to +\infty$. Then by (A.3.29) and part 6) of Proposition A.3.1 there exists an operator $A \in S_2(H)$ such that

$\lim\limits_{n \to \infty} \|A_n - A\|_2 = 0$. To verify that $A \in S_1(H)$, let $A = US$ be the polar decomposition of A and put $C_n = U^* A_n$. Obviously

$$\lim_{n \to \infty} \|C_n - S\|_2 = 0, \tag{A.3.39}$$

$$\lim_{m,\,n \to \infty} \|C_n - C_m\|_1 = 0. \tag{A.3.40}$$

Let $\{e_\alpha\}$ be an orthonormal eigenbasis of S with eigenvalues s_α. From (A.3.39) it follows that

$$s_\alpha = (Se_\alpha, e_\alpha) = \lim_{n \to \infty} (C_n e_\alpha, e_\alpha). \tag{A.3.41}$$

To prove that $A \in S_1(H)$ it suffices to verify that $\sum\limits_\alpha s_\alpha < +\infty$. This, in turn, follows from (A.3.41) and the estimate

$$\sup_n \sum_\alpha |(C_n e_\alpha, e_\alpha)| < +\infty, \tag{A.3.42}$$

which holds in view of the inequality

$$\sum_\alpha |(Ce_\alpha, e_\alpha)| \leqq \|C\|_1, \qquad C \in S_1(H), \tag{A.3.43}$$

which is easily derived from (A.3.33).

It remains to prove that

$$\lim_{n \to \infty} \|A_n - A\|_1 = 0.$$

Let $\varepsilon > 0$ and select N so large that

$$\|A_m - A_n\|_1 \leqq \varepsilon, \qquad m, n \geqq N. \tag{A.3.44}$$

Let us prove that

$$\|A_n - A\|_1 \leqq \varepsilon \qquad n \geqq N. \tag{A.3.45}$$

From (A.3.43) it follows that

$$\lim_{n \to \infty} \mathrm{Sp}(BA_n) = \mathrm{Sp}(BA) \text{ for any } B \in \mathscr{L}(H).$$

But by (A.3.44), this gives

$$|\mathrm{Sp}(BA_n) - \mathrm{Sp}(BA)| \leqq \varepsilon \qquad \text{for } n \geqq N,$$

if $\|B\| \leqq 1$. It only remains to use (A.3.33) to arrive at the desired (A.3.45). $\quad\square$

A.3.5 Expressing the trace in terms of the operator kernel

To express the trace in terms of the operator kernel is very simple from the formal point of view although difficult technically (it is difficult to justify an easily discovered formal formula). In this section, we introduce a formal scheme which should be justified in detail in each concrete case.

Let X be a space with measure dx and K a trace class operator on $L^2(X)$ with kernel $K(x, y)$. We would like to write K in the form

$$K = L_1 L_2, \qquad L_1, L_2 \in S_2(L^2(X)) \tag{A.3.46}$$

(this can for instance be done as follows: let $K = US$ be the polar decomposition, and take $L_1 = U\sqrt{S}$, $L_2 = \sqrt{S}$). Denote by $L_1(x, y)$ and $L_2(x, y)$ the kernels of L_1 and L_2. Formally, we have

$$K(x, y) = \int L_1(x, z)\, L_2(z, y)\, dz. \tag{A.3.47}$$

Since we may also write $K = (L_1^*)^* L_2$, where L_1^* has kernel $L_1^*(x, y) = \overline{L_1(y, x)}$, then by Propositions A.3.3 (formula (A.3.14)) and A.3.2 justifying

$$(K_1, K_2)_2 = \int K_1(x, y)\, \overline{K_2(x, y)}\, dx\, dy \tag{A.3.48}$$

(where $K_j \in S_2(L^2(X))$, $K_j(x, y)$ is the kernel of K_j) we have

$$\mathrm{Sp}\, K = (L_2, L_1^*)_2 = \int L_1(y, x)\, L_2(x, y)\, dx\, dy = \int L_1(x, z)\, L_2(z, x)\, dz\, dx, \tag{A.3.49}$$

or

$$\mathrm{Sp}\, K = \int K(x, x)\, dx, \tag{A.3.50}$$

Actually, a justification of the formal calculations (A.3.47)–(A.3.50) is possible, for instance, when X is a compact manifold with boundary, dx is a measure on X determined by a positive smooth density and the kernel $K(x, y)$ is continuous. An example of these kind of arguments based on the Mercer's theorem is carried out in §13 in proving Theorem 13.2.

Note the basic difficulty in the justification: the kernel $K(x, y)$ is only defined up to a set of measure zero in $X \times X$, but in (A.3.50) there enters the restriction of $K(x, y)$ on the diagonal in $X \times X$ – which is a set of measure 0.

Another version of this kind of argument is to try and obtain the integral (A.3.50) as a limit of integrals on a small neighbourhood of the diagonal appropriately normalized (cf. Gohberg and Krein [1], chapter III, §10). We omit the details, since this is not used in the main text of this book.

Exercise A.3.1. Show that

$$\|AB\|_1 \leq \|A\|_2 \|B\|_2, \qquad A, B \in S_2(H).$$

Exercise A.3.2. Show that

$$\|B\| = \sup_{\substack{A \in S_1(H) \\ \|A\|_1 \leqq 1}} |\operatorname{Sp}(AB)|.$$

Exercise A.3.3. Let J be a two-sided ideal in $\mathscr{L}(H)$ and assume $0 \leqq A \leqq B$ with $B \in J$. Show that $A \in J$.

Hint: Show that $A^{1/2} = CB^{1/2}$, where $C \in \mathscr{L}(H)$ and $\|C\| \leqq 1$.

A Short Guide to the Literature

Here we only mention the works most closely connected with the material covered in the book. We make no claims whatsoever on bibliographical completeness. I have tried as far as possible to avoid referring to short communications, so that mostly monographs and detailed papers or survey articles are referred to.

Chapter I

The concept of a pseudodifferential operator (ΨDO) originated in the theory of multidimensional singular integral operators (cf. Michlin [1] and the references therein). Subsequently, singular integro-differential operators emerged (cf. Agranovich [1] and references therein). The term "pseudodifferential operator" was coined by Friedrichs and Lax [1]. In the present form ΨDO appeared basically in the work by Kohn and Nirenberg [1] (where the ΨDO which we call "classical" were considered). The symbol classes $S^m_{\varrho,\delta}$ and the corresponding operator classes were introduced by Hörmander [2]. The theorem on the invariance of the class of ΨDO under changes of variables is also due to him (cf. Hörmander [1], [2]). The proof of this theorem given here is based on ideas of Kuranishi.

The Fourier integral operators (FIO) were introduced and systematically studied in Hörmander [6]. The closely related concept of a canonical operator had been studied earlier by Maslov in connection with various asymptotic problems (cf. Maslov [1], [2], Maslov and Fedoryuk [1], and Duistermaat [3]). Hörmander's work was also preceeded by the works of Eskin [1], [2], Egorov [1], [2], containing the ideas developed by Hörmander. In the works by Nirenberg and Treves [1] and Egorov [1–4] ΨDO and the simplest FIO were used to study local solvability and regularity of solutions for general operators of principal type.

In the exposition of the theory of oscillatory integrals, FIO, and in the construction of algebras of ΨDO, I basically follow Hörmander [6]. Hypoellipticity in §5, is presented in the spirit of Hörmander [2]. In the same work there is given described here in detail a sketch of the theory of Sobolev spaces. More complete information on Sobolev spaces can be found in Hörmander [7], Nikolskii [1], Sobolev [1], Besov, Il'in and Nikolskii [1], Lions and Magenes [1]. An elementary presentation of the basic facts in the theory of ΨDO can be found in the book by Wells [1].

Note that what we call FIO are usually called local FIO. To avoid an excessive increase of the volume of the book, I deliberately left out the considerably more subtle and complex theory of global FIO. Global FIO or their equivalent, the canonical Maslov operator, are however necessary in a series of applications (for instance in the theory of hyperbolic equations or in spectral theory). Therefore, after the reader has acquired familiarity with local FIO and their applications, he has to learn about global FIO (this might be done for instance from the papers by Hörmander [6] and Duistermaat and Hörmander [1] or from the lectures of Duistermaat [1]) and Maslov canonical operator theory (for instance, from the book by Maslov and Fedoryuk [1]).

Concerning other questions of the theory of ΨDO and FIO, we refer the reader to the monographs by Friedrichs [1], Eskin [3], Taylor [1], Duistermaat [1], Tréves [1], Grushin [2] and the papers by Agranovich [1], Kumano-go [1], [2], Beals and Fefferman [1], Beals [1], [2], [3], and Calderón [1].

Let us mention the important results found by Calderón and Vaillancourt [1] making more precise the boundedness theorem (for example they considered operators of the class $L^0_{\lambda,\delta}$ with $0 \le \varrho = \delta < 1$) (cf. also Watanabe [1] and Kumano-go [3]). As for the action of ΨDO on L^p-spaces, see Muramatu [1] and Illner [1], for the action on Hölder classes, see Durand [1], and for the action on Gevrey classes and classes of analytic functions see Volevič [1], Du-Chateau and Tréves [1], and Baouendi and Goulaouic [1].

Operators with complex phase-function were considered by Kučerenko [1], Miščenko, Sternin and Shatalov [1], and by Melin and Sjostrand [1].

Research on the index problem, for which ΨDO provided an essential tool, excerted a strong influence on the development of the theory ΨDO: cf. Atiyah and Singer [1], Fedosov [1], Atiyah, Bott and Patodi [1], Hörmander [5], Atiyah [1], [2]. The technique of ΨDO was used in the work by Atiyah and Bott [1] on the Lefschetz theorem.

Important concrete applications of ΨDO to classical problems in the theory of partial differential equations can be found in Oleinik and Radkievič [1], and Maz'ya and Paneyah [1].

Chapter II

The description of the structure of complex powers of elliptic operators in terms of ΨDO and the theorem on meromorphic continuation of the kernel of complex powers and of the ζ-function are due to Seeley [1], [2], [3]. Variations and generalizations of this theory can be found in Nagase and Shinkai [1], Hayakawa and Kumano-go [1], Kumano-go and Tsutsumi [1], Smagin [1], [2]. The Tauberian technique was first exploited in the classical work of Carleman [1]. A survey of several variants of the Tauberian technique is given in Hörmander [4]. The proof of the Ikehara Tauberian theorem given here, is close to the one given in the book by Lang [1].

Let us mention an important application of the results of Seeley, namely, the already mentioned work of Atiyah, Bott and Patodi [1], where a new presentation of index theory is given. For a self-adjoint, non-semibounded operator A, Atiyah, Patodi and Singer [1] studied the function

$$\eta_A(z) = \sum_j (\text{sign } \lambda_j) \, |\lambda_j|^z.$$

We mention also the papers by Ray and Singer [1], [2] on analytic torsion close to this circle of ideas.

The papers by Seeley [2], [3] also contain a study of the ζ-function for an operator corresponding to an elliptic boundary problem. We did not touch upon the spectral theory of boundary problems, for this the reader is referred to the monograph of Berezanskij [1].

The meromorphic continuation of the ζ-function is intimately related with the asymptotic behaviour of the trace of the resolvent (as $\lambda \to \infty$) and with the asymptotic behaviour of the θ-function

$$\theta(t) = \sum_j e^{-\lambda_j t} \quad \text{as } t \to +0$$

(cf. for instance Duistermaat and Guillemin [1]). These questions are considered in the papers by Fujiwara [1] and Greiner [1].

Pseudodifferential systems, elliptic in the sense of Douglis and Nirenberg are studied from this view point by Koževnikov [1]. The asymptotic behaviour of $N(\lambda)$ as $\lambda \to +\infty$ (without an estimate of the remainder) for hypoelliptic operators on a compact manifold with boundary was obtained by Moscatelli and Thomson [1].

A discussion of the theory of pseudodifferential boundary problems with parameters can be found in Agranovich [2].

As we have noted in the main text, the theorem on the continuation of the ζ-function does not allow us to obtain any substantial information concerning the eigenvalues of non-self-adjoint operators. A survey of different questions of the spectral theory of non-selfadjoint differential and pseudodifferential can be found in Agranovich [3].

Chapter III

Theorem 16.1., which is due to Hörmander, is proved in [4]. Chapter III is an extensive presentation of this work, supplemented with an exposition of all the indispensable auxiliary facts. The work by Hörmander [4] also contains some results bearing on the case of a non-compact manifold and some results on Riesz' means (concerning this, see also Hörmander [3]).

The description of the structure of the operator $\exp(-it A)$ as an FIO, is essentially based on ideas from geometric optics (concerning this, see the book by Babič and Buldyrev [1]). These ideas were exploited by Lax [1] to construct asymptotic solutions and a parametrix of hyperbolic systems. Hörmander [4] developed the method of Lax and arrived at the indicated description of the structure of the operator $\exp(-it A)$, which is essentially equivalent to a description of the singularities of the fundamental solution of a pseudodifferential hyperbolic equation. Note also that for large t, the operator $\exp(-itA)$ already is a global FIO (cf. the literature guide to Chapter I.). The possibility of representing the exponential operator $\exp(-tA)$ and its kernel in terms of Feynman type integrals is studied in a series of works (cf. e.g. Maslov and Shishmarev [1]).

Notice, that the method of obtaining the asymptotics behaviour of spectral function by considering a hyperbolic equation, was first employed by Levitan [1]. Let us also mention Levitan [2], which contains important supplements to the work of Hörmander [4].

Further results on the asymptotic behaviour of eigenvalues, which connect this question with the geometry of the bicharacteristics, can be found in the papers of Colin de Verdière [1], Chazarain [1], Duistermaat [2], Weinstein [1], Shnirelman [1]. Note in particular the work by Duistermaat and Guillemin [1], where under certain assumptions, the second in the asymptotics of $N(\lambda)$ as $\lambda \to \infty$ is obtained for an selfadjoint elliptic operator on a closed manifold. Using FIO Rozenblyum [1] got very sharp results on asymptotic behaviour of eigenvalues for operators on a circle.

Concerning the geometry of the spectrum cf. also the book by Berger, Gauduchon and Mazet [1], an article by Molčanov [1] and interesting papers by Gilkey [1], [2], [3].

We did not touch on questions connected with the spectral asymptotics of non-smooth or degenerate operators and boundary problems. Regarding this, we refer the reader to the lectures of Birman and Solomyak [1] and their survey [2], where an extensive bibliography on spectral asymptotics can also be found. A survey of a number of important results concerning eigenfunction expansions can be found in the article by Alimov, Il' in and Nikišin [1].

Chapter IV

Essentially the theory of ΨDO in $\mathbb{R}^n$ emerged long ago in connection with mathematical questions of quantum mechanics. (cf. e.g. Berezin [1], Berezin and Shubin [1], [2]). Several versions of this theory can be found in the works of Rabinovič [1], Kumano-go [1], [2], Grušin [1], Shubin [1], [5], Beals [2], Feigin [2].

In Beals [3] and Shubin [5], there is discussed the structure of operators, which are functions of ΨDO in $\mathbb{R}^n$ with uniform (in x) estimates of the symbols (such as in Kumano-go [1]).

Various questions, related to the Fredholm properties of ΨDO on non-compact manifolds, are considered by Cordes and McOwen [1]. Recently a number of papers have appeared devoted to ΨDO on nilpotent Lie groups (in particular on the Heisenberg group). Cf. e.g. Rothschild and Stein [1].

The construction given here of the algebra of ΨDO in $\mathbb{R}^n$ is close to the one in Shubin [1]. The concept of the anti-Wick symbol was introduced by Berezin [2] and is a variation of Friedrich's construction [1] (see also Kumano-go [1], [2]).

Concerning applications of the Wick and anti-Wick symbol and also more general symbols, see the papers by Berezin [2], [3], Berezin and Shubin [1], and Shubin [2], [3]. With the help of inequalities for Sp $\exp(-itA)$, the asymptotic behaviour of the eigenvalues is obtained in the work Berezin [2] (without remainder estimate). The results of §25 and §26 are essentially contained in the work of Shubin [1] (see also [4]) – The method of approximate spectral projection and all the results in §28–30 are contained in Tulovskij and Shubin [1]. A significant modification of this method was offered by Feigin [1], [2].

We mention that the method of approximate spectral projection is essentially a variational method. Variational methods began with the classical work of H. Weyl [1], [2]. To find the asymptotic behaviour of eigenvalues for operators in $\mathbb{R}^n$, one can also apply the Tauberian method, cf. the work by Kostyučenko [1]. A survey of all results on the spectral asymptotics for operators in $\mathbb{R}^n$ can be found in the already cited work by Birman and Solomyak [2].

Appendix 1

On the basis of the earlier concept of singular support of a hyperfunction, due to Sato [1], Hörmander [6] introduced the concept of a wave-front for a distribution. The theorem on propagation of singularities in the form given here is due to Duistermaat and Hörmander [1] (see also Hörmander [8]). The proof given here is due to Tulovski. Another presentation can be found in the lectures by Nirenberg [1]. In a number of later works more subtle questions connected to the propagation of wave fronts have been considered (see e.g. the work by Ivrii [1] and references therein).

Appendix 2

In this appendix the results of Roitburd [1] are presented. A closely related result but without an estimate of the remainder term was obtained by Berezin [2]. Other information on quasi-classical asymptotic formulae and references to the literature, can be found in the monograph by Maslov and Fedoryuk [1].

Bibliography

Agranovich, M. S.
[1] Elliptic singular integro-differential operators, Uspehi Mat. Nauk **20**, No. 5, 3–120 (1965) (Russian), also in Russian Math. Surveys **20**, 1–122 (1965)
[2] Boundary value problems for systems with a parameter, Mat. Sb. **84**(126), 27–65 (1971) (Russian), also in Math. USSR Sb. **13**, 25–64 (1971)
[3] Spectral properties of diffraction problems. In: Voitovich, N. N., Katsenelenbaum, B. Z., Sivov, A. N.: Generalized method of eigenvibrations in diffraction theory, Moscow, Nauka, 1977, p. 289–416 (Russian). See also: Agranovich, M. S., Katsenelenbaum, B. Z., Sivov, A. N., Voitovich, N. N. Generalized method of eigenoscillations in diffraction theory. Translated from the Russian manuscript by Vladimir Nazaikinskii.
WILEY-VCH Verlag Berlin GmbH, Berlin, 1999. 377 pp.

Alimov, Š. A., Il'in, V. A., Nikišin, E. M.
[1] Questions on the convergence of multiple trigonometric series and spectral expansions. 1, II. Uspehi Mat. Nauk **31**, No. 6 (192), 28–83 (1976), **32**, No. 1 (193), 107–130, 271 (1977) (Russian), also in Russian Math. Surveys **31**(1976), No. 6, **32** (1977), No. 1

Arnold, V.I.
[1] Mathematical methods of classical mechanics. Graduate Texts in Math., vol. 60, Springer-Verlag, Berlin, New York 1978, second edition 1989

Atiyah, M.
[1] Elliptic operators and compact groups. Lect. Notes Math. **401**, 1–93 (1974)
[2] Elliptic operators, discrete groups and von Neumann algebras. Astérisque **32–33**, 43–72 (1976)

Atiyah, M., Bott, R.
[1] A Lefschetz fixed point formula for elliptic complexes. I. Ann. of Math., Ser. 2, 374 407 (1967); II. Applications, Ann. of Math., Ser. 2, **88**, 451–491 (1968)

Atiyah, M., Bott, R., Patodi, V. K.
[1] On the heat equation and the index theorem. Invent. Math. **19**, 279–330 (1973); Errata. Invent. Math. **28**, 277–288 (1975)

Atiyah, M., Patodi, V. K., Singer, I. M.
[1] Spectral asymmetry and Riemannian geometry. I, Math. Proc. Cambridge Phil. Soc. **77**, 43–69 (1975); II, **78**, 405–432 (1975); III, **79**, 71–99 (1976)

Atiyah, M., Singer, I. M.
[1] The index of elliptic operator. I, Ann. of Math. **87**, 484–530 (1968)

Babič, V. M., Buldyrev, V. S.
[1] Asymptotic methods in short wave diffraction problems. Vol. 1. The method of canonical problems. Nauka, Moscow 1972, 456 pp. (Russian). See also: Babič, V. M., Buldyrev, V. S. Short-wavelength diffraction theory. Asymptotic methods. Translated

from the 1972 Russian original by E. F. Kuester. Springer Series on Wave Phenomena, 4. Springer-Verlag, Berlin, 1991, xi+445 pp.

Baouendi, M. S., Goulaouic, C.
[1] Problèmes de Cauchy pseudo-différentiels analytiques non linéaires. (French) Séminaire Goulaouic-Schwartz 1975/1976: Équations aux dérivèes partielles et analyse fonctionnelle, Exp. No. 13, 11 pp. Centre Math., École Polytech., Palaiseau, 1976

Beals, R.
[1] Spatially inhomogeneous pseudodifferential operators. II. Comm. Pure Appl. Math. 27, 161–205 (1974)
[2] A general calculus of pseudodifferential operators. Duke Math. J. 42, 1–42 (1975)
[3] Characterization of pseudodifferential operators and applications. Duke Math. J. 44, 45–58 (1977)

Beals, R., Fefferman, C.
[1] Spatially inhomogeneous pseudodifferential operators. I. Comm. Pure Appl. Math. 27, 1–24 (1974)

Berezanskii, Yu. M.
[1] Expansions in eigenfunctions of selfadjoint operators, Akad. Nauk Ukrain. SSR. Institut Matematiki. Naukova Dumka, Kiev 1965, 798 pp.; also American Mathematical Society. Providence, R.I., 1968, 809 pp.

Berezin, F. A.
[1] Representation of operators by means of functionals. Trudy Moscow. Mat. Obšč. 17, 117–196 (1967) (Russian), also in Trans. Moscow Math. Soc. 129–217 (1967)
[2] Wick and anti-Wick Operator symbols. Mat. Sbornik, 86, No. 4, 578–610 (1971) (Russian), also in Math. USSR Sbornik 15, No. 4, 577–606 (1971)
[3] Covariant and contravariant symbols of operators. Izv. Akad. Nauk SSSR Ser. Mat. 36, No. 5, 1134–1167 (1972) (Russian), also in Math. USSR Izvestija 6, No. 5, (1972) 1117–1151

Berezin, F. A., Shubin, M. A.
[1] Symbols of operators and quantization. Colloquia mathematica Societatis Janos Bolyai 5, Hilbert space operators, Tihany (Hungary), 21–52 (1970)
[2] Lectures on quantum mechanics. Moscow State University, 1972 (Russian). See also: The Schrödinger equation. Kluwer Acad. Publishers, Dordrecht, 1991, 555 pp. See also: The Schrödinger equation. Translated from the 1983 Russian edition by Yu. Rajabov, D. A. Leites and N. A. Sakharova and revised by Shubin. With contributions by G. L. Litvinov and D. A. Leites. Mathematics and its Applications (Soviet Series), 66. Kluwer Academic Publishers Group, Dordrecht, 1991. xviii+555 pp.

Berger, M., Gauduchon. P., Mazet. E.
[1] Le spectre d'une varieté Riemannienne. Lect. Notes Math. 194 (1971), 251 pp.

Besov, O. V., Ilin, V. P., Nikolskii, S. M.
[1] Integral representations of functions and imbedding theorems. Vol. I. Translated from the Russian. Scripta Series in Mathematics. Edited by Mitchell H. Taibleson. V. H. Winston & Sons, Washington, D.C.; Halsted Press [John Wiley & Sons], New York-Toronto, Ont.-London, 1978. viii+345 pp.

Birman, M. S., Solomjak, M. Z.
[1] Quantitative analysis in Sobolev's imbedding theorems and applications to spectral theory (Russian). Tenth Mathematical School (Summer School, Kaciveli/Nalchik, 1972), pp. 5–189. Inst. Mat. Akad. Nauk Ukrain. SSR, Kiev 1974. See also: Quantitative analysis in Sobolev imbedding theorems and applications to spectral theory. Translated from the Russian by F. A. Cezus. American Mathematical Society Transla-

tions, Series 2, 114. American Mathematical Society, Providence, R.I., 1980. viii+132 pp.

[2] Asymptotic properties of the spectrum of differential equations (Russian), Mathematical analysis, vol. 14, pp. 5–58, Akad. Nauk SSSR, Vsesojuz. Inst. Naučn. i Tehn. Informacii, Moscow 1977

Calderon, A. P.

[1] Lecture notes on pseudo-differential operators and elliptic boundary value problems. I. Cursos de Matemática, No. 1. [Courses in Mathematics, No. 1] Consejo Nacional de Investigaciones Cientificas y Tecnicas, Instituto Argentino de Matemática, Buenos Aires, 1976. 89 pp.

Calderon, A. P., Vaillancourt, R.

[1] A class of bounded pseudodifferential operators. Proc. Nat. Acad. Sci. USA **69**, 1185–1187 (1972)

Carleman, T.

[1] Propriétés asymptotiques des functions fondamentales des membranes vibrantes. C. R. 8-ème Congr. des Math. Scand. Stockholm (1934), Lund, 34–44 (1935)

Chazarain, J.

[1] Formule de Poisson pour les variétés riemanniennes, Invent. Math. **24**, 65–82 (1974)

Chern, S. S.

[1] Complex manifolds. Lectures in Univ. of Chicago, Autumn 1955 – Winter 1956. See also: Complex manifolds. Textos de Matemática, No. 5, Instituto de Física e Matemática, Universidade do Recife, 1959, v+181 pp.

Colin de Verdier, Y.

[1] Spectre du laplacien et longueurs des géodésique périodiques. II. Comp. Math. **27**, 159–184 (1973)

Cordes, H. O., McOwen, R. C.

[1] The C^*-algebra of a singular elliptic problem on a noncompact Riemannian manifold. Math. Z. **153**, No. 2, 101–116 (1977)

Du-Chateau, P., Tréves, F.

[1] An abstract Cauchy-Kovalevski theorem in scale of Gevrey classes. Symp. Mathematica **7**, 135–163 (1971)

Duistermaat, J. J.

[1] Fourier integral operators. Translated from Dutch notes of a course given at Nijmegen University, February 1970 to December 1971. Courant Institute of Mathematical Sciences, New York University, New York, 1973. iii+190 pp. See also: Fourier integral operators. Progress in Mathematics, 130. Birkhäuser Boston, Inc., Boston, MA, 1996. x+142 pp.

[2] The spectrum and periodic geodesics. Lecture on the AMS Summer Institute on differential geometry, Stanford, August 1973

[3] Oscillatory integrals, Lagrange immersions and unfolding of singularities. Comm. Pure Appl. Math. **27**, No. 2, 207–231 (1974)

Duistermaat, J. J., Guillemin, V. W.

[1] The spectrum of positive elliptic operators and periodic bicharacteristics. Invent. Math. **29**, 39–79 (1975)

Duistermaat, J. J., Hörmander, L.

[1] Fourier integral operators II, Acta Math. **128**, 183–269 (1972)

Durand, M.

[1] Paramétrix d'opérateurs elliptiques de classe C^μ. Bull. Soc. Math. France **103**, 21–63 (1975)

Egorov, Ju. V.
[1] Canonical transformations and pseudodifferential operators. Trudy Moscow. Mat.
     Obšč. **24** (1971), 3–28 (Russian), also in Trans. Moscow Math. Soc. **24** (1971), 1–28
     (1974)
[2] Necessary conditions for the solvability of pseudodifferential equations of principal
     type. Trudy Moscow. Mat. Obšč **24**, 29–41 (1971) (Russian), also in Trans. Moscow
     Math. Soc. **24** (1971), 29–42 (1974)
[3] Subelliptic operators. Uspehi Math. Nauk **30**:2, 57–114 (1975), 57–114 (Russian),
     also in Russian Math. Surveys **30**:2, 59–118 (1975)
[4] On subelliptic operators. Uspehi Mat. Nauk **30**: 3, 57–104 (1975) (Russian), also in
     Russian Math. Surveys **30**:3, 55–105 (1975)
[5] Linear differential equations of principal type. Translated from the Russian by Dang
     Prem Kumar. Contemporary Soviet Mathematics. Consultants Bureau, New York,
     1986. viii+301 pp.

Eskin, G. I.
[1] The Cauchy problem for hyperbolic systems in convolutions. Mat. Sbornik (N. S.) **74**
     (116), No. 2, 262–297 (1967) (Russian), also in Math. USSR Sbornik, v. 3 (1967),
     303–332 (1969)
[2] Systems of pseudodifferential equations with simple real characteristics. Mat. Sbornik
     (N. S.) **77** (119), No. 2, 174–200 (1968) (Russian), also in Math. USSR, Sbornik v. 6
     (1968), 159–183 (1970)
[3] Boundary value problems for elliptic pseudodifferential equations. Translated from the
     Russian by S. Smith. Translations of Mathematical Monographs, 52. American Math-
     ematical Society, Providence, R.I., 1981. xi+375 pp.

Fedosov, B. V.
[1] Analytic formulas for the index of elliptic operators. Trudy Moscow. Mat. Obšč. **30**.
     159–242 (1974) (Russian), also in Trans. Moscow Math. Soc. **30**, 159–240 (1974)

Feigin, V. I.
[1] Asymptotic distribution of eigenvalues for hypoelliptic systems in IR^n. Mat. Sb. (N.
     S.) **99** (141) (1976), No. 4, 594–614 (Russian), also in Math. USSR Sb. **28** (1976),
     No. 4, 533–552 (1978)
[2] Some classes of pseudodifferential operators in IR^n and some applications. Trudy
     Moscow. Mat. Obšč **36**, 155–194 (1978) (Russian), also in Trans. Moscow Math. Soc.
     36 (1978)

Friedrichs, K. O.
[1] Pseudo-differential operators. An introduction. Notes prepared with the assistance of
     R. Vaillancourt. Revised edition. Courant Institute of Mathematical Sciences, New
     York University, New York 1970, vi+279 pp.

Friedrichs, K. O., Lax, P. D.
[1] Boundary value problems for first order operators. Comm. Pure Appl. Math. **18**, No.
     1–2, 355–388 (1965)

Fujiwara, D.
[1] On the asymptotic formula for the Green operators for elliptic operators on compact
     manifolds. J. Fac. Sci. Univ. Tokyo. Sec. 1, **14**, No. 2, 251–283 (1967)

Gelfand, I. M., Šilov, G. E.
[1] Generalized functions. Volume 1: Properties and operations. Volume 2: Spaces of fun-
     damental and generalized functions. Volume 3: Some questions in the theory of dif-
     ferential equations. Moscow, Fizmatgiz, 1958–1959 (Russian), also Academic Press,
     New York, London 1964, 1968

Gilkey P. B.
[1] Curvature and the eigenvalues of the Laplacian for elliptic complexes. Adv. Math. **10**, 344–382 (1973)
[2] Curvature and the eigenvalues of the Dolbeault complex for Kaehler manifolds. Adv. Math. **11**, 311–325 (1973)
[3] Spectral geometry and the Kaehler condition for complex manifolds. Invent. Math. **26**, 231–258 (1974)

Gohberg, I. C., Krein, M. G.
[1] Introduction to the theory of linear nonselfadjoint operators. Nauka, Moscow 1965 (Russian). See also: Introduction to the theory of linear nonselfadjoint operators. Translated from the Russian by A. Feinstein. Translations of Mathematical Monographs, Vol. 18. American Mathematical Society, Providence, R.I. 1969, xv+378 pp.

Greiner, P.
[1] An asymptotic expansion for the heat equation. Arch. Ration. Mech. and Anal. **41**, No. 1, 163–218 (1971)

Grušin, V. V.
[1] Pseudodifferential operators in IR with bounded symbols. Funktional. Anal. i Priložen. **4**, No. 3, 37–50 (1970) (Russian), also in Functional Analysis Appl. 4 (1970), 202–212 (1971)
[2] Pseudodifferential operators. Moscow Institute of Electronic Engineering, Moscow 1975 (Russian)

Hayakawa, K., Kumano-go, H.
[1] Complex powers of a system of pseudo-differential operators. Proc. Jap. Acad. **47**, 359–364 (1971)

Hörmander, L.
[1] Pseudo-differential operators. Comm. Pure Appl. Math. **18**, 501–517 (1965)
[2] Pseudo-differential operators and hypoelliptic equations. Singular integrals (Proc. Sympos. Pure Math., Vol. X, Chicago, Ill., 1966), pp. 138–183. Amer. Math. Soc., Providence, R.I., 1967
[3] On the Riesz means of spectral functions and eigenfunction expansions for elliptic differential operators. (Proc. Annual Sci. Conf., Belfer Grad. School Sci., Yeshiva Univ., New York, 1965–1966). Some Recent Advances in the Basic Sciences, Vol. 2, Belfer Graduate School of Science, Yeshiva Univ., New York, pp. 155–202, 1969
[4] The spectral function of an elliptic operator. Acta Math. **121**, 193–218 (1968)
[5] On the index of pseudodifferential operators, Schriftenr. Inst. Math. Dtsch. Akad. Wiss. Berlin, A, No. 8, 127–146 (1971)
[6] Fourier integral operators, I, Acta Math. **127**, 79–183 (1971)
[7] Linear partial differential operators. Springer-Verlag, Berlin New York, 1976, vii+285 pp.
[8] On the existence and the regularity of solutions of linear pseudo-differential equations, L'Enseignement mathematique **17**: 2, 99–163 (1971)

Illner, R.
[1] On algebra of pseudo-differential operators in $L^p(\mathrm{IR}^n)$. Comm. Part. Diff. Eq. 2, No. 4, 359–394 (1977)

Ivrii, V. Ja.
[1] Wave fronts of the solutions of certain pseudo-differential equations. Functional. Anal. i Priložen. **10**: 2, 71–72 (1976) (Russian), also in Functional Analysis Appl. **10**, 141–142 (1976)

Kohn, J. J., Nirenberg, L.
[1] An algebra of pseudo-differential operators Comm. Pure Appl. Math. **18**, 269–305 (1965)

Kostjučenko, A. G.
[1] Asymptotic behavior of the spectral function of selfadjoint elliptic operators. Fourth Math. Summer School (Kaciveli, 1966), pp. 42–117. Naukova Dumka, Kiev 1968 (Russian)

Koževnikov, A. N.
[1] Spectral problems for pseudodifferential systems elliptic in the Douglis-Nirenberg sense and their applications. Mat. Sb. **92**, No. 1, 60–88 (1973) (Russian), also in Math. USSR Sb. **21**, 63–90 (1973)

Kučerenko, V. V.
[1] Asymptotic solution of the Cauchy problem for equations with complex characteristics, Itogi Nauki i Tekhniki, Sovremennye Problemy Matematiki **8**, 41–136, (1976) Moscow (Russian), also in J. Soviet Math. **13**, 24–80 (1980)

Kumano-go, H.
[1] Remarks on pseudo-differential operators. J. Math. Soc. Japan **21**, 413–439 (1969)
[2] Algebras of pseudo-differential operators. J. Fac. Sci. Univ. Tokyo, Sec. 1A, **17**, 31–50 (1970)
[3] Pseudo-differential operators of multiple symbol and the Calderon-Vaillancourt theorem. J. Math. Soc. Japan **27**, No. 1, 113–119 (1975)

Kumano-go, H., Tsutsumi, C.
[1] Complex powers of hypoelliptic pseudo-differential operators with applications. Osaka J. Math. **10**, 147–174 (1973)

Lang, S.
[1] Algebraic numbers. Addison-Wesley Publishing Co., Inc., Reading, Mass.-Palo Alto-London 1964, ix+163 pp.

Lax, P. D.
[1] Asymptotic solutions of oscillatory initial value problems. Duke Math. J. **24**, 627–646 (1957)

Levitan, B. M.
[1] On the asymptotic behavior of the spectral function of a self-adjoint second order differential equation. Izvestiya Akad. Nauk SSSR, Ser. Mat. **16**, 325–352 (1952) (Russian), also in Amer. Math. Soc. Trans. (2), **101**, 192–221 (1973)
[2] Asymptotic behavior of the spectral function of an elliptic equation. Uspehi Mat. Nauk **26**:6, 151–212 (1971) (Russian), also in Russian Math. Surveys **26**(1971), No. 6, 165–232 (1972)

Lions, J.-L., Magenes, E.
[1] Problemes aux limites non homogènes et applications, v. **1**, Dunod, Paris 1968, 372 pp.

Maslov, V. P.
[1] Perturbation theory and asymptotic methods. Moscow, 1965 (Russian)
[2] Operational methods. Mir Publishers, Moscow 1976

Maslov, V. P., Fedorjuk, M. V.
[1] Quasiclassical approximation for the equations of quantum mechanics. Nauka, Moscow 1976, 296 pp. (Russian), English translation Mathematical Physics and Applied Mathematics, 7. D. Reidel Publishing Co., Dordrecht-Boston, Mass., 1981

Maslov, V. P., Shishmarev, I. A.
[1] T-product of hypoelliptic operators. Itogi Nauki Tekh., Ser. Sovrem. Probl. Mat. **8**, 137–197 (1977) (Russian), also in J. Sov. Math. **13**, 81–118 (1980)

Mazja, V. G., Panejah, B. P.
[1] Degenerate elliptic pseudodifferential operators and the oblique derivative problem. Trudy Moskov. Mat. Obšč. **31**, 237–295 (1974) (Russian), English translation in: Trans. Moscow Mathematical Society for the year 1974 (vol. 31), Amer. Math. Soc. Providence, R.I., 1976, 247–305

Melin, A., Sjöstrand, J.
[1] Fourier integral operators with complex-valued phase functions. Lect. Notes Math. **459**, 121–223 (1975)

Mikhlin, S. G.
[1] Multidimensional singular integrals and integral equations (Russian), Fizmatgiz, Moscow 1962. English translation, Pergamon Press, Oxford-New York-Paris, 1965, XI+255 pp.

Mishchenko, A. S., Sternin, B. Yu., Shatalov, V. E.
[1] Geometry of Lagrangian manifolds and the canonical Maslov operator in complex phase space. Itogi Nauki Tekh., Ser. Sovrem. Probl. Mat. **8**, 5–39 (1977) (Russian), also in J. Sov. Math. **13**, 1–23 (1980)

Molčanov, S. A.
[1] Diffusion processes and Riemannian geometry. Uspehi Mat. Nauk **30**:1, 3–59 (1975) (Russian), also in Russian Math. Surveys **30**, No. 1, 1–63 (1975)

Moscatelli, V. B., Thompson, M.
[1] Distribution asymptotique des valeurs propres d'opérateurs pseudo-differentiels sur des variétés compacts. C. R. Acad. Sci. **284**, No. 6, A373–A375 (1977)

Muramatu, T.
[1] On the boundedness of a class of operator-valued pseudo-differential operators in L^p-space. Proc. Japan Acad. 49, 94–99 (1973)

Nagase, M., Shikai, K.
[1] Complex powers of non-elliptic operators. Proc. Jap. Acad. **46**, 779–783 (1970)

Nikol'skii, S. M.
[1] Approximation of functions of several variables and imbedding theorems. Nauka, Moscow 1969, 480 pp. (Russian), English translation, Die Grundlehren der mathematischen Wissenschaften **205**, Berlin-Heidelberg-New York: Springer-Verlag, VIII, 420 pp.

Nirenberg, L.
[1] Lectures on linear partial differential equations. Conference board of the mathematical sciences, Regional Conference series in mathematics. Amer. Math Soc. No. 1 (1973)

Nirenberg, L., Treves, F.
[1] On local solvability of linear partial differential equations. Part I: Necessary conditions. Comm. Pure Appl. Math. **23**, 1–38 (1970). Part II: Sufficient conditions. Comm. Pure Appl. Math. **23**, 459–509 (1970)

Olejnik, O. A., Radkevič, E. V.
[1] Second order equations with non-negative characteristic form. In Matem. Anal. 1969, ed. R. V. Gamkrelidze, Moscow 1971 (Russian). See also: Second order equations with nonnegative characteristic form. Translated from the Russian by Paul C. Fife. Plenum Press, New York-London, 1973. vii+259 pp.

Rabinovič, V. S.
[1] Pseudodifferential equations in unbounded domains with conical structure at infinity. Mat. Sb. (N. S.) **80** (122), 77–97 (1969) (Russian), also in Math. USSR Sb. **9**, 73–92 (1969)

Ray, D., Singer, I. M.
[1] R-Torsion and the Laplacian on Riemannian manifolds. Adv. Math. **7**, 145–210 (1971)
[2] Analytic torsion for complex manifolds. Ann. of Math. **98**, 154–177 (1973)

Riesz, F., Sz.-Nagy, B.
[1] Functional analysis. Translated from the second French edition by Leo F. Boron. Reprint of the 1955 original. Dover Books on Advanced Mathematics. Dover Publications, Inc., New York, 1990. xii+504 pp.

Roitburd, V. L.
[1] The quasiclassical asymptotic behavior of the spectrum of a pseudodifferential operator. Uspehi Mat. Nauk **31**:4, 275–276 (1976) (Russian)

Rotschild, L. P., Stein, E. M.
[1] Hypoelliptic differential operators and nilpotent groups. Acta Math. **137**, No. 3–4, 247–320 (1976)

Rozenbljum, G. V.
[1] Near-similarity of pseudodifferential systems on the circle. Dokl. Akad. Nauk SSSR **223**:3, 569–571 (1975) (Russian), also in Soviet Math. Dokl. **16** (1975), No. 4, 959–962 (1976)

Rudin, W.
[1] Functional analysis. Second edition. International Series in Pure and Applied Mathematics. McGraw-Hill, Inc., New York, 1991. xviii+424 pp.

Sato, M.
[1] Regularity of hyperfunction solutions of partial differential equations. Actes Congr. int. mathématiciens, 1970, v. 2, Paris, 785–794 (1971)

Seeley, R. T.
[1] Complex powers of an elliptic operator. Proc. Symp. in are Math. **10**, 288–307 (1967)
[2] The resolvent of an elliptic boundary problem. Amer. J. Math. **91**, 889–920 (1969)
[3] Analytic extension of the trace associated with elliptic boundary problems. Amer. J. Math. **91**, 963–983 (1969)

Shnirel'man, A. 1.
[1] The asymptotic multiplicity of the spectrum of the Laplace operator. Uspehi Mat. Nauk **30**, No. 4 (184), 265–266 (1975) (Russian)

Shubin, M. A.
[1] Pseudodifferential operators in $\mathbb{R}^n$. Dokl. Akad. Nauk SSSR **196**, No. 2, 316–319 (1971) (Russian), also in Soviet Math. Dokl. **12**, No. 1, 147–151 (1971)
[2] Certain properties of pseudo-differential operators with nonsmooth symbols. Dokl. Akad. Nauk SSSR **207**, No. 3, 551–553 (1972) (Russian), also in Soviet Math. Dokl. **13**, 1586–1589 (1972)
[3] Spectral properties of operators with covariant and contravariant symbol and a variational principle. Vestnik Moskov. Univ., Ser. 1, Math. Mech. **28**, No. 3, 51–57 (1973) (Russian)
[4] The essential selfadjointness of uniformly hypoelliptic operators. Vestnik Moskov. Univ. Ser. I Mat. Mech. **30**, No. 2, 91–94 (1975) (Russian), also in Moskow Univ. Math. Bull. **30**, No. 1–2, 147–150 (1975)
[5] Pseudodifferential almost-periodic operators and von Neumann algebras. Trudy Moscow. Mat. Obšč. **35**, 103–164 (1976) (Russian), also in Trans. Moscow Math. Soc., Issue 1. 103–166 (1979)

Smagin, S. A.
[1] Fractional powers of a hypoelliptic operator in $\mathbb{R}^n$, Dokl. Akad. Nauk SSSR **209**, 1033–1035 (1973) (Russian), also in Soviet Math. Dokl. **14**, 585–588 (1973)

[2] Complex powers of hypoelliptic systems in $\mathbb{R}^n$. Mat. Sb. (N. S.), **99** (141), No. 3, 331–341, 479 (1976) (Russian), also in Math. USSR Sbornik **28**, 291–300 (1976)

Sobolev, S. L.
[1] Introduction to the theory of cubature formulae. Nauka, Moscow 1974, 808 pp. (Russian). See also: Cubature formulas and modern analysis. An introduction. Translated from the 1988 Russian edition. Gordon and Breach Science Publishers, Montreux, 1992. xvi+379 pp.

Taylor, M.
[1] Pseudo differential operators. Lect. Notes in Math. **416**. Springer-Verlag, Berlin New York, 1974, iv+155 pp.

Treves, F.
[1] An introduction to pseudo-differential operators and Fourier integral operators. Universidade Federal de Pernambuco, Institute de Matematika, Editora Universitaria, Recife 1973

Tulovskiĭ, V. N., Shubin, M. A.
[1] On asymptotic distribution of eigenvalues of pseudodifferential operators in $\mathbb{R}^n$. Mat. Sbornik **92**, No. 4, 571–588 (1973) (Russian), also in Math. USSR Sbornik **21**, No. 4, 565–588 (1973)

Volevič, L. R.
[1] Pseudodifferential operators with holomorphic symbols and Gevrey classes, v. 24, 1971, p. 43–68 (Russian), also in Trans. Moscow Math. Soc. **24**, 43–72 (1971)

Watanabe, K.
[1] On the boundedness of pseudodifferential operators of type $\varrho,\ \delta$ with $0 \leq \varrho = \delta < 1$, Tohoku Math. J. **25**, No. 3, 339–345 (1973)

Weinstein, A.
[1] Fourier integral operators, quantization, and the spectra of Riemannian manifolds. (English. French summary.) With questions by W. Klingenberg and K. Bleuler and replies by the author. Géométrie symplectique et physique mathématique (Colloq. Internat. CNRS, No. 237, Aix-en-Provence, 1974), pp. 289–298. Éditions Centre Nat. Recherche Sci., Paris, 1975

Wells, R. O.
[1] Differential analysis on complex manifolds. Second edition. Graduate Texts in Mathematics, 65. Springer-Verlag, New York Berlin, 1980, x+260 pp.

Weyl, H.
[1] Das asymptotische Verteilungsgesetz der Eigenwerte linearer partieller Differentialgleichungen. Math. Ann. **71**, 441–479 (1912)
[2] Über die Abhängigkeit der Eigenschwingungen einer Membran von der Begrenzung. J. reine angew. Math. **141**, 1–11 (1912)

<h1 style="text-align:center">Index of Notation</h1>

Subject Index